**PAUL DRIBEN**

**HARVEY HERSTEIN**

# portrait of humankind

## AN INTRODUCTION
## TO HUMAN BIOLOGY
## AND PREHISTORIC CULTURES

**Canadian Cataloguing in Publication Data**

Driben, Paul, 1946–
    Portrait of humankind: an introduction to
human biology and prehistoric cultures

ISBN 0–13–064015–8

1. Anthropology. 2. Culture. 3. Man, Prehistoric.
I. Herstein, Harvey H. II. Title.

GN25.D75 1994        301        C93-095110-7

Prentice-Hall, Inc., Englewood Cliffs, New Jersey
Prentice-Hall International (UK) Limited, London
Prentice-Hall of Australia, Pty. Limited, Sydney
Prentice-Hall Hispanoamericana, S.A., Mexico City
Prentice-Hall of India Private Limited, New Delhi
Prentice-Hall of Japan, Inc., Tokyo
Simon & Schuster Asia Private Limited, Singapore
Editora Prentice-Hall do Brasil, Ltda., Rio de Janeiro

ISBN 0–13–064015–8

Acquisitions Editor: Michael Bickerstaff
Developmental Editor: Maurice Esses
Production Editor: Imogen Brian
Production Coordinator: Deborah Starks
Permissions and Photo Research: Karen Taylor
Cover and Internal Design: Olena Serbyn
Cover Image: First Light Associated Photographers
Page Layout and Illustrations: Jansom/Janette Thompson


  2 3 4 5    RRD     98 97 96 95

 This book is printed on recycled paper.

Printed and bound in the United States of America

Every reasonable effort has been made to obtain permissions for all articles and data used in this
edition. If errors or omissions have occurred, they will be corrected in future editions provided written
notification has been received by the publisher.

IN MEMORY OF ENID TRILLER
AND I.N. HERSTEIN

# BRIEF CONTENTS

# CONTENTS

*Chapter 10*  ## MESOLITHIC, NEOLITHIC, AND CIVILIZED
CULTURES  339

*Chapter 11*   ## RACE AND RACISM    381

# PREFACE

In Inuktitut, the aboriginal language of the Inuit in Siberia, Alaska, Canada, and Greenland, there are many words for *snow*. None correspond exactly with the English word for *snow*. Instead, there are separate and distinct terms for falling snow, snow on the ground, snow that can cause avalanches, and so on. This detailed elaboration not only reflects the fact that knowing about snow is vital to the Inuit way of life, but also makes it possible for the Inuit to "see" their Arctic environment in ways that others cannot.

A similar phenomenon occurs in anthropology. Like other disciplines, anthropology has its own technical terms. These allow anthropologists to "see" humankind in a unique way, from a perspective that is both fascinating and exciting. Whereas many subjects tend to take things apart, anthropology is one of the few disciplines that tries to put things together, to present a universal portrait of humankind from its biological beginnings through to the appearance of the cultures of the ancient and the modern world.

The focus of this book is on biological anthropology and archaeology, and in order to help students master the language of these branches of the discipline, we have **highlighted** and defined key terms and concepts in the body of the text. We have also included a list of key terms and concepts (with a pronunciation guide) at the conclusion of each chapter, and placed them in a glossary that appears at the end of the text. These devices should make learning the language of anthropology all the easier.

Two other pedagogical aids also deserve special mention. All too often students find it difficult to locate reading materials on specific topics they may wish to pursue, either to expand their knowledge or to prepare term papers and essays. In order to alleviate this problem we have included an extensive list of selected readings at the end of each chapter, and have arranged the list in a way that we hope will help students secure the specialized information they are seeking. Say, for instance, students want to learn more about the origin of bipedalism — the ability to walk upright. They can begin by consulting the bibliographic entries listed under the main heading "The Origin of Bipedalism," in the selected readings that appear at the end of Chapter Six.

The other pedagogical aid, found throughout the text, is the use of "quotable quotes" — statements selected from a wide range of books and articles that allow students to hear first-hand from scholars in Canada and abroad who have devoted their lives to the ongoing study of humankind. The quotations, which are accompanied by photographs, maps, tables, and diagrams, are included both for their intrinsic interest and to add to the material presented. An example of a "quotable quote" is presented on the next page. It is followed by two tables: Table P.1, which contains information about recent Canadian market trends and job prospects for new graduates in Anthropology and related disciplines, and Table P.2, which shows the principal areas and fields of employment among those who have graduated with a specialized degree in Anthropology.

# P.1 BECOMING AN ANTHROPOLOGIST

Morton Fried recalls the reactions to his decision to become an anthropologist.

•  •  •

When I told my parents that I was thinking of going into anthropology as a career, my father was pleased, but I could tell that my mother was stunned and unhappy . . . Partly because of her desire to please my father, partly because she was always extremely indulgent with her "one and only," she strove to conceal her deep dismay and anxiety. It was not that she opposed an intellectual career; her own mother had battled through life to produce two lawyer sons. Indeed, she hoped I would follow my uncles' road or go into medicine, or perhaps become a school teacher. But an anthropologist? What was that? One thing was sure. You couldn't make a living at it. She worried about what she would tell her "family," the numerous brothers and sisters to whom she often seemed most attached.

Years later, when I was a graduate student, my wife was asked by the grocer what her husband did. Overlooking the fact that it would still be several years before I became a true professional, my wife replied, "He's an anthropologist." "What in hell's that?" asked the grocer. "Can he make a living at it?"

Source: Morton H. Fried, *The Study of Anthropology* (New York: Thomas Y. Crowell Company, Inc., 1972), pp. 3–4.

---

**TABLE P.1**  Recent Canadian market trends and job prospects for new graduates in Anthropology and related disciplines

| EMPLOYMENT TRENDS | NUMBER OF JOBS 1989 | AVERAGE ANNUAL GROWTH RATES ACTUAL 1981–1989 | (%) PROJECTED 1989–1995 | NUMBER OF JOB OPENINGS 1989–1995 |
|---|---|---|---|---|
| This occupation | 1251 | 0.2 | 2.2 | 821 |
| All occupations | 12 434 282 | 1.5 | 1.5 | 8 081 262 |

Source: COPS (Canadian Occupational Projection System). *Job Futures: An Occupational Outlook to 1995*, Vol. 1 (Ottawa: Minister of Supply and Services, 1990), p. 100.

---

**TABLE P.2**  Principal areas and fields of employment among those who have graduated with a specialized degree in anthropology and related disciplines (based on 1986 census data)

| AREAS OF EMPLOYMENT | FIELDS OF EMPLOYMENT | % EMPLOYED |
|---|---|---|
| Services: | education | 21 |
| | business | 14 |
| | health and welfare — non-hospital | 10 |
| Public Administration: | provincial | 19 |
| | federal | 14 |
| | municipal | 6 |
| Other: | | 16 |

Source: COPS (Canadian Occupational Projection System). *Job Futures: An Occupational Outlook to 1995*, Vol. 1 (Ottawa: Minister of Supply and Services, 1990), p. 100.

In preparing this book we thought it best to divide the text into eleven chapters. Chapter One explains what anthropology is, what makes it unique, and why it is relevant; Chapters Two through Eight focus on the story of humankind as told by biological anthropologists; Chapters Nine and Ten concentrate on what archaeologists have learned about prehistoric cultures; and Chapter Eleven looks at the concept of race and the phenomenon of racism. Although this arrangement serves our purpose, others may find it useful to "mix and match." Our hope is that the book will be enjoyed in whatever order the chapters are read.

# ACKNOWLEDGEMENTS

Much of the credit for this book must go to the scores of scholars who have devoted themselves to the study of humankind; we hope that we have presented their data and ideas in a way that will encourage a new generation of students to choose anthropology as a career.

We would also like to acknowledge the manuscript reviewers, whose critical comments not only forced us to clarify vague passages in the working manuscript, but also directed us to many of the sources consulted in the preparation of the book. We are likewise indebted to the library staffs at Lakehead University, the University of Manitoba, and the University of Winnipeg for their assistance in securing research materials.

On a more personal level, we would like to thank the following friends and colleagues for their insights and help: Miguel Bombin, Luke Dalla Bona, Scott Hamilton, Gladys Hector, Dave Kemp, El Molto, Mark Nisenholt, Joe Stewart, and Nick van Eeden. Lisa Driben and Bonnie Frost deserve special mention in this connection as well.

Thanks are also due to the team from the College Division of Prentice Hall Canada who worked with us during the production of the book. We are especially indebted to Acquisitions Editor, Mike Bickerstaff, for his faith in the project; to Senior Developmental Editor, Maurice Esses, for his constant encouragement and support; and to Production Editor, Imogen Brian, for her wonderful editorial suggestions.

Finally, and most of all, we would like to thank our spouses, Carol Driben and Rose Herstein. Without their love, help, and inspiration, this book would have been impossible.

*Paul Driben*
*Harvey Herstein*

# KEY TO PRONUNCIATION GUIDES

| VOWEL SOUNDS | | CONSONANT SOUNDS | | SYLLABIFICATION INDICATED BY | ACCENT SYLLABLES INDICATED BY |
|---|---|---|---|---|---|
| a | as in back | b | as in boy | – | |
| ah | as in father | ch | as in chip | | |
| ahr | as in car | d | as in dog | | |
| air | as in hair | f | as in fat | | |
| ay | as in hay | g | as in get | | |
| aw | as in raw | h | as in hope | | |
| a̲ | as in ago | hw | as in wheat | | |
| | | j | as in jug | | |
| e | as in yes | k | as in cup | | |
| ee | as in see | l | as in leg | | |
| eer | as in beer | m | as in my | | |
| eye | as in eyelid | n | as in no | | |
| e̲ | as in taken | ng | as in sing | | |
| | | p | as in pen | | |
| i | as in hit | r | as in red | | |
| i̲ | as in direct | s | as in sit | | |
| | | sh | as in she | | |
| oh | as in go | t | as in top | | |
| oo | as in do | th | as in thin | | |
| or | as in door | th̲ | as in there | | |
| oor | as in poor | v | as in very | | |
| ow | as in cow | w | as in will | | |
| oy | as in oil | y | as in you | | |
| o̲ | as in connect | z | as in zoo | | |
| | | zh | as in vision | | |
| uh | as in upper | | | | |
| ur | as in turn | | | | |
| uu | as in full | | | | |
| u̲ | as in focus | | | | |
| y̲ | as in by | | | | |

Source: Adapted from Charles Harrington Elster. *There Is No Zoo in Zoology and Other Beastly Mispronunciations* (New York: Macmillan Publishing Company, 1988), pp. xvii-xx.

# CHAPTER ONE

## WHAT IS ANTHROPOLOGY?

*contents at a glance*

# INTRODUCTION

**Anthropology** is the study of humankind, all over the globe, in the past as well as the present. It is a young discipline, scarcely one hundred years old, but its contribution to our knowledge has been spectacular. Since the closing decades of the nineteenth century, when the idea of a unified study of humankind was first proposed, anthropologists have shed more light on what it means to be human than anyone could have imagined. Were it not for anthropologists we would know far less about our biological history, the identity of our close animal relatives, and the significance of our "racial" traits. Nor would we know as much about the drama of our ancestors' early experiments with culture, the special connections that exist between communication and culture, or the panorama of customs and traditions that make every culture unique. The study of humankind also has a practical side; anthropologists have been called upon to help solve problems ranging from how to identify deceased persons from their skeletal remains to how to salvage ancient cultural materials that are in danger of being destroyed. Anthropology is consequently diverse and enlightening, and this is why, however imperfect the discipline, students almost never see themselves or those around them in the same way as they did before they studied the subject.

# BIOLOGICAL ANTHROPOLOGY

Partly because of its scope, but also because being human is a biocultural experience, anthropologists have found it useful to divide their discipline into two complementary parts: biological (or physical) anthropology, and cultural (or social) anthropology. **Biological anthropology** is concerned with human biology in the broadest possible sense, and this is reflected in the wide range of subjects studied by its practitioners.

## PALAEOANTHROPOLOGY

Some biological anthropologists focus their attention on the antecedents of humankind. They are called **palaeoanthropologists**, and their aim is to reconstruct the biological history of our species. In order to accomplish this goal palaeoanthropologists pay special attention to the **fossil record** — what remains of organisms that are no longer alive. Above all they are concerned with human and human-like fossils; seeking out locations where these are likely to be found (such as Olduvai Gorge in Tanzania — see Figure 1.1), obtaining permission to search for the fossils, acquiring financial support to undertake the work, and writing up the results of their research in a way that illuminates the evolutionary history of our kind.

The story that they have pieced together is fascinating. Although there are gaps in the fossil record, particularly when it comes to human and

Palaeoanthropologist Louis S.B. Leakey (1903–1972) searching for the fossils of our ancestors in Olduvai Gorge, Tanzania

human-like fossils, those that have been discovered confirm that our biological history spans millions of years. About 70 million years ago (mya), when the dinosaurs became extinct, a new group of much smaller animals — the prosimians (whose name means monkey-like) — began to flourish. Their descendants, which are called **primates**, include living **prosimians** such as lemurs, indriids, lorises, and tarsiers; modern **anthropoids** or monkeys, apes, and humans; and the extinct forms that gave rise to the prosimians and anthropoids that are alive today. Since our biological history is part and parcel of the biological history of the primates, the evolution of our kind is best understood by telling the entire story, by looking at human evolution in the context of primate evolution as a whole. Palaeoanthropologists are consequently interested not only in human evolution, but also in the evolution of our primate relations, whose biological histories are related to but different from our own. See QQ 1.1.

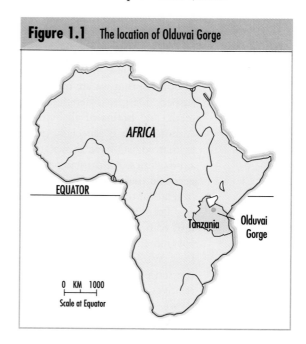

**Figure 1.1**  The location of Olduvai Gorge

AFRICA

EQUATOR

Tanzania

Olduvai Gorge

0  KM  1000
Scale at Equator

## PRIMATOLOGY

Other biological anthropologists investigate the similarities and the differences between humankind and the more than 200 other species of surviving primates. These anthropologists are called **primatologists**, and they have discovered that we are more like our non-human relatives than many people suspect. Despite differences in appearance, the overall structure of the limbs, teeth, and brains of all living primates is based on the same general plan. Although this is especially true with respect to monkeys, apes, and humans, primatologists have shown that prosimians are patterned along remarkably similar lines. All primates tend to have free-moving limbs with clawless fingers and toes, a relatively small number of teeth, and a relatively large brain in relation to the overall size of the body.

Primates are also alike in another respect. By studying the social behaviour of prosimians, monkeys, and apes in the wild, primatologists have discovered that many species of living non-human primates have rich social lives. While this finding is interesting in and of itself, it is doubly exciting in the sense that information about the social behaviour of our closest living relatives may help us to better understand how we ourselves behave, provided we appreciate that their behavioural patterns are neither direct nor simple analogues of our own. See QQ 1.2.

## 1.2 THE IMPORTANCE OF PRIMATOLOGY

Linda Marie Fedigan, Professor of Anthropology at the University of Alberta, explains how studying the behaviour of non-human primates contributes to our understanding of ourselves.

• • •

Why . . . study the behavior of primates? There are so many possible answers to this question that I will make my answer a personal one. I study these animals firstly because they fascinate and delight me, and I think this may be true of all primatologists, and secondly because I am greatly interested in how and why creatures live in social groups, which may not be true of all primatologists. Had I been trained as a zoologist rather than an anthropologist I might have studied other socially-living animals such as elephants or herring gulls; however, the study of primates and not the study of proboscids [or elephants] is one of the subdisciplines of anthropology. For primates do constitute the order of animals to which humans also have been taxonomically consigned and the study of humankind's agreement with, and divergence from, other animals is one of the definitions of anthropology. Thus, it is reasonable to ask what we learn about *ourselves* from the study of these animals. While I have said that our primate relatives are not simplified versions of our own species, that sort of reductionism does not rule out other approaches to the relevance of nonhuman to human primates. So I would suggest that a third reason for studying the behavior of prosimians, monkeys and apes, is to enrich our perception and understanding of human behavioral patterns in the expanded perspective of the vast array of patterns existing in other primates, animals whose evolutionary paths separated from our own more recently than those of other mammals.

Source: Linda Marie Fedigan, *Primate Paradigms: Sex Roles and Social Bonds.* (Montréal: Eden Press, Inc., 1982), p. 4.

## POPULATION BIOLOGY

Still other biological anthropologists study the outward appearance and the genetic characteristics of contemporary human populations. They are known as **population biologists**, and one of their most important discoveries is that no so-called race of people is biologically superior to another. Instead, what they have determined is that people everywhere are capable of equally impressive achievements, and this is hardly surprising considering the biological unity of humankind. Whatever our differences, modern humankind is a single **species** — a biological population whose members can interbreed with reproductive success — and given this fact, the idea that we all possess the same biocultural potential makes eminent sense.

Population biologists have also challenged a more fundamental idea about race. In the past many scholars believed that humankind could be divided into a fixed number of races on the basis of features such as skin colour, height, and so on. Some people still believe this to be true. The idea is false. Certainly there are differences between us, and the differences are worth exploring, at least to better understand human diversity and its potential adaptive significance. However, while population biologists agree that human variation should be studied, their research has shown that races are notoriously hard to identify when objective criteria are used. See QQ 1.3.

# 1.3 RACIAL IDENTIFICATION

Boyce Rensberger comments on the problem of identifying races on the basis of outward appearance.

• • •

There is a . . . conclusion about race that is often misunderstood. Despite our . . . constant everyday references to race, no one has ever discovered a reliable way of distinguishing one race from another. While it is possible to classify a great many people on the basis of certain physical features, there are no known feature[s] or groups of features that will do the job in all cases.

Skin color won't work. Yes, most Africans from south of the Sahara and their descendants around the world have skin that is darker than that of most Europeans. But there are millions of people in India, classified by some anthropologists as members of the Caucasoid, or "white," race who have darker skins than most Americans who call themselves black. And there are many Africans living in sub-Saharan Africa today whose skins are no darker than the skins of many Spaniards, Italians, Greeks or Lebanese.

What about stature as a racial trait? Because they are quite short, on the average, African Pygmies have been considered racially distinct from other dark-skinned Africans. If stature, then, is a racial criterion, would one include in the same race the tall African Watusi and the Scandinavians of similar stature?

The little web of skin that distinguishes Oriental eyes is said to be a particular feature of the Mongoloid race. How, then, can it be argued that the American Indian, who lacks this epicanthic fold, is Mongoloid?

Even more hopeless as racial markers are hair color, eye color, hair form, the shapes of noses and lips or any of the other traits put forward as typical of one race or another.

Source: Boyce Rensberger, "Racial Odyssey," *Science Digest*, Vol. 89, No. 1 (1981), p. 52.

Some of the peoples of the modern world

United Nations

# CULTURAL ANTHROPOLOGY

**Cultural anthropology** picks up the story of humankind where biological anthropology leaves off, at a point in the remote past when our ancestors developed the capacity for culture — a capacity that allows people everywhere to behave in ways that are not biologically inherited but that are learned from other people. Cultural anthropologists consequently focus their attention on humankind's cultural achievements, which developed over time and were and are expressed in different ways in different parts of the world.

## ARCHAEOLOGY

**Archaeologists** concentrate on the cultural achievements of the past. Although some study historic cultures, the majority are interested in how people lived in **prehistoric** times — from the point at which humankind developed the capacity for culture to the beginning of written history, which was inaugurated about 5 500 years ago in the Near East and later in other parts of the world. In order to reconstruct extinct cultures archaeologists unearth and analyze the remains that prehistoric peoples left behind, including their implements and the ruins, if any, of their buildings (see QQ 1.4). They also pay special attention to the natural resources that prehistoric peoples depended upon for food, shelter, and clothing, which are recovered from sites such as Head Smashed-In, Alberta, which is one of the oldest "buffalo jumps" in North America (see Figure 1.2 and Figure 1.3). Taken together, these materials provide archaeologists with the clues that they use to reconstruct the cultures of the past, ranging from those whose tool-makers fashioned the first stone implements to the ones whose civilizations formerly dominated the world. On a more general level, archaeologists are also interested in the rise and fall of prehistoric cultures and in how such cultures influenced those that prevail today.

 **1.4 PREHISTORIC ARCHAEOLOGY**

James Wright, former Head of the Scientific Section, Archaeological Survey of Canada, Canadian Museum of Civilization, describes the goals of prehistoric archaeology.

• • •

The archaeologist interested in prehistory studies people of the past who left no written records. Because these people lived before writing was invented or used in their part of the world, the story of their life and times is usually called *prehistory* rather than *history*. This absence of written records is one of the major differences in the kinds of data available to historians and prehistorians. While the former reconstruct history largely from documentary evidence left by past civilizations, the latter must rely on the use of physical objects that have survived the centuries in order to tell the story of long-vanished peoples. These objects may be artifacts, such as discarded or lost stone and bone tools; features, like the vague outline in the soil where an ancient house stood; and material, like the butchered bones and burnt plant remains that represent the animals and plants that were eaten . . . .

As more local prehistories become known, they in turn can be compared to each other and provide a general picture of the overall prehistory of a larger region, country, or continent. The process is continually being built upon. As new sites are excavated, producing a growing body of information, old interpretations are discarded or modified to permit an increasingly accurate reconstruction of prehistory . . . .

Source: James V. Wright, *Six Chapters of Canada's Prehistory*. (Toronto: Van Nostrand Reinhold Ltd., 1976), pp. 11–23.

**Figure 1.2**   An artist's rendering of bison plunging over the sandstone cliffs at Head-Smashed-In

Drawing by Judith Nickol from *The First Albertans* by Gail Helgason, Lone Pine Publishing. ©1987. Used by permission of Lone Pine Publishing, Edmonton, Alberta.

**Figure 1.3**   The location of the Head-Smashed-In site

## LINGUISTIC ANTHROPOLOGY

**Linguistic anthropologists** have a different orientation; they are concerned with the special relationship that exists between communication and culture. The relationship between language and culture is an outstanding example. Without **language**, which consists of words and rules for their use, culture would not and could not exist. Realizing this, the first generation of linguistic anthropologists began to wonder about the origin of language — how and when our ancestors learned to speak. Since then, linguistic anthropologists have also become interested in how the speakers of different languages "see" the world, how they combine sounds into words and words into sentences, and how languages, which are grouped into families (see Table 1.1 on page 10), change over time. At one time linguistic anthropologists were concerned almost exclusively with language. Today they are also interested in how sociocultural conditions influence speech, in how people communicate through gestures and tones of voice, and in how non-human animals communicate. See QQ 1.5.

## QQ 1.5 COMMUNICATION AND CULTURE

Alfred Smith describes the concerns of linguistic anthropologists.

• • •

The linguistic anthropologist, like the . . . [archaeologist] and the physical [or biological] anthropologist, studies the various peoples of the world: how they resemble and differ from one another, and how they have changed and developed. The linguistic anthropologist differs from the[se] other anthropologists in being less concerned with human biology and technology . . . and being more concerned with the languages of the people of the world. Each language is a code of human interaction. Some of these codes involve words and sentences. Some of these codes are vocal but nonverbal and involve tones of voice. Still other codes are nonverbal and nonvocal and involve gestures and other kinds of action and conduct.

While the mathematician [who also studies communication] is concerned with electronic signals, the linguistic anthropologist studies human signals. While the social psychologist is generally concerned with human communication in his own culture in his own time, the linguistic anthropologist works with a worldwide range of data, with all the codes of the Greeks, Bantu, and the Samoans. This broad range of data provides a necessary test of generalizations that may have been derived from too small a sample, from studying only one's own way of behavior.

Source: Excerpt from *Communication and Culture: Readings in the Codes of Human Interaction* by Alfred G. Smith, copyright © 1966 by Holt, Rinehart and Winston, Inc. Reprinted by permission of the publisher.

## ETHNOLOGY

Finally there are **ethnologists**, whose principal concern is with contemporary cultures and those of the recent past. Every culture — and there are thousands — has a kinship system that determines whom people count as their relatives, an economic system that provides them with the means to earn a livelihood, a political system that helps them to settle disputes, and a magical-religious system that provides a set of sacred beliefs that allows them to communicate with the supernatural world. But beyond these general similarities, cultures are remarkably varied. The Arunta of Australia and the Crow Indians of the American Plains do not have the same kinship system; the Bushmen in southern Africa and the Inca Indians of Peru earn their

## TABLE 1.1 Canada's indigenous language families and languages

| LANGUAGE FAMILY | LANGUAGES | LANGUAGE FAMILY | LANGUAGES |
|---|---|---|---|
| Algonquian | Abenaki<br>Delaware<br>Potawatomi<br>Malecite<br>Micmac<br>Blackfoot<br>Montagnais-Naskapi<br>Ojibwa<br>Cree | Kutenai<br><br>Salishan | Kutenai<br><br>Sechelt<br>Squamish<br>Straits<br>Bella Coola<br>Comox<br>Halkomelem<br>Lillooet<br>Okanagan<br>Thompson<br>Shuswap |
| Athapaskan | Tagish<br>Sarcee<br>Han<br>Sekani<br>Kaska<br>Beaver<br>Hare<br>Dogrib<br>Chilcotin<br>Tahltan<br>Tutchone<br>Slave<br>Kutchin<br>Carrier<br>Chipewyan | Siouan<br><br>Tlingit<br><br>Tsimshian<br><br><br><br>Wakashan | Dakota<br><br>Tlingit<br><br>Southern Tsimshian<br>Nass-Gitksan and<br>Coastal Tsimshian<br><br>Haisla<br>Heiltsuk<br>Kwakwala<br>Nitinat<br>Nootka |
| Eskimo-Aleut | Inuktitut | | |
| Haida | Haida | | |
| Iroquoian | Tuscarora<br>Onondaga<br>Oneida<br>Cayuga<br>Seneca<br>Mohawk | | |

Source: Michael Foster, *Indigenous Languages in Canada* (Ottawa: Commissioner of Official Languages), 1982.

Ethnologist Paul Driben interviewing an Ojibwa hunter in the Lansdowne House Indian Reserve in Ontario, Canada

**Figure 1.4** The location of the Lansdowne House Indian Reserve

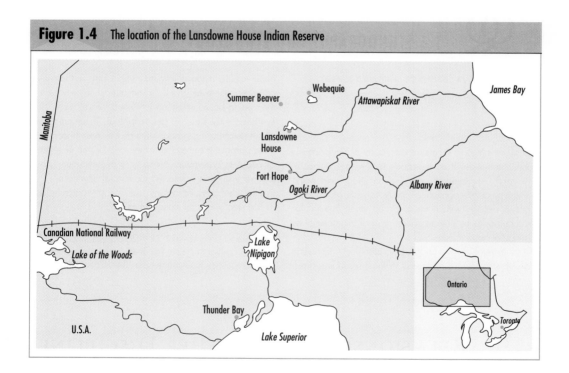

livelihoods in different ways; the citizens of Russia and the Ojibwa Indians of Canada use different methods to settle disputes; and New York's Hasidic Jews and the Trobriand Islanders of Melanesia hold different religious beliefs.

To determine the extent of the variation, ethnologists consult the **ethnographic record**, which contains descriptions of cultures around the world. The descriptions are called **ethnographies**, and the research on which these are based is generally conducted in one of two ways. Sometimes information about the customs and traditions of a group of people is best collected by examining the contents of historical records. This approach, which is known as **ethnohistorical research**, has proven to be particularly effective in studying the cultures of the recent past. But whenever a culture has living representatives, **ethnographic research** is preferred. This involves observing people on a first-hand basis, questioning them about their values and beliefs, and, whenever possible, participating with them as they go about their everyday affairs. Taken together, these methods have produced the information that ethnologists use to compare how the capacity for culture is expressed in different parts of the world. Ethnologists also rely on the ethnographic record to develop and test theories about the impacts of cultures on one another, about the process of culture change, and about the development of specific customs such as marriage, gift-giving, warfare, and so on. See QQ 1.6.

## 1.6 ETHNOGRAPHY AND ETHNOLOGY

Claude Lévi-Strauss discusses the relationship between ethnography and ethnology.

. . .

In all countries, it seems, ethnography is interpreted in the same way: It corresponds to the first stages in research — observation and description, field work. The typical ethnographical study consists of a monograph dealing with a social group small enough for the author to be able to collect most of his material by personal observation . . . .

In relation to ethnography, ethnology represents a . . . step toward synthesis. Without excluding direct observation, it leads toward conclusions sufficiently comprehensive to preclude, or almost to preclude, their being based solely on first-hand information. The synthesis may be . . . geographical, if information about neighboring groups is to be collated; historical, if the purpose is to reconstruct the past of one or several peoples; systematic, if one type of technique, custom, or institution is selected for special attention . . . . In all these cases, ethnology includes ethnography as its first step and is an extension of it.

Source: Selected excerpt from page 352 of *Structural Anthropology* by Claude Lévi-Strauss Copyright © 1963 by Basic Books, Inc. Reprinted by permission of Basic Books, Inc., a division of HarperCollins, Publishers, Inc.

## RELATIONSHIPS WITH OTHER DISCIPLINES

Like experts in other fields, anthropologists work closely with scholars in other disciplines. The Dakhleh Oasis project is a good example. In 1969, Geoffrey Freeman, an amateur archaeologist from Canada, formed an organization known as the Society for the Study of Egyptian Antiquities (SSEA), which was subsequently awarded the right to excavate the Dakhleh Oasis site in Egypt's Great Western Desert. See Figure 1.5.

Twelve years later the Royal Ontario Museum (ROM) became involved in the project, and, since then, under the direction of Anthony Mills, the museum's assistant curator of Egyptology, an international multidisciplinary team composed of archaeologists, geologists, botanists, zoologists, and palaeoanthropologists has reached a major conclusion. Thanks to their efforts, 8300 square kilometres of territory that was once described as an "infinite space of mystery and terror" is now known to have been occupied continuously for the past 200 000 years (see QQ 1.7). What remains to be determined is why, where stone age peoples once hunted and later emperors ruled, there are now scarcely 40 000 people scattered over a vast desert whose landscape is distinguished only by date trees, farms, and little villages.

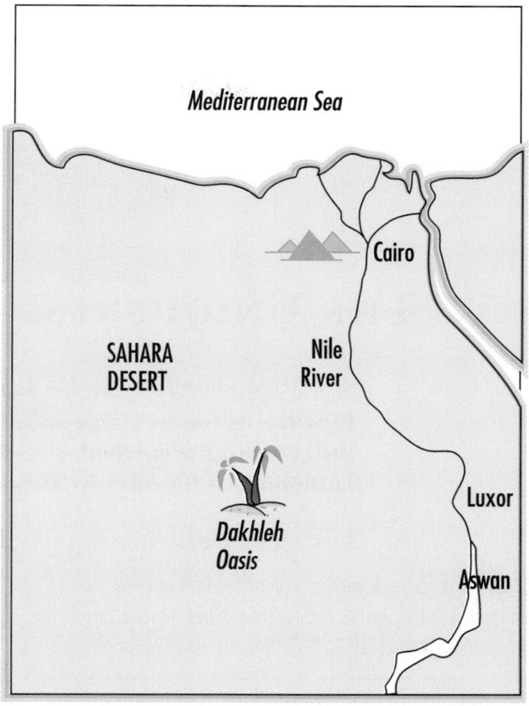

The Dakhleh Oasis project is one of many examples of multidisciplinary research in which anthropologists have collaborated with scholars in other disciplines, and this is to be expected considering that researchers other than anthropologists are also concerned with humankind. But while anthropologists may study the same subject matter as experts in other fields, their approach is unique.

**Figure 1.5** The location of the Dakhleh Oasis

## QQ 1.7 MULTIDISCIPLINARY RESEARCH AT THE DAKHLEH OASIS, EGYPT

Harry Thurston explains why Geoffrey Freeman originally decided to study the Dakhleh Oasis from a multidisciplinary point of view and the outcome of his decision.

• • •

From the beginning, Freeman believed that because of the distinct nature of the oasis and the antiquity of its human archaeological record, the Dakhleh project should be a multidisciplinary effort, involving natural scientists as well as archaeologists. "I thought, if you're going to study something like the Dakhleh Oasis, beginning 200 000 years ago, you might as well study everything about it and not just the rocks and the tombs and the temples and the hieroglyphs." Freeman wanted to know the origin of the oasis people and how and why they came to the oasis in the first place. Similarly, he wanted to know how the flora and fauna arrived at the oasis and, subsequently, why many species disappeared.

Today . . . Freeman's original vision has given the project a unique place in Egyptology . . . . [Although] Freeman, stricken with blindness in 1982, has not returned to the field . . . [under the direction of Anthony Mills], there are now 25 major scientists . . . [working on the project] from institutions in Canada, the United States, Australia and Poland . . . .

Source: Harry Thurston, "Everlasting Oasis," *Equinox*, Vol. VI: 5, No. 35 (1987), pp. 32–35.

# THE UNIQUENESS OF ANTHROPOLOGY

## THE COMPARATIVE METHOD

One special feature of the anthropological approach is the widespread use of the **comparative method**, a research strategy that is based on analyzing similarities and differences in a comprehensive way. Palaeoanthropologists, for example, rely on comparison to interpret the fossil record, taking note of the similarities and differences between ourselves and the forerunners of our species in order to explain the evolution of our kind. By comparing our own biological and social make-up with those of living prosimians, monkeys, and apes, primatologists have been able to determine what we have in common with our primate relations and what sets us apart (see Figure 1.6). And were it not for the fact that population biologists have compared the physical features of tens of thousands of people from hundreds of locations, the ideas that races can be ranked and identified with ease might still prevail today.

Cultural anthropologists make use of the comparative method as well. By comparing prehistoric cultures, archaeologists have been able to reconstruct the main lines of variation of world prehistory; by comparing languages, linguistic

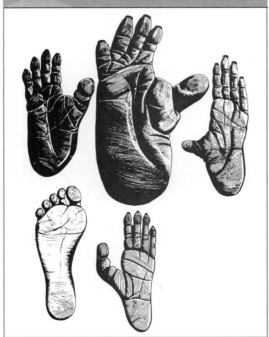

**Figure 1.6**   A comparison of the feet of apes and humans (top left to right: chimpanzee, gorilla, and orangutan; bottom left to right: human and siamang)

From THE APES by Vernon Reynolds. Copyright © 1967 by Vernon Reynolds. Used by permission of the publisher, Dutton, an imprint of New American Library, a division of Penguin books USA Inc.

anthropologists have been able to identify the common components of language and to determine which languages are related; and, by following the same strategy, ethnologists have been able to explain not only why cultures are different, but also why certain customs and traditions are universal. The comparative method is consequently important in each and every branch of the discipline. See Figure 1.7 and QQ 1.8.

**Figure 1.7** The cultural areas of the Indian and Inuit peoples of Canada. Boundaries are determined by comparing the cultural attributes of Indian and Inuit peoples at the time of contact with Europeans.

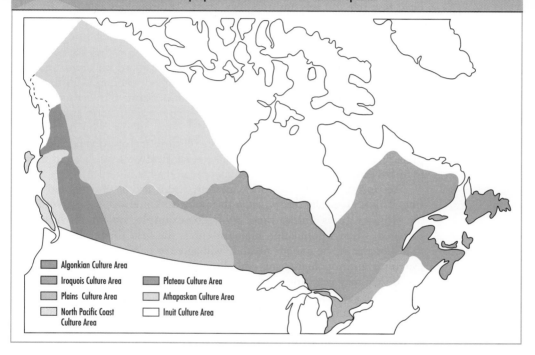

Algonkian Culture Area
Iroquois Culture Area
Plains Culture Area
North Pacific Coast Culture Area
Plateau Culture Area
Athapaskan Culture Area
Inuit Culture Area

## 1.8 ANTHROPOLOGY AND THE COMPARATIVE METHOD

George Peter Murdock discusses the importance of the comparative method in cultural anthropology.

• • •

[There] can never be any generally valid science of man which is not specifically adapted to, and tested with reference to, the diverse manifestations of human behavior encountered in the thousands of human societies differing from our own that are known to history and ethnography. Whatever other methods of investigation may be employed — and they are numerous — the comparative method is indispensable. Without it, no combination of other methods can achieve scientific results of universal application. At the most they can only produce culture-bound generalizations, approximately valid for a particular group of related societies during a particular segment of their history, but incapable of generalization to other societies except as highly tentative working hypotheses, and equally incapable of predicting future developments in periods of rapid social change or even of comprehending them after they have occurred.

Source: George Peter Murdock, "Anthropology as a Comparative Science," *Behavioral Science*, Vol. 2 (1957), p. 249.

## FIELDWORK

Field research also sets anthropology apart. Anthropology is not a dry subject; it is vibrant and alive, in part because of the strong emphasis that its practitioners place upon **fieldwork**, a research method that brings anthropologists into direct contact with the subjects and objects they study.

Palaeoanthropologists do not simply speculate about the origin and evolution of humankind. In search of clues about our biological history they seek out and excavate fossils all over the world. Primatologists likewise do not study the behaviour of non-human primates only in zoos. Instead, they observe our close animal relatives in the wild, in the jungles of South America, in the forests of Asia, and in the interior of Africa. And while preparing for such expeditions may take years, fieldwork can be, and usually is, an exciting and rewarding experience. See QQ 1.9.

Yet as romantic as fieldwork may seem, it can also be difficult and upsetting. This is especially true when anthropologists come face-to-face with people from cultures unlike their own. For example, no matter how well prepared, when a population biologist enters the field there is no guarantee that those he or she intends to study will agree to participate in the research. Linguistic anthropologists and ethnologists may suffer a similar fate; however much they may want to make observations, conduct interviews, and participate in the everyday lives of people in order to better understand their language and culture, they too may be regarded as unwelcome visitors. And even when anthropologists are accepted they still may experience **culture shock** — the feeling of alienation that is associated with being a foreigner, a stranger in a strange land — and this can be depressing, so much so that fieldwork is sometimes brought to an abrupt halt. See QQ 1.10 on page 18.

## HOLISM

The anthropological approach is also unique because it deals with humankind in a **holistic** way, paying attention to virtually every aspect of the human experience. To be sure, many anthropologists study highly specialized topics. A look at the research interests of anthropologists who teach at Canadian universities attests to the fact. Consider the research that the members of the Département d'anthropologie at Université Laval in Québec City have pursued. There, since the sixties, faculty members have conducted ethnographic research in Québec, Latin America, Africa, and Melanesia, and in those locations they have studied subjects ranging from land use and occupancy among the Inuit of Arctic

# 1.9 THE REWARDS OF FIELDWORK

Jane Goodall, who has studied chimpanzees in the Gombe National Park in western Tanzania for more than thirty years, talks about her first success in the field.

. . .

[In] 1960, after four long, difficult months in the field, I made my first really exciting observation — I saw a chimp fashion and use crude tools!

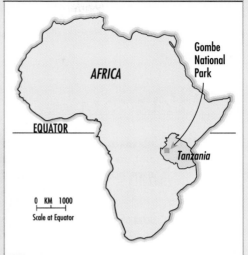

The location of the Gombe National Park

AFRICA

Gombe National Park

EQUATOR

Tanzania

0   KM   1000
Scale at Equator

Animals Animals © Stewart D. Halperin, #M-14032

. . . I saw David Greybeard [the chimp's name] and as I watched him I could hardly believe my eyes. He was carefully trimming the edges from a wide blade of sword grass!

I gazed, scarcely daring to breathe, as he pushed the modified stem into the [termite] nest. He left it for a moment, then pulled it out and picked off . . . [the insects] with his lips. The chimp continued probing with the stem until it bent double. He then discarded it and reached out to pick a length of vine. With a sweeping movement of one hand, he stripped the leaves from the vine, bit a piece from one end, and set to work again with his newly prepared tool.

For an hour I watched . . . Finally, after trying each of the holes in turn, he dropped the piece of vine and wandered away.

▼

A chimpanzee "termiting"

▲

Source: Jane Goodall, *My Friends the Wild Chimpanzee.* (Washington, D. C.: National Geographic Society, 1967), p. 31.

Ethnologist Napoleon Chagnon, who spent nineteen months among the Yąnomamö Indians of southern Venezuela and northern Brazil, explains the difficulties of living among people whose language and customs are unlike one's own.

• • •

### The homeland of the Yąnomamö

Map from YANOMAMÖ: THE FIERCE PEOPLE, Second Edition by Napoleon Chagnon, copyright © 1977 by Holt, Rinehart and Winston, Inc., reprinted by permission of the publisher

My first day in the field illustrated to me what my teachers meant when they spoke of "culture shock." I had traveled in a small, aluminum rowboat propelled by a large outboard motor for two and a half days. This took me . . . deep into Yąnomamö country . . .

We arrived at the village, Bisaai-teri, about 2:00 PM . . . In just a few moments I was to meet my first Yąnomamö, my first primitive man. What would it be like . . .

As we walked down the path . . . I pondered the wisdom of having decided to spend a year and a half with this tribe before I had even seen what they were like. I am not ashamed to admit, either, that had there been a diplomatic way out, I would have ended my fieldwork then and there. I did not look forward to the next day when I would be left alone with the Indians; I did not speak a word of their language, and they were decidedly different from what I had imagined them to be. The whole situation was depressing, and I wondered why I ever decided to switch from civil engineering to anthropology in the first place. I had not eaten all day, I was soaking wet from perspiration, the gnats were biting me, and I was covered with red pigment, the result of a dozen or so complete examinations I had been given by as many burly Indians. These examinations capped an otherwise grim day. The Indians would blow their noses into their hands, flick as much of the mucus off that would separate in a snap of the wrist, wipe the residue into their hair, and then carefully examine my face, arms, legs, hair, and the contents of my pockets . . .

So much for my discovery that primitive man is not the picture of nobility and sanitation I had conceived him to be. I soon discovered that it was an enormously time-consuming task to maintain my own body in the manner to which it had grown accustomed in the relatively antiseptic environment of the northern United States. Either I could be relatively well fed and relatively comfortable in a fresh change of clothes and do very little fieldwork, or, I could do considerably more fieldwork and be less well fed and less comfortable.

Napoleon Chagnon/Anthro-Photo, #0771

▼

Napoleon Chagnon dressed as a Yąnomamö

▲

Source: Excerpt from YANOMAMÖ: THE FIERCE PEOPLE, Second Edition by Napoleon Chagnon, copyright © 1977 by Holt, Rinehart and Winston, Inc., reprinted by permission of the publisher.

Québec to urban ethnicity in Papua New Guinea. Meanwhile, anthropologists elsewhere in the country (see Table 1.2) have matched their achievements, studying an equally diverse range of highly specialized topics.

On the surface, such specialized interests may seem to contradict the claim that anthropology is holistic. But, when taken together, the detailed studies conducted by anthropologists from Canada and abroad form the basis for a holistic view of our species. While biological anthropologists tell us about our evolutionary history and the development of our capacity for culture, cultural anthropologists make us aware of our early experiments with technology, communication, and culture and the customs and traditions that people cherish today. Thus, instead of dwelling on one era or aspect of the human experience, anthropology focuses on the broad history of our species from its origin to the present. This makes anthropology more than just a biological science, more than just a humanity, and more than just a social science; it is all these and more. See QQ 1.11.

## 1.11 THE HOLISTIC NATURE OF ANTHROPOLOGY

**Margaret Mead looks at holism in anthropology.**

• • •

Anthropology is a uniquely situated discipline, related in diverse ways to many other disciplines, each of which, in specializing, has also inadvertently helped to fragment the mind of modern man. Anthropology is a humanity . . . concerned with the arts of language and with the versions that human cultures have given of the definition of man and of man's relationship to the universe; anthropology is a science, concerned with discovering and ordering the behavior of man-in-culture; anthropology is a biological science, concerned with the physical nature of man, with man's place in evolution, with the way genetic and racial differences, ecological adaptations, growth and maturation, and [biological] differences are implicated in man's culture and achievements; anthropology is a historical discipline, concerned with reading the record of man's far past and establishing the links which unite . . . the threads between the preliterate and the literate world wherever the sequence occurs, in Egypt, in China, in Crete, or in a modern African state. Anthropology is a social science, although never only a social science, because in anthropology man, as part of the natural world, as a biological creature, is not separated from man as a consumer or pro-

Margaret Mead (1901–1978)

ducer, member of a group, or possessor of certain psychological faculties. Anthropology is an art. The research skills that go into good field work are as complex as the skills of a musician or a surgeon . . . .

Source: Margaret Mead, "Anthropology and an Education for the Future," In David G. Mandelbaum, Gabriel W. Lasker, and Ethel M. Albert, eds. *The Teaching of Anthropology* © 1963 The Regents of the University of California. (Los Angeles: University of California Press), p. 596.

**Table 1.2**  The number of full-time anthropology faculty in anthropology, archaeology, anthropology/archaeology, and anthropology/sociology departments in Canadian universities in 1992–1993

| INSTITUTION | TYPE OF DEPARTMENT | NUMBER OF FULL-TIME ANTHROPOLOGY FACULTY |
|---|---|---|
| U Alberta | anthropology | 24 |
| Brandon U | anthropology/sociology | 2 |
| U British Columbia | anthropology/sociology | 25 |
| U Calgary | archaeology | 22 |
| U Calgary | anthropology | 11 |
| Carleton U | anthropology/sociology | 14 |
| Concordia U* | anthropology/sociology | 7 |
| Dalhousie U | anthropology/sociology | 6 |
| U Guelph | anthropology/sociology | 3 |
| Lakehead U | anthropology | 5 |
| Laurentian U | anthropology/sociology | 3 |
| U Laval | anthropology | 21 |
| U Lethbridge | anthropology | 7 |
| U Manitoba | anthropology | 16 |
| McGill U | anthropology | 19 |
| McMaster U | anthropology | 18 |
| Memorial U of Newfoundland | anthropology | 16 |
| U Montréal | anthropology | 23 |
| Mount Allison U* | anthropology/sociology | 2 |
| Mount Saint Vincent U* | anthropology/sociology | 2 |
| U New Brunswick | anthropology | 7 |
| U Prince Edward Island* | anthropology/sociology | 3 |
| U Regina | anthropology | 4 |
| Saint Francis Xavier U* | anthropology/sociology | 1 |
| Saint Mary's U | anthropology | 5 |
| U Saskatchewan | anthropology/archaeology | 7 |
| Simon Fraser U | archaeology | 10 |
| Simon Fraser U | anthropology/sociology | 9 |
| U Toronto | anthropology | 40 |
| Trent U | anthropology | 17 |
| U Victoria | anthropology | 11 |
| U Waterloo | anthropology | 6 |
| U Western Ontario | anthropology | 14 |
| Wilfrid Laurier U | anthropology/sociology | 4 |
| U Windsor | anthropology/sociology | 6 |
| U Winnipeg | anthropology | 6 |
| York U | anthropology | 14 |
| | Total | 410 |

Sources: American Anthropological Association. *A Guide to Departments/A Directory of Members: aaa Guide* (Washington D C: American Anthropological Association, 1992), pp. 1–213; Reimer, Bill, ed. *1990–1991 Guide to Departments: Anthropology, Archaeology, Sociology in Universities and Museums in Canada.* Ottawa: Canadian Sociology and Anthropology Association/Canadian Anthropology Society, 1992; and personal communication with those institutions followed by an asterisk.

# TECHNICAL VOCABULARY

Finally, anthropology is unique because of its **technical vocabulary**. Like other academic disciplines, anthropology has developed a specialized language of its own. This language makes it possible for anthropologists to discuss human beings and their behaviour in a unique fashion (see QQ 1.12). Certainly there are words in anthropology that are familiar to everyone; "evolution" and "culture" are two outstanding examples. In anthropology, however, these terms and others like them are defined in a special way.

For example, when anthropologists speak about evolution they usually have one of three meanings in mind. Depending on the context, **evolution** may mean biological change through time, the process by which new species arise from old ones, or may refer to the theory that Charles Darwin proposed. The same is true with respect to the concept of culture. Depending on the context,

Page from an Anthropological Dictionary

---

## R

**race.** The common use of the word in English is to refer to a group of persons who share common physical characteristics and form a discrete and separable population unit has no scientific validity, since evolutionary theory and PHYSICAL ANTHROPOLOGY have long since demonstrated that there are no fixed or discrete racial groups in human populations. Instead, human groups constantly change and interact, to such an extent that modern population genetics focuses on CLINES or patterns of the distribution of specific genes rather than on artificially created racial categories. However as a folk concept in Western and non-Western societies the concept of race is a powerful and important one, which is employed in order to classify and systematically exclude members of given groups from full participation in the social system controlled by the dominant group. As a folk concept, race is employed to attribute not only physical characteristics but also psychological and moral ones to members of given categories, thus justifying or naturalizing a discriminatory social system. (*See* RACISM.)

**race relations.** The relations between persons and groups conceived of according to folk categories as being of different RACES. There is not always a clear distinction between race relations and ETHNIC relations, since groups conceived of as racially separate often also possess an ethnic identity (or may be divided internally into different ethnic minorities) which confuses this distinction. The first major contribution to the theory of race relations in the social sciences was that of Park and the Chicago school. He held that race relations were the product of racial consciousness created by competition between populations attempting to occupy the same ecological niche (1950). The Chicago school tended to view racial conflicts as stages in the process of integration which

would eventually result from population migration and contact. This view was challenged by social stratification theory which argued that race was an expression or extension of social CLASS, and should be analysed as an aspect of social stratification in general. According to stratificationists, the Chicago school's emphasis on a movement towards integration disguised the fundamental nature of race relations, which are the product not of competition but of the exploitation of non-European by European races. (*See* PLURAL SOCIETY.)

**racism, racialism.** Doctrines of or beliefs in racial superiority, including the belief that race determines intelligence, cultural characteristics and moral attributes. Racism includes both racial PREJUDICE and racial discrimination, and thus is also employed describe social systems which systematically discriminate against given racial categories. Many writers employ the term 'institutionalized racism' to refer to the social structural aspect of racism and the manner in which specific racial prejudices and stereotypes are incorporated into legal, administrative and social systems. Institutionalized racism may be analysed as a product of class interests and class ideology, as well as at an international level as a product of COLONIALIST and IMPERIALIST strategies employing racism as an important element in the justification and maintenance of relations of exploitation and unequal exchange with subordinate populations who happen to be physically different. Students of racism have pointed out how the rise and fall of racial stereotypes and racial prejudice is closely related to the changing historical relations between different populations and above all to the interests of dominant groups (*see* APARTHEID; SLAVERY). However there has been little systematic anthropological study of racism and the forms which it takes in colonial and

238

---

## ⵕⵕ 1.12 TECHNICAL TERMS

Philosopher May Brodbeck explains why technical terms are indispensable in academic disciplines such as anthropology.

• • •

Some features of the world stand out, almost begging for names. Concepts of clouds, thunder, table, dog, wealth, hunger, color, shape, and the like, name differentiated slices of reality that impinge willy-nilly on all of us. The terms of commonsense name these obtru-

sive daily experiences. Other features of the world have to be cut out as it were. They are discerned only by a more subtle and devious examination of nature, man, and society than is made in everyday life. These more covert aspects of experience are named by . . . concepts [that are] in principle not . . . [those] of common speech [but of academic disciplines].

Source: May Brodbeck, "General Introduction," *Readings in the Philosophy of the Social Sciences.* (New York: Macmillan Company, 1968), pp. 3–5.

**culture** may be used to call attention to the learned patterns of thought and behaviour that are characteristic of either humankind as a whole, the members of a particular society such as the Arunta or the Ojibwa, or the members of a particular subgroup in a society such as Francophone Canadians, Evangelical Christians, and so on. Thus, as these examples show, in order to understand the world from an anthropological perspective, the technical terms of the discipline must be mastered.

# THE VALUE OF ANTHROPOLOGY

## APPLIED ANTHROPOLOGY

The study of anthropology is worthwhile. As already mentioned, anthropological research can be used to help solve practical problems. In Canada, for example, Aboriginal peoples and governments frequently rely on the knowledge that ethnologists acquire in order to help resolve disputes over access to lands and natural resources. Ethnologists have also been called upon to help improve the delivery of health care services in remote locations and to immigrant and disadvantaged groups.

Research undertaken by linguistic anthropologists may likewise be directed towards practical ends, particularly when it comes to designing and implementing programs that are intended to protect minority language rights and promote the use of minority languages. Archaeologists, too, often act in an applied capacity. While some have helped to reconstruct the historic forts and trading posts that are now open to the public, others have conducted salvage operations in order to help retrieve and preserve prehistoric materials that are threatened by industrial development.

Meanwhile, in addition to using their expertise to analyze skeletal remains for theoretical purposes, some biological anthropologists have been asked to use their skills in skeletal analysis to help law enforcement personnel establish the identity of a deceased person and the manner and cause of death; this branch of the discipline is known as forensic anthropology. In many Canadian universities these and other applied topics are dealt with in advanced undergraduate and graduate courses in anthropology. See QQ 1.13.

 **1.13 FORENSIC ANTHROPOLOGY**

The following is a description of a course in forensic anthropology offered in the Department of Archaeology at Simon Fraser University by Associate Professor Mark Skinner.

• • •

The course is designed to familiarize participants with current methods of human skeletal analysis. It is of particular interest to medico-legal investigators, law enforcement personnel, students of prehistoric and fossil human skeletal biology and archaeologists.

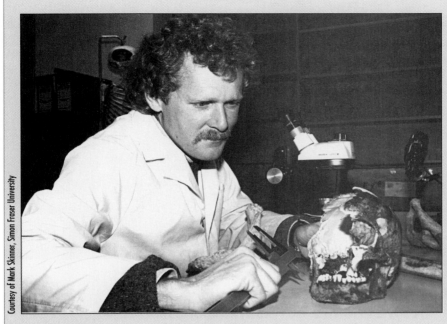

Courtesy of Mark Skinner, Simon Fraser University

▼
Mark Skinner
▲

It will show how the specialist, trained in the analysis of human skeletal material, assists law enforcement agencies in the derivation of maximum personal information from bones, teeth, associated materials and their contextual relationships to establish the identity of the person(s) concerned, cause of death, time elapsed since death and other relevant information.

Source: Department of Archaeology, Simon Fraser University. *Forensic Anthropology*, (Burnaby, British Columbia: Simon Fraser University), 1982.

## COMBATTING RACISM

The study of anthropology is also worthwhile because of the light that it sheds on the ways individuals and groups interact. Studying the biological and cultural characteristics of humankind makes people appreciate that they have much more in common with each other than they usually realize. Consider our biological features. While these certainly vary, it is clear that the biological significance of the variation is minimal. Despite differences in appearance we are all *Homo sapiens* — a species that has been evolving for millions of years.

Anthropologists rely on this fact to speak out against the notion that there is a scientific basis for **racism** — the belief that people whose physical features are allegedly or actually different from our own are biologically inferior and consequently can be mistreated. And this is a lesson that anthropologists are uniquely equipped to teach because of what they have discovered to be true and false about the biological diversity of our kind. See QQ 1.14.

Demonstrators
protesting against
racism in Germany

Bahide Arzlan
50 Jahre
von Neonazis verbrannt
am 22.11.92

Yilliz Arzlan
10 Jahre
von Neonazis
verbrannt am 22.11.92

Ayse Yilmaz
15 Jahre
von Neonazis verbrannt
am 22.11.92

## QQ 1.14 SPEAKING OUT AGAINST RACISM

Morton Fried discusses the importance of anthropology with respect to combatting racism.

• • •

. . . I would argue that every student should be exposed to at least one course in general anthropology . . .

It is only in such a course that students can deal with race and racism. The former, of course, is a concept that applies to biology and requires knowledge of the physical and chemical bases of life . . . And, too, while the phenomenon of racism has very little and sometimes nothing to do with the phenomenon of race, this message cannot be properly delivered until

a minimum understanding of the biological facts has been appreciated. Today no justification is needed for insisting on the broadest treatment of problems of race and racism in classrooms at various levels. The presence of anthropology in a curriculum can be compared only to the acquisition of such basic skills as the control of literacy and the fundamentals of calculation. There may be a time when these basic messages will be so well disseminated at lower levels in the curriculum that there will be no need for them in higher education, but that day is still to come.

Source: Selected excerpt from *The Study of Anthropology* by Morton H. Fried. © 1972 by Thomas Y. Crowell Company, New York. Reprinted by permission of HarperCollins Publishers, Inc.

## COMBATTING ETHNOCENTRISM

Anthropology also encourages people not to make superficial judgements about cultures. In our age of almost instant communication we frequently hear about "strange" and "unusual" customs, often considering them to be either "humorous" or "barbaric" — Inuit seal hunting in the Canadian Arctic is a good example.

In the opinion of many animal-rights and animal-welfare activists, sealing is not only ecologically dangerous, but also an affront to the dignity of both the pursuers and the pursued. Anti-sealing spokespersons from organizations such as Greenpeace, the International Fund for Animal Welfare (IFAW), and I KARE Wildlife Coalition have consequently lobbied long and hard for the abolition of sealing. Although their initial concern was to stop the commercial whitecoat ("baby seal") hunt in the Atlantic region, by the mid-1970s Inuit hunters in the Canadian Arctic had also been identified as culprits. The ensuing campaign against sealing by the aforementioned organizations and others like them was massive and unrelenting, and it produced the desired result. See QQ 1.15 and Table 1.3.

## QQ 1.15 THE CAMPAIGN AGAINST SEALING

The Royal Commission on Seals and the Sealing Industry in Canada looks at one protester's view of the campaign against sealing.

• • •

Public statements by [Steven] Best, formerly of IFAW (Canada) and now with I KARE Wildlife Coalition, provide considerable insight into . . . [the philosophy and operations] of those environmental groups for which the seal issue has been particularly important:

*I feel quite content* [Best told the Commission] *to create an industry based on trying to protect wildlife, as big if not bigger than the ones based on exploiting wildlife . . . .*

*And if Inuit and Newfoundlanders have a right to make a living out of killing seals and make it a part of their culture, do I not have a right to make a living saving them and make that part of my culture?*

It is implicit in Best's remarks that the seal protest movement is an industry in its own right, and that . . . [w]hile the policies of the protest groups are determined by the broader issues of conservation, cruelty or ethics, these groups depend on financial support by the public, and therefore must focus some of their attention on those aspects of these issues that can attract broad public interest, such as the clubbing of whitecoats . . . .

According to Best, this industry will achieve its aims by continuing the controversy and obtaining political power, and some peoples and cultures will suffer from these activities:

*Controversy and continuing controversy* [he said] *is what will protect wildlife . . . . I intend to continue the conflict going in all issues, constantly, because what destroys the markets and what prevents people from taking wildlife is the fact that someone is going to be always around screaming at them for doing it . . . .*

*What it really boils down to is that you want to live one way, I want to live another way. We have a problem, and whoever gets the most amount of political power gets to say what's going to happen . . . .*

Source: Royal Commission on Seals and the Sealing Industry in Canada. *Seals and Sealing in Canada*, Vol. 2. (Ottawa: Department of Fisheries and Oceans, pp. 87–88. Reproduced with the permission of the Minister of Supply and Services Canada, 1993

Yet seen in the context of Inuit culture, the pursuit of seals is anything but a senseless slaughter of defenceless animals undertaken by people who are simply responding to international market demands. In fact, when the European Community banned the importation of harp seal pup skins (whitecoats) and

| Table 1.3 | Major events in the anti-sealing campaign |
| --- | --- |

| 1955 | Observers report on the inhuman killing of harp seals. |
| --- | --- |
| 1964 | The film *Les Phoques* is aired in Europe. |
| 1967 | A 'Save the Seals' campaign is launched. |
| 1971 | Canada imposes harp seal quotas. |
| 1972 | The US passes the Marine Mammal Protection Act. |
| 1977 | Greenpeace's seal policy explicitly encompasses the Inuit. |
| 1983 | The European Community agrees to a binding two-year ban on harp and hooded seal imports. |
| 1984 | Canada forms a Royal Commission to investigate the sealing controversy. |
| 1985 | The European Community renews its boycott. The Home Rule government of Greenland asks Greenpeace to differentiate publicly between Inuit and commercial sealing. Greenpeace declines. |
| 1989 | The EC votes for an indefinite boycott of sealskins. |

Source: George Wenzel, *Animal Rights, Human Rights: Ecology, Economy and Ideology in the Canadian Arctic.* (Toronto: University of Toronto Press, 1991), p. 46. Reprinted by permission of the publisher.

hooded seal pup skins (bluebacks) in 1983, extended the ban in 1985, and then, in 1989, banned the importation of all seal products for an indefinite period, the Inuit were appalled. And this was not just because the income that they earned from sealing dropped by almost 85 percent during this interval; much more was at stake than the accumulation of wealth. As one Inuit hunter told the Royal Commission on Seals and the Sealing Industry in Canada: "To tell you in few minutes about the significance of seals to our people is much like you having to explain to us the significance of agriculture for your civilization."

▼ Inuit eating fresh seal ▲

Lyn Hancock

More than a century of ethnographic research has confirmed this view. While the Inuit certainly hunt seals to provide themselves with food and trade goods, these are not the only or even the primary reasons for the hunt. The fact of the matter is that the Inuit pursue seals in order to reinforce their social ties with one another, and, equally important, to maintain the intimate relationship that has always existed between themselves and seals. This unique relationship is based on the Inuit belief that seals (and other animals) are persons in their own right. Seals make themselves available to Inuit hunters, but only if they are treated with the dignity that the Inuit say such non-human persons rightly deserve. It is these underlying social and ideological principles rather than commerce that governs Inuit sealing (see QQ 1.16). Thus, in order to avoid making a superficial judgement about Inuit sealing, the custom must be examined not in terms of foreign standards, but in the context of the culture in which it occurs.

## 1.16 SUBSISTENCE HUNTING AND INUIT CULTURE

George Wenzel, an anthropologist and geographer who teaches at McGill University, discusses the relationship between Inuit subsistence and Inuit culture.

• • •

Inuit subsistence [hunting] has . . . its own social institutions, principles, and rules. For Inuit, the basis of secure, successful subsistence is the social relatedness of one person to another, rather than individual prowess or special equipment . . . . As a result, subsistence is more than a means of survival. It is a set of culturally established responsibilities, rights and obligations that affect every man, woman and child each day . . . .

Inuit subsistence [also] has its own ideology. It comes in part from the belief that many animals were human in times past . . . but it also arises from the social relatedness of its daily practice. Inuit do not segregate the qualities enjoyed by human beings from those enjoyed by animals. Animals share with

humans a common state of being that includes kinship and family relations, sentience, and intelligence. The rights and obligations that pertain among people extend to other members of the natural world. People, seals, polar bear, birds, and caribou are joined in a single community in which animals give men food and receive acknowledgment and revival.

In all subsistence societies, it is these social and ideological considerations . . . and their meaning, that are primary. For Inuit, the non-material component of life is central. Harvesting involves the relating of society to the environment. This cultural adaptation is integrated in its pattern and goals. All living things are part of one system of reciprocal rights and responsibilities. In this cultural system, harvesting is the point of articulation between hunter and animal, society and environment.

Source: George Wenzel, *Animal Rights, Human Rights: Ecology, Economy and Ideology in the Canadian Arctic.* (Toronto: University of Toronto Press, 1991), pp. 60–61. Reprinted by permission of the publisher.

What is true about Inuit sealing is also true with respect to other patterns of culture. Evaluating such patterns on the basis of standards that prevail elsewhere is known as **ethnocentrism**, and one of the great lessons that anthropology teaches is that ethnocentrism inevitably leads to misunderstanding and intolerance, both of which can have severe adverse impacts on the social solidarity of multicultural societies and on the relationships that exist between nations with different cultural traditions. In order to avoid this pitfall, anthropologists insist that the customs and traditions of the world's cultures must be studied in an objective way. To this end they analyze cultural differences not in emotional terms, but in terms of the more neutral concepts of their discipline.

This is what some anthropologists have in mind when they speak of **cultural relativism**. Others regard cultural relativism as a directive that warns them to avoid making any derogatory value judgements about cultures, except where racist or ethnocentric practices are involved; in such cases anthropologists can and do speak out against oppression. No matter which approach is adopted, an objective study of culture can be truly enlightening. See QQ 1.17.

## QQ 1.17 IS ANTHROPOLOGY WORTH STUDYING?

Franz Boas, one of the founders of modern anthropology, talks about the benefits of the discipline.

• • •

Anthropology is often considered a collection of curious facts, telling about the particular appearance of exotic people and describing their strange customs and beliefs. It is looked upon as an entertaining diversion, apparently without any bearing upon the conduct of life of civilized communities.

This opinion is mistaken . . . . [A] clear understanding of the principles of anthropology illuminates the social processes of our own times and may show us, if we are ready to listen to its teachings, what to do and what to avoid.

Source: Franz Boas, *Anthropology and Modern Life.* (New York: W. W. Norton & Company, Inc., 1962), p. 11.

Franz Boas (1858–1942)

The Bettmann Archive

## KEY TERMS AND CONCEPTS

ANTHROPOLOGY

BIOLOGICAL ANTHROPOLOGY

PALAEOANTHROPOLOGISTS (*pay'-lee-oh + anthropologists*)

FOSSIL RECORD

PRIMATES (*pry'-mayts*)

PROSIMIANS (*proh-sim'-ee-anz*)

ANTHROPOIDS (*anth'-roh-poydz*)

PRIMATOLOGISTS

POPULATION BIOLOGISTS

SPECIES

CULTURAL ANTHROPOLOGY

ARCHAEOLOGISTS (*ahr'-kee-ahl'-uh-jists*)

PREHISTORIC

LINGUISTIC ANTHROPOLOGISTS

LANGUAGE

ETHNOLOGISTS (*eth-nawl'-uh-jists*)

ETHNOGRAPHIC RECORD
(*eth'-naw-gra-fik* + *record*)

ETHNOGRAPHIES (*eth-naw'-gra-feez*)

ETHNOHISTORICAL RESEARCH

ETHNOGRAPHIC RESEARCH

COMPARATIVE METHOD

FIELDWORK

CULTURE SHOCK

HOLISTIC (*hoh-lis'-tik*)

TECHNICAL VOCABULARY

EVOLUTION

CULTURE

RACISM

ETHNOCENTRISM

CULTURAL RELATIVISM

# SELECTED READINGS

## INTRODUCTION

Darnell, R., ed. *Readings in the History of Anthropology*. New York: Harper & Row, 1974.

Preston, R. and Tremblay, M-A. "Anthropology." *The Canadian Encyclopedia* (revised edition), Vol. 1 (1988), pp. 80–83.

## BIOLOGICAL ANTHROPOLOGY

Banton, M. *Racial Consciousness*. New York: Longman, 1988.

Boaz, N. "History of American Paleoanthropological Research on Early Hominidae, 1925–1980." *American Journal of Physical Anthropology*, Vol. 56 (1980), pp. 397–405.

Cole, S. *Leakey's Luck*. New York: Harcourt Brace Jovanovich, 1975.

Gilmore, H. "From Radcliffe-Brown to Sociobiology: Some Aspects of the Rise of Primatology within Physical Anthropology." *American Journal of Physical Anthropology*, Vol. 56 (1981), pp. 387–392.

Milner, R. and Prost, J. "The Significance of Primate Behavior for Anthropology." In Korn, N., and Thompson, F., eds. *Human Evolution: Readings in Physical Anthropology* (second edition). New York: Holt, Rinehart and Winston, 1967, pp. 125–136.

Spencer, F., ed. *A History of American Physical Anthropology, 1930–1980*. New York: Academic Press, 1982.

Szathmary, E. "Anthropology, Physical." *The Canadian Encyclopedia* (revised edition), Vol. 1 (1988), pp. 83–84.

## CULTURAL ANTHROPOLOGY

Bernard, H. *Research Methods in Cultural Anthropology*. Newbury Park, California: Sage, 1988.

Darnell, R. "Anthropology, Linguistic." *The Canadian Encyclopedia* (revised edition), Vol. 1 (1988), p. 83.

Deetz, J. *Invitation to Archaeology*. New York: Natural History Press, 1967.

Fladmark, *A Guide to Basic Archaeological Field Procedures*. Department of Archaeology, Simon Fraser University, Publication Number 4, 1978.

Forbis, R. and Noble, W. "Archaeology." *The Canadian Encyclopedia* (revised edition), Vol. 1 (1988), pp. 91–94.

Foster, Michael. *Indigenous Languages in Canada*. Ottawa: Commissioner of Official Languages, 1982.

Greenberg, J. *Anthropological Linguistics: An Introduction*. New York: Random House, 1968.

Hammersley, M. and Atkinson, P. *Ethnography: Principles in Practice*. London: Tavistock, 1983.

Jochim, M. "Archaeology as Long-Term Ethnography." *American Anthropologist*, Vol. 93 (1991), pp. 308–321.

Potter, P., Jr. "The 'What' and 'Why' of Public Relations for Archaeology: A Postscript to Decicco's Public Relations Primer." *American Antiquity*, Vol. 55 (1990), pp. 608–613.

Sapir, E. *Language*. New York: Harcourt Brace, 1921.

## RELATIONSHIPS WITH OTHER DISCIPLINES

Castillos, J. *A Reappraisal of the Published Evidence on Egyptian Predynastic and Early Dynastic Cemeteries*. Toronto: Society for the Study of Egyptian Antiquities, SSEA Studies, Number 1, 1982.

Johnson, N. "Anthropology and the Humanities: A Reconsideration." *Anthropology and Humanism Quarterly*, Vol. 14 (1989), pp. 82–89.

Melbye, F. "Human Remains from a Roman Period Tomb in the Dakhleh Oasis, Egypt." *Journal of the Society for the Study of Egyptian Antiquities*, Vol. 13 (1983), pp. 193–201.

Molto, J. "Dakhleh Oasis Project: Human Skeletal Remains from the Dakhleh Oasis, Egypt." *Journal of the Society for the Study of Egyptian Antiquities*, Vol. 16 (1986), pp. 119–127.

## THE UNIQUENESS OF ANTHROPOLOGY

Berreman, G. *Behind Many Masks: Ethnography and Impression Management in a Himalayan Village*. Ithaca, New York: Society for Applied Anthropology, Monograph Number 4, 1962.

Cole, D. "The Origins of Canadian Anthropology, 1850–1910." *Journal of Canadian Studies*, 1973, Vol. 7 (1973), pp. 33–45.

Darnell, R. "History of Anthropology in Historical Perspective." *Annual Review of Anthropology*, Vol. 6 (1977), pp. 399–417.

DeVita, P., ed. *The Naked Anthropologist: Tales from Around the World*. Belmont, California, Wadsworth, 1992.

Gold, G. and Tremblay, M-A. "After the Quiet Revolution: Quebec Anthropology and the Study of Quebec." *Ethnos*, Vols. 1 and 2 (1982), pp. 103–132.

Goodall, J. *Through a Window: My Thirty Years with the Chimpanzees of Gombe*. Boston: Houghton-Mifflin, 1990.

Holy, L., ed. *Comparative Anthropology*. New York: Basil Blackwell, 1987.

Hunter, D. and Whitten, P. *Encyclopedia of Anthropology*. New York: Harper & Row, 1976.

Kuper, A. "Anthropologists and the History of Anthropology." *Critique of Anthropology*, Vol. 11 (1991), pp. 125–142.

Maybury-Lewis, D. *The Savage and the Innocent* (second edition). Boston: Beacon Press, 1988.

Nakane, C. "Becoming an Anthropologist." In Richter, D., ed. *Women Scientists: The Road to Liberation*. London: Macmillan, 1982, pp. 45–60.

Rodman, W. and Rodman M. "To Die on Ambae: On the Possibility of Doing Fieldwork Forever." *Anthropologica*, Vol. 31 (1989), pp. 25–43.

Seymour-Smith, C. *Dictionary of Anthropology*. Boston: G. K. Hall, 1986.

Stevenson, J. *Dictionary of Concepts in Physical Anthropology*. New York: Greenwood Press, 1991.

Tremblay, M-A. "Anthropological Research and Intervention at Laval University Since the Sixties." *Western Canadian Anthropologist*, Vol. 15 (1988), pp. 3–28.

## THE VALUE OF ANTHROPOLOGY

Barzun, J. *Race: A Study in Superstition*. New York: Harper & Row, 1965.

Bidney, D., ed. *The Concept of Freedom in Anthropology*. The Hague, Mouton, 1963.

Boas, F. *Race, Language and Culture*. New York: Macmillan, 1940.

Burley, D. *Nipawin Reservoir Heritage Study: Resource Evaluations, Impacts and Mitigation* (Vol. 1). Saskatoon: Saskatchewan Research Council, 1982.

Chambers, E. *Applied Anthropology: A Practical Guide*. Englewood Cliffs, New Jersey: Prentice-Hall, 1985.

Gould, S. *The Mismeasure of Man*. New York: Norton, 1981.

Jarvie, I. "Recent Work in the History of Anthropology and its Historiographic Problems." *Philosophy of Social Science*, Vol. 19 (1989), pp. 345–375.

Salisbury, R. "Anthropology, Applied." *The Canadian Encyclopedia* (revised edition). Vol. 1 (1988), p. 83.

Skinner, M. and Lazenby, R. *Found! Human Remains: A Field Manual for the Recovery of the Recent Human Skeleton*. Burnaby, British Columbia: Archaeology Press (Department of Archaeology, Simon Fraser University), 1983.

Stocking, G. *Race, Culture, and Evolution: Essays in the History of Anthropology*. New York: Free Press, 1968.

Ubelaker, D. and Scammell, H. *Bones: A Forensic Detective's Casebook*. New York: HarperCollins, 1992.

United Nations Educational, Scientific and Cultural Organization. *The Race Concept: Results of an Inquiry*. Paris: UNESCO, 1952.

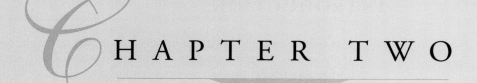

# CHAPTER TWO

## OUR PLACE IN NATURE

# INTRODUCTION

It is only during the last three hundred years that scientists have come to appreciate that humans are part of the natural world. Before then there was a widespread belief that our species occupied a special place in the universe; humankind was thought to be a unique form of life.

At the time, those who promoted the idea did not know that our **anatomy** (the organs of the body) and our **physiology** (the operation of the organs) are similar to those of other animals, particularly prosimians, monkeys, and apes. Nor did they know that, like humans, many other primate species have fascinating social lives. However, between the start of the fifteenth century and the end of the seventeenth century opinions about humankind's uniqueness were revised. Major improvements in maritime navigation and in the quality of ships made it possible for explorers to travel the world. On their voyages they discovered hundreds of new varieties of plants and animals, and this set the stage for scientists to unravel the mystery of our place in nature.

# EDWARD TYSON (1650–1708)

One of the first to tackle the problem was an English physician named Edward Tyson. In 1698 Tyson acquired the corpse of an immature male chimpanzee that had died as a result of a badly infected jaw shortly after it had been brought by ship from Angola to London. Tyson, having briefly seen the chimpanzee alive, took the cadaver home where he dissected it and compared its anatomical features with those of monkeys and humans. Based on this pioneering research Tyson concluded that his specimen, which he called a "Pygmie," shared some traits with monkeys and others with humans, a conclusion that later research confirmed. See QQ 2.1.

## 2.1 TYSON'S "PYGMIE"

**Vernon Reynolds summarizes Tyson's research.**

• • •

Tyson compared his specimen, point by point, at each stage of the anatomical dissection, with monkeys on the one hand, and with man on the other. In this way he drew up two lists. The first consisted of forty-seven characters in which his "Pygmie" resembled man more closely than it resembled any monkey, and the second consisted of thirty-four characters in which the "Pygmie" resembled a monkey more closely than it resembled man. He concluded that the "Pygmie" was an intermediate type, wholly distinct from both man and the monkeys. In his own words: "In this *Chain* of the *Creation,* as an intermediate link between an *Ape* [that is, monkey] and a *Man,* I would place our *Pygmie.*"

Source: Vernon Reynolds, *The Apes: The Gorilla, Chimpanzee, Orangutan, and Gibbon — Their History and Their World* (New York: Harper & Row, Publishers, 1967), p. 47.

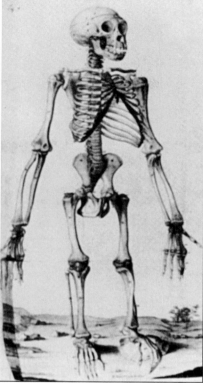

Tyson's "Pygmie" and its skeleton

Tyson, however, was confused about the identity of his specimen. He suggested that his "Pygmie" represented a fabled race of tiny people, in his words, "a puny race of mankind." He was mistaken; the "puny race" he had in mind were the Pygmies who live in and on the edge of the tropical rain forest in Zaïre, in central Africa. At the time of Tyson's research Pygmies had not yet been contacted by Europeans; they were known only through garbled, second-hand accounts. In ensuing years Tyson's opinion that chimpanzees were diminutive humans was roundly criticized by anatomists who were better informed. But it was not until two hundred years after his initial pronouncement, when Europeans came into direct contact with Pygmy people, that it became clear that Tyson's specimen was not a Pygmy but an ape.

# LINNAEUS (1707–1778)

Another scientist who questioned our relationship to other species was the great Swedish naturalist, Carl von Linné. Better known as Linnaeus, his research was more comprehensive than Tyson's, so much so that he now occu-

Linnaeus
(1707–1778)

The Bethmann Archive

pies a special place in the history of science. Linnaeus looked at the anatomy and the physiology of all living things. Then, on the basis of what he had learned about their features, Linnaeus assigned them a position in a **taxonomy** — a system of biological classification in which organisms are arranged first into large groups and then into increasingly smaller units on the basis of a rigorous and scrupulous analysis of the similarities and differences in their biological traits. See QQ 2.2.

 **2.2 LINNAEUS' SCHEDULE**

In a letter written in 1761, Linnaeus explained how he occupied himself while he organized his information.

● ● ●

I lecture every day for an hour in public and afterwards give private instruction to a number of pupils. Then comes an hour with some Danes and two Russians. Having thus talked for five hours before lunch, in the afternoon I correct work, prepare my manuscript for the printers and write letters to my botanical friends, visit the garden and deal with people who want to consult me, and look after my little property — with the result that often I hardly have a moment to eat . . . . If the Almighty spares me for a few more years I will free the aging horse from the yoke so that it does not break down and finally become ridiculous. If I can then succeed in growing a few rare plants in my garden I shall be happy.

Source: Linnaeus quoted in Wilfrid Blunt and William T. Stern. *The Compleat Naturalist: A Life of Linnaeus.* (London: William Collins, Sons & Company Limited, 1971), pp. 167–168.

In his book *Systema Naturae*, originally published in 1735 and revised many times thereafter, Linnaeus presented his findings. These paved the way for subsequent generations of scientists to develop an increasingly sophisticated taxonomy of the world's species, which is now based not only on similarities and differences in anatomy and physiology, but also on similarities and differences in genetic make-up and patterns of growth. Susceptibility to disease is also often taken into account: organisms that suffer from the same infectious or inherited diseases are more closely related than those that are prone to different diseases.

Taking the abovementioned factors into consideration, taxonomists currently divide life as we know it into large groups called kingdoms — the plant kingdom and the animal kingdom being familiar examples — and then proceed to subdivide each kingdom into successively smaller units whose principal names, in descending order, are phylum, class, order, family, genus, and species. By adding prefixes such as "sub," "infra," and "super" to these designations, the system becomes all the more refined. Table 2.1 shows how humankind is classified in an abbreviated version of a modern taxonomy of animal life, which is based on Linnaeus' original plan.

| TABLE 2.1 | The place of humankind in an abbreviated version of a modern taxonomy of animal life | | | |
|-----------|---------|-------------|-----------|----------|
| KINGDOM | PHYLUM | SUBPHYLUM | CLASS | ORDER |
| Animalia | Chordata | Vertebrata | Mammalia | Primates |

# HUMANS AND OTHER ANIMALS

Careful consideration of Table 2.1 reveals that humans have a great deal in common with other species. Since we are members of the **Kingdom Animalia** we are ultimately related to all organisms that possess a nervous system, whose nourishment is acquired by ingesting their food, and whose anatomy and physiology make them capable of moving about — in other words, to all forms of animal life. Within this broad category we are next placed into a special, less inclusive phylum of animals known as **chordata,** whose distinguishing feature is that their bodies are supported, at some stage in their lives, by a **notochord** — a flexible rod of cartilage that runs lengthwise along the back from the rump to the neck where it leads into a hollow nerve cord that is connected to the brain. For greater precision we are then assigned membership in the subphylum **vertebrata**, which includes only those chordates that have a true spine or backbone. Thereafter we are placed into a special class of vertebrates known as **mammalia**, whose members have hair, are warm blooded, and suckle their young. Our closest taxonomic relatives, however, are the primates, the mammalian order that includes prosimians, monkeys, and apes.

# THE PRIMATE PATTERN

## SOCIAL CONSIDERATIONS

Although primates do not possess a single conspicuous specialization, such as the trunks of Proboscidia (elephants), they tend to differ from other mammals in a number of important respects. One is their social lives. For instance, unlike

A female mountain gorilla and her offspring

Animals Animals © Bruce Davidson, #608105-M

most mammals, primates are not characterized by a breeding season that occurs once or twice each year. Instead, many species are capable of engaging in sexual intercourse throughout the year, and this usually encourages more sustained social contact between the sexes. In addition, because primate lifespans are comparatively long, social relationships in many species can and do become more intimate over time. Primate females also bear fewer offspring at birth than most mammals, and because the offspring are born in an immature state and often take years to mature, almost all primate mothers care for their young for much longer periods of time than other mammalian mothers. In fact, the mother-offspring relationship is one of the most important features of primate social life. It is by controlling access to her newborn offspring and by coordinating the behaviour of her infant with her own that a primate mother helps to ensure that her offspring will survive, and it is primarily by observing and interacting with its mother that an immature primate learns how to adapt to its animate and inanimate surroundings as it matures. See QQ 2.3.

## QQ 2.3 PRIMATE MOTHERS AND THEIR OFFSPRING

**Carol Berman comments on mother-infant relationships among nonhuman primates.**

. . .

In most species of nonhuman primate, an infant's relationship with its mother is its first, most intense, and in many cases, most enduring social relationship. Among cercopithecines [Old World monkeys], mothers and infants initially spend nearly all of their time in ventro-ventral [abdomen-to-abdomen] contact and on the nipple . . . . At first the mother takes primary responsibility for maintaining contact and proximity to the infant, frequently initiating contact, rarely refusing the infant's demands for contact, and frequently restraining it from leaving. Gradually, the pair spend more time out of contact and at a distance. The infant grad-

ually takes on primary responsibility for maintaining contact and proximity and is rejected more frequently by the mother . . . [who] may use rejections to assess an infant's readiness for independence . . . .

Mothers also regulate their infants' interaction with group companions . . . . They can encourage interaction by leading or carrying the infant near certain individuals and by discouraging its clinging. They can [also] discourage interaction by restraining the infant, by carrying it away from certain individuals, or by directly threatening, restraining, or distracting the infant's companion.

Source: Carol M. Berman, "Variation in Mother-Infant Relationships: Traditional and Nontraditional Factors," in Meredith F. Small, ed. *Female Primates: Studies by Women Primatologists.* (New York: Alan R. Liss, Inc., 1984), p. 18.

# BIOLOGICAL CONSIDERATIONS

The anatomy and the physiology of primates likewise help to distinguish the order. For example, in contrast to most other mammals, primates generally have a small snout and a poor sense of smell. They also have a comparatively small number of teeth that are best suited to a diet of fruit, leaves, flowers, and insects. But while primates lack the large snouts and fangs that most mammals possess, their senses of sight and of touch are especially keen, and their brains are more fully developed.

Consider the sense of sight. All living primates have eyes that are protected by a bony ring called the **postorbital bar**, most can see in colour, and many are capable of **stereoscopic vision** or depth perception. This ability to see objects in three dimensions is made possible partly because primate eyes tend to face forward rather than to the sides, and partly because of the unique way in which the primate brain receives and interprets visual information. Because the eyes tend to face forward, the visual fields overlap. And since the optic fibres from each eye are connected to both hemispheres of the brain rather than only to one hemisphere, the visual information that the brain receives can be processed in a way that produces a three-dimensional image. See Figure 2.1.

**FIGURE 2.1** Stereoscopic vision

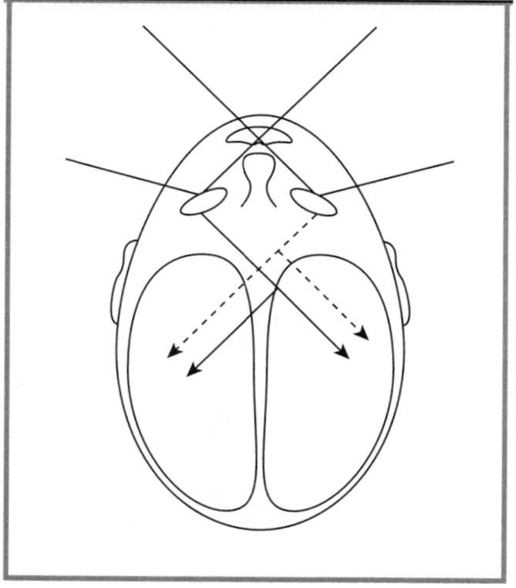

The primates' sense of touch complements their visual acuity. Instead of immobile hoofs or paws, primates have versatile hands and feet that are usually equipped with highly mobile and sensitive **digits** (fingers and toes). And instead of terminating in curved claws, their digits are almost always equipped with flat nails. Many species also have an **opposable thumb** that can be placed against the palm of the hand, and an **opposable big toe** that can be placed against the sole of the foot. Taken together, these features enable the members of most primate species to grip objects either with power or precision depending upon the task. See QQ 2.4.

The primate brain also merits attention. One of the most outstanding features of the primates is that they are equipped with a relatively large and heavy brain in relation to the overall size of the body (see Figure 2.2). Another is that the cerebral hemispheres, which coordinate mental functions, are exceptionally well developed. Although the behavioural implica-

# 2.4 POWER AND PRECISION

John Napier explains how human beings use their hands to grip objects with power and precision.

• • •

In the hand at rest — with the fingers slightly curled, the thumb lying in the plane of the index finger, the poise of the whole reflecting the balanced tension of opposing groups of muscles — one can see something of its potential capacity. From the position of rest, with a minimum of physical effort, the hand can assume either of its two prehensile working postures. The two postures are demonstrated in sequence by the employment of a screw driver to remove a screw solidly embedded in a block of wood . . . . The hand first grips the tool between flexed fingers and the palm with the thumb reinforcing the pressure of the fingers; this is the "power grip." As the screw comes loose, the hand grasps the tool between one or more fingers and the thumb, with the pulps, or inner surfaces, of the finger and thumb tips fully opposed to one another; this is the "precision grip." Invariably it is the nature of the task to be performed, and not the shape of the tool or object grasped, that dictates which posture is employed. The power grip is the grip of choice when the full strength of the hand must be applied and the need for precision is subordinate; the precision grip comes into play when the need for power is secondary to the demand for fine control.

Source: From John Napier, "The Evolution of the Hand," *Scientific American*, Vol. 207, No. 6, pp. 57–58. Copyright © 1962 by Scientific American Inc.

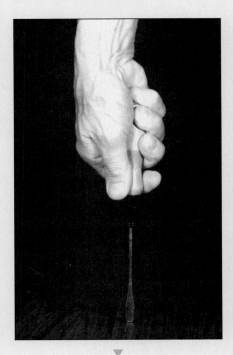

Power grip

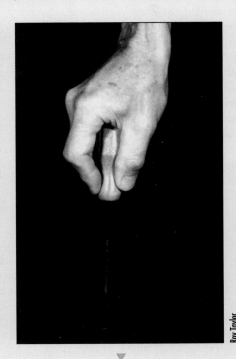

Precision grip

Roy Taylor

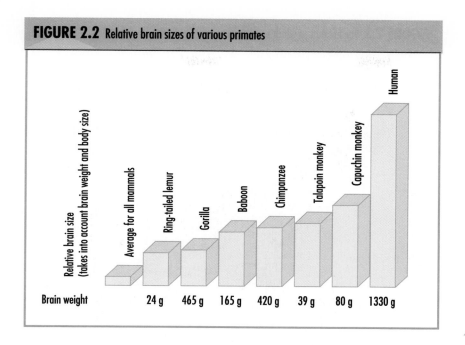

**FIGURE 2.2** Relative brain sizes of various primates

Relative brain size
(takes into account brain weight and body size)

Average for all mammals

Ring-tailed lemur

Gorilla

Baboon

Chimpanzee

Talapoin monkey

Capuchin monkey

Human

Brain weight     24 g    465 g    165 g    420 g    39 g    80 g    1330 g

tions of these traits have yet to be fully explored, it is widely believed that they enhance conscious thought and make it possible for primates, more than any other mammals, to learn and to accumulate knowledge.

# PRIMATE HABITATS

The abovementioned features represent the **primate pattern**. This amalgam of social and biological traits on the whole, is especially well suited to **arboreal habitats** — a variable group of tropical forest environments in which survival is enhanced by hands and feet that can grasp branches and limbs; acute vision; good hand-eye coordination; and a brain that facilitates the ability to learn, store, and act on information that is acquired from others. Although a few primate species live elsewhere than in arboreal habitats — some baboons, for example, live on the tropical grasslands, or **savanna** of Africa — it is in arboreal settings that the vast majority of primates are currently found, mostly in the tropical forests of Asia, Africa, and South America. There some reside in long-established **primary rain forests** with fully mature trees, dense canopies, and dark understoreys; some in regenerating **secondary rain forests** that contain maturing trees, discontinuous canopies, and green understoreys; others in **woodland forests** bordered by tropical grasslands that feature deciduous trees interspersed with grasses and bushes; and still others in relatively luxuriant **gallery forests** that line the banks of waterways (see Figure 2.3). Rather than

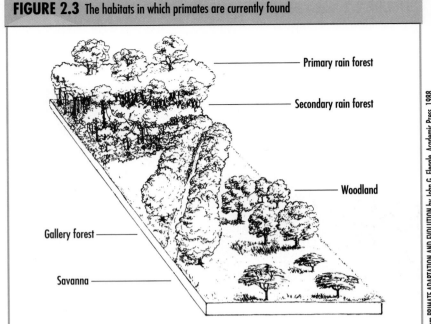

**FIGURE 2.3** The habitats in which primates are currently found

Primary rain forest

Secondary rain forest

Woodland

Gallery forest

Savanna

From PRIMATE ADAPTATION AND EVOLUTION by John G. Fleagle, Academic Press, 1988

range freely throughout these ever-changing tropical forest environments, however, the primate species that are currently found there occupy specific ecological niches to which they have become adapted during the course of their biological histories. The territory they generally travel within and defend is called a **home range**. See QQ 2.5.

## 2.5 PRIMATE HABITATS

John Fleagle explains how occupying specific ecological niches enables primates to better cope with their tropical forest environments.

• • •

Primates live in a complex environment with many constantly changing variables. One way groups of primates deal with this complexity is to restrict their activities to a limited area of forest that they know well. Thus

we find that primates are very conscious of real estate. In contrast with many birds or other mammals, most primates spend their days, years, and often their entire lives in a single, relatively small patch of forest. To exploit this patch effectively, they must know many things about it — the different food trees and their seasonal cycles, the best pathways for moving, the best water sources, and the safest places for sleeping.

Source: John G. Fleagle, *Primate Adaptation and Evolution* (San Diego: Academic Press, Inc., 1988), p. 51.

Of course, if we look closer at the more than 200 recognized species of living primates, along with differences in their habitats, we see other differences among them, and these are reflected in the way that the members of the order are classified.

As a rule, taxonomists divide the primates into two suborders: **Prosimii** — the prosimians or lower primates; and **Anthropoidea** — the higher primates or monkeys, apes, and humans. Thereafter, as the contents of Table 2.2 indicate, increasingly finer distinctions are made.

| **TABLE 2.2** A simplified classification of living primates | | | | | |
|---|---|---|---|---|---|
| ORDER | SUBORDER | INFRAORDER | SUPERFAMILY | FAMILY | SUBFAMILY |
| Primates | Prosimii (prosimians) | Lemuriformes Tarsiiformes | | | |
| | Anthropoidea (monkeys, apes, and humans) | Platyrrhini (New World monkeys) | | Callitrichidae Cebidae | |
| | | Catarrhini (Old World monkeys, apes, and humans) | Cercopithecoidea (Old World monkeys) | | Cercopithecinae Colobinae |
| | | | Hominoidea (apes and humans) | Hylobatidae (gibbons and siamangs) Pongidae (orangutans, gorillas, and chimpanzees) Hominidae (humans) | |

# PROSIMII

Many of us have seen lemurs, indriids, lorises, and tarsiers in zoos — these are prosimians. Tree shrews, which are members of the family **Tupaiidae**, were likewise once classified as prosimians, but this is no longer the case. Although the brain, limbs, and teeth of these fascinating squirrel look-alikes from Southeast Asia and adjacent islands (see Figure 2.4) resemble those of the prosimians, they have claws on all of their digits, a relatively long snout, and a comparatively good sense of smell. In addition, because their eyes are located on either side of the snout and face sideways rather than forward, tree shrews lack stereoscopic vision. They also live close to the ground in thickets and shrubs rather than in trees. For these reasons, the reassignment of the Tupaiidae into a non-primate taxonomic category makes good sense. See QQ 2.6.

**FIGURE 2.4** The present geographic range of Tupaiidae

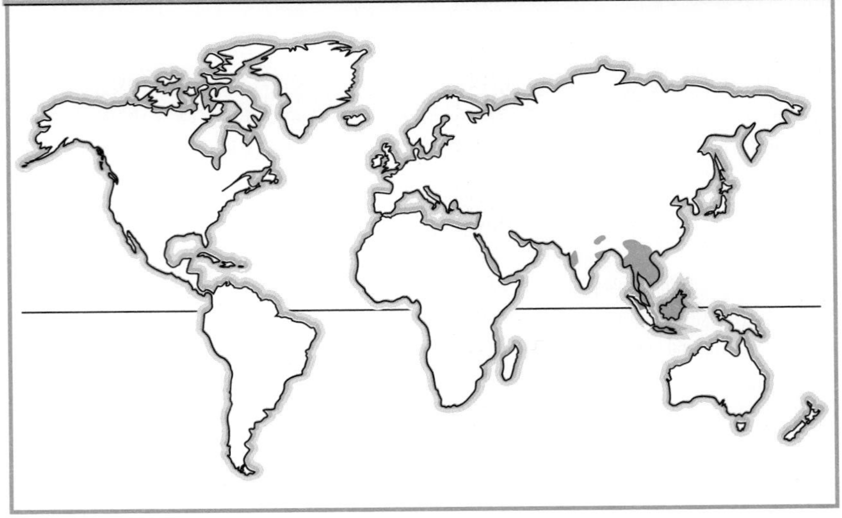

## QQ 2.6 TREE SHREWS

W. Patrick Luckett looks at the modern taxonomic position of tree shrews.

• • •

During the past 200 years, studies of tupaiid systematic relationships have resulted in several alternative hypotheses for their possible affinities with . . . [many orders including] primates . . . . In most cases, these suggested relationships have been based on concepts of overall (phyletic) similarity . . . . [But careful] analysis indicates that . . . [tree shrews] share few, if any, uniquely derived character states with primates . . . .

This apparently remote phyletic position provides support for Butler's . . . suggestion that Tupaiidae should be allocated to a separate order Scandentia. Alternative taxonomic schemes which would be consistent with current hypotheses . . . [are the inclusion of Tupaiidae within . . . [the] order Insectivora [which includes living shrews, moles, hedgehogs, and their extinct ancestors], or else to list them simply as [placental mammals with an uncertain affinity].

Source: W. Patrick Luckett, "The Suggested Evolutionary Relationships and Classification of Tree Shrews," *Comparative Biology and Evolutionary Relationships of Tree Shrews* (New York: Plenum Press, 1980), p. 28.

▼

Common

tree shrew

▲

## LEMURIFORMES

The species that represent the **Lemuriformes** — the prosimian infraorder that includes the lemurs and the indriids of Madagascar and the lorises of Africa and Asia (see Figure 2.5) — are much less ambiguously primates than are tree shrews. Although they too are small furry animals, their inclusion in our order makes good taxonomic sense.

Like all primates, Lemuriformes have agile limbs, a well-developed sense of sight, and a comparatively large brain. They also have a relatively small number of teeth, oftentimes arranged in a 3:1:3:3/3:1:3:3 pattern; that is, 40 teeth in all, including 3 incisors, 1 canine, 3 premolars, and 3 molars in each upper quadrant of the mouth, and 3 incisors, 1 canine, 3 premolars, and 3 molars in each lower quadrant. And even though Lemuriformes are unable to manipulate objects with precision because their fingers work as a unit, they can grasp

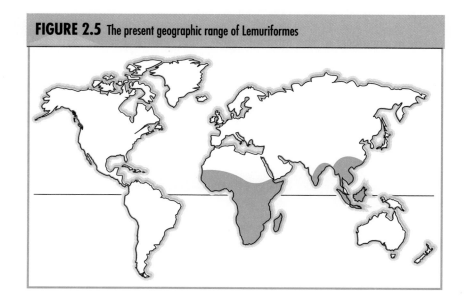

**FIGURE 2.5** The present geographic range of Lemuriformes

objects with power since they have nails on all of their digits except for the second toes on the hind feet; these toes are equipped with claws that are used for scratching and grooming. Also like most primates, Lemuriformes live in arboreal habitats where they feed chiefly on fruit, leaves, flowers, and insects.

Lemuriformes are all the more interesting to primatologists because of the differences among them. For example, while some are **arboreal quadrupeds** that move slowly on all fours up the trunks of trees and then scurry likewise but more rapidly along branches, others are **vertical clingers and leapers** that hop short distances and make highly precise leaps from time to time between distantly spaced branches or trees (see Figure 2.6). Lemuriformes also vary in size. Whereas one of the smallest, the Brown lesser mouse lemur of Madagascar (*Microcebus rufus*), weighs about 60 grams and has a combined head and body length of about 12 centimetres, the Indri (*Indri indri*) from the same island is at the opposite end of the spectrum; it has a combined head and body length that ranges between 57 centimetres and 70 centimetres and can weigh 10 kilograms or more.

**FIGURE 2.6** An artist's rendering of vertical clinging and leaping

The members of the infraorder also differ with respect to their daily routines: **diurnal** ones move about by day, **nocturnal** ones are active at night, and **crepuscular** species are active at dawn and again at twilight. And while some Lemuriformes spend most of their time alone, others live in groups of up to 20 individuals composed of several adults and their immature offspring. These differences enable Lemuriformes to survive and breed successfully in the tropical forest in multitudinous ways. See QQ 2.7.

## QQ 2.7 THE SOCIAL ORGANIZATION OF LEMURS

G. Mitchell compares the social structure of the Mongoose lemur (*Lemur mongoz*) and the Ruffed lemur (*Varecia vareigatus*).

• • •

The social structures [of lemurs] are quite varied . . . . [Take the case of] *Lemur mongoz* [the mongoose lemur from northwestern Madagascar] and *Varecia vareigatus* . . . [the ruffed lemur from the east coast of

Mongoose lemur     Ruffed lemur

Animals Animals © Michael Dick, #647073M

© Zoological Society of San Diego 1993

Madagascar, both of which are crepuscular species].

The *Lemur mongoz* species spend ten times as much time in direct contact as do *Varecia vareigatus* . . . . The contact includes higher levels of grooming [in which one animal picks dirt and parasites from another's hair with its hands and teeth to initiate social contact] and huddling both of which are almost absent in the ruffed lemur. Ruffed lemurs are solitary and come together only for the breeding season whereas *Lemur mongoz* . . . are apparently monogamous.

Source: G. Mitchell, *Behavioural Sex Differences in Nonhuman Primates* (New York: Van Nostrand Reinhold Company, 1979), pp. 220–221.

## TARSIIFORMES

The **Tarsiiformes**, too, are unambiguously primates. This prosimian infraorder is represented exclusively by four species of tarsiers that live in the primary and secondary tropical rain forests of Southeast Asia. See Figure 2.7.

Compared to other primates Tarsiiformes possess two extraordinary features. One is their unusually large eyes, which they rely on to spot the insects that they hunt at dusk; each eye is larger than the tarsiers' brain. In fact, the

**FIGURE 2.7** The present Geographic range of Tarsiiformes

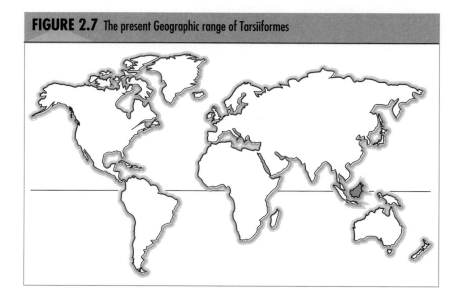

eyes are so massive that it is easier for tarsiers to rotate their heads than to move their eyes to see to the sides. The other special feature of Tarsiiformes is their exceptionally long ankle or **tarsal bones**. These bones, from which the name tarsiers was derived, provide these relatively small primates with so much leverage when they jump that they can cover a distance of up to two metres when they leap from branch to branch or tree to tree, which they typically do to escape from predators or to capture insect prey.

An adult tarsier

© Zoological Society of San Diego 1993

Given these specialized features it may be hard to believe that we are related to Tarsiiformes. Yet we are. If anything, we are even more closely related to Tarsiiformes than we are to Lemuriformes. Based on the structure of the tarsiers' brain, nose, eyes, and reproductive system, some taxonomists have suggested that Tarsiiformes should be classified as anthropoids rather than as prosimians. In either event, despite differences in our appearance, our anatomy and physiology are based on the same overall plan, one that stresses the importance of the senses of sight and of touch over the sense of smell. It is these senses, more than any others, that have enabled the small-bodied Tarsiiformes to survive in their arboreal habitat. See QQ 2.8.

## QQ 2.8 PROSIMIAN ADAPTATION

Sarel Emierl and Irven DeVore discuss the adaptive significance of the eyes and snouts of prosimians.

• • •

To watch a prosimian is, in a sense, to be transported back to the world of 50 million years ago. For it requires only a little imagination to perceive in these furry, bushy-tailed, transitional little animals the early forms of adaptation to arboreal life. Consider . . . their eyes and their snouts, the vehicles of sight and smell. To a considerable degree, these two senses are complementary. Both are methods of obtaining information, and the more use an animal makes of one, the less it will depend on the other. The balance is dictated by the way it exploits its environment. To an animal that lives on the ground and is active at night, smell can be extremely useful: it can identify objects not by looking at them but by sniffing them. To one that lives away from the ground and is active by day, the value of the sense of smell is lessened and that of vision is increased: this is seen particularly well in birds, whose sense of smell is as poor as their vision is keen. Even in the trees, the sense of smell is far less valuable than it is on the ground, and vision is far more; for it helps its possessor to avoid possibly fatal falls and also to identify food amid the rich and colorful foliage.

Source: From *Life Nature Library: The Primates* by Sarel Emierl, Irven DeVore and the Editors of Time-Life Books © 1965 Time-Life Books Inc., New York, p. 12.

# MONKEYS

Like most prosimians, virtually all monkeys have tails. Apes and humans do not, and in this sense monkeys are more like prosimians than they are like apes and humans. But since monkeys have more features in common with apes and humans than they do with prosimians, taxonomists group monkeys with the higher primates or anthropoids.

Like other anthropoids, virtually all monkeys have a relatively flat face that is capable of a variety of expressions, a relatively small snout, and a poor sense of smell. They also have eyes that face directly forward, opposable thumbs and big toes, and flexible digits that can grasp objects with power and precision. In addition, monkeys have relatively large and well-developed brains. But since not all monkeys are the same, taxonomists distinguish between them, especially between those in the New World and those in the Old World.

## NEW WORLD MONKEYS (INFRAORDER PLATYRRHINI)

New World monkeys, which are found in the equatorial rain forest and woodland savanna stretching from Mexico down through South America to Paraguay, northern Argentina, and southern Brazil, belong to the infraorder **Platyrrhini** (see Figure 2.8). The platyrrhines are named after the shape of their nose. Almost all New World monkeys have a **platyrrhine nose**, which features widely separated nostrils that open to the side. Upon closer examination, however, two families can be distinguished. These are the **Callitrichidae**, which are otherwise referred to as marmosets and tamarins, and the **Cebidae**, which includes squirrel monkeys, capuchins, howler monkeys, woolly monkeys, spider monkeys, and woolly spider monkeys.

Although the members of both families are almost completely arboreal, there are some interesting differences between them. One is that cebids are larger than callitrichids. For example, whereas some cebids — male howlers are an example — have a combined head and body length of about 60 centimetres and weigh upwards of 10 kilograms, the combined head and body length of callitrichids is generally less than 20 centimetres and none weighs more than one kilogram. Another difference is that cebids have more teeth than callitrichids — 36 arranged in a 2:1:3:3/2:1:3:3 pattern as opposed to

**FIGURE 2.8** The present geographic range of Platyrrhini

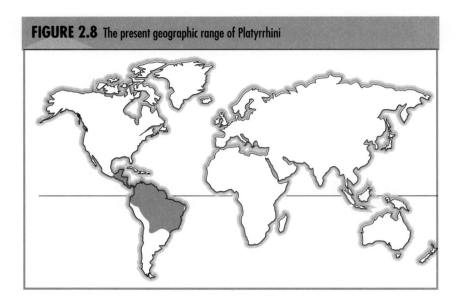

Adult howler monkey and infant

© Zoological Society of San Diego 1993

32 arranged in a 2:1:3:2/2:1:3:2 pattern. Their digits are also different. Whereas cebids have opposable thumbs and nails on all of their digits, callitrichids have non-opposable thumbs and claws on all of their digits save for the big toes, which are equipped with nails. Yet another difference between the two families is that several cebids have a grasping or **prehensile tail** that they use like a hand, a feature that is not present among callitrichids.

The most striking differences among the platyrrhines, however, are found between the individual species that are included in the infraorder. This is especially true with respect to their social systems and diets. For example, whereas the members of some species live together in small family units that contain a single male-female adult couple and their dependent offspring, others live in extended families, and still others in loosely-knit social aggregates or communities that contain between 10 and 20 individuals. Such primate communities are called **troops**. And while the members of some species are **insectivores** that feed mainly on insects, others are **folivores** that feed mainly on leaves, and still others are **frugivores** that feed mainly on fruit. There is even one cebid, the night or owl monkey (*Aotus trivirgatus*), that is nocturnal; it is found in various forest habitats stretching from western Panama south to northern Argentina (see QQ 2.9). Thus, like the Lemuriformes, the Platyrrhines sustain themselves in the tropical forest in a wide variety of ways.

## 2.9 *Aotus trivirgatus*

Patricia Wright describes how the night monkey (*Aotus trivirgatus*) has adapted to a nocturnal niche.

• • •

The problems that all nocturnal mammals must overcome to be night active are: (1) colder temperatures; (2) low light levels; (3) differences in predators and competitors and (4) differences in availability of certain food items. By integrating knowledge gained from the laboratory on physiology and neurobiology with information from new field studies of *Aotus*, we can better understand the nocturnal monkey niche, and the adaptations that better equip this monkey to live at night.

The comparative approach to the behavioral ecology of the night monkey has revealed several important differences between the diurnal *Callicebus* [*moloch*, the dusky titi,] and the nocturnal *Aotus*. In contrast to the day monkey, ranging patterns are often circular and chosen routes are habitual and rigid, one or two sleeptrees are used for many years. These habitually travelled and scent-marked routes would make orientation easier in the dark. Sleep sites are hidden in tree holes or dense vine tangles. This suggests *Aotus* needs to hide from day-time predators, but is not threatened by night time predators who could easily memorize habitual travel routes.

[In addition, l]ow basal metabolic rate and dense fur may be important in energy conservation during the colder nights. Small body size allows limited ranging patterns and a more insectivorous diet, which may also be important to the energetics of being nocturnal.

Source: Patricia C. Wright, "The Nocturnal Primate Niche in the New World," *Journal of Human Evolution*, Vol. 18, No. 7, (1989), pp. 653–654.

# OLD WORLD MONKEYS
## (INFRAORDER CATARRHINI)

Unlike New World monkeys, Old World monkeys belong to the infraorder **Catarrhini**. So do apes and humans. That we are closely related is indicated by our noses and our teeth. Like apes and humans, all Old World monkeys possess a **catarrhine nose** with nostrils that are close together and open downward. All catarrhines also have 32 teeth arranged in a 2:1:2:3/2:1:2:3 pattern. Old World monkeys, which are found in Africa, Asia, and Gibraltar (see Figure 2.9), are consequently more like humans than are their Central and South American counterparts.

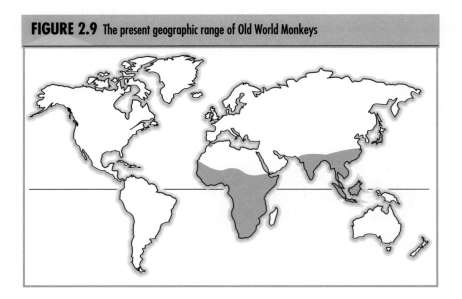

**FIGURE 2.9** The present geographic range of Old World Monkeys

# SUBFAMILIES
## (CERCOPITHECINAE AND COLOBINAE)

Although all Old World monkeys are grouped together in the superfamily **Cercopithecoidea**, two subfamilies can be distinguished: the **Cercopithecinae**, which are predominantly frugivores, and the **Colobinae**, which are predominantly folivores. In turn, both subfamilies can be divided into a number of smaller, subsidiary groups. As Table 2.3 indicates, the Cercopithecinae include the guenons and the mangabeys that are both found in Africa; the baboons and the baboon-like drills and mandrills that live in Africa and in the Arabian Peninsula; and the macaques from Africa, Asia, and Gibraltar. Meanwhile, the Colobinae include the African guerezas and the langurs from Asia. But while some of these catarrhine monkeys are small arboreal animals, others such as baboons are large **terrestrial quadrupeds**, which habitually move about with all fours on the ground. Thus, on the surface, there seems to be no clear-cut blueprint for catarrhine monkeys; they inhabit a wide variety of habitats and vary in size and appearance.

It is important to remember that Old World monkeys are our close relatives. In fact, we are so closely related that medical researchers have begun to explore the possibility of using the organs of catarrhine monkeys to extend human life. Such was the case with Baby Fae, a human infant who made medical history in 1984 when surgeons

▼
Macaque
▲

© Zoological Society of San Diego 1993

**TABLE 2.3** The main groups of Old World monkeys (Cercopithecoidea)

| SUBFAMILY | SUBSIDIARY GROUP | GEOGRAPHICAL RANGE | DESCRIPTION |
|---|---|---|---|
| Cercopithecinae | Guenons | Sub-Saharan Africa | Primarily arboreal monkeys with long tails; numerous species are found in forested areas |
| | Mangabeys | Sub-Saharan Africa | A small group of semiarboreal monkeys usually found in gallery forests |
| | Baboons, drills mandrills | Sub-Saharan Africa and Arabian Peninsula | Large terrestrial monkeys with long faces; some are found in forests while others live in arid open country |
| | Macaques | Gibraltar, North Africa, and Asia from Afghanistan to Indonesia | A very adaptable and widespread group of primarily terrestrial monkeys, usually rather large and stocky |
| Colobinae | Guerezas | Sub-Saharan Africa | Arboreal leaf-eating monkeys with reduced thumbs; restricted mainly to forested areas |
| | Langurs | Asia from India to Indonesia | Primarily large but lightly built arboreal monkeys. Numerous species are known |

Source: adapted from Elizabeth S. Watts, *Biology of the Living Primates* (Dubuque, Iowa: Wm. C. Brown Company Publishers, 1975), p. 18. Used with the permission of Elizabeth S. Watts.

removed her defective heart and replaced it with one from a young baboon. Although Baby Fae died less than a month after the operation, cross-species organ transplants, which are called **xenografts**, may eventually become more common. Research along these lines is continuing; on 8 December 1992, for example, medical researchers at the University of Western Ontario Hospital announced to the press that they were about to launch a research project "to establish a safe, effective drug therapy for baboon-to-human transplants, so that in the future, patients who would die because a human organ is not available might be successfully transplanted."

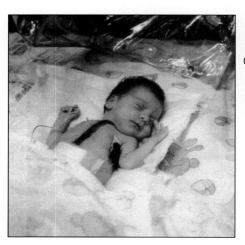

▼

Baby Fae
and baboon
donor

▲

© 1985 Discover Magazine

## THE "BARBARY APE"

In conjunction with their biological features, the social systems of catarrhine monkeys also indicate that they are close relatives of humankind. Despite the fact that these systems vary, Old World monkeys are, like ourselves, highly gregarious. The so-called "Barbary ape" (*Macaca sylvanus*), a macaque — not an ape — from northern Algeria and Morocco that was introduced to Gibraltar, is one of many outstanding examples.

Like the majority of higher primates, Barbary apes live together in troops, which, in their case, contain 50 or more individuals that arrange themselves into smaller social units that forage for fruit, leaves, bark, and roots on scrublands, on rocky cliffs, and in mid- and high-altitude forests. Although several related females that remain in the same natal group throughout their lives are the mainstay of the troop, these females are accompanied by adult males that typically move from one group to another after they reach adolescence. Within such troops, which are formed in much the same way among most higher primates, the complex social relations that characterize *Macaca sylvanus* are established; with adult males and females both contributing to the socialization of the immature. See QQ 2.10.

## 2.10 THE SOCIALIZATION OF *Macaca sylvana*

Frances Burton, Professor of Anthropology at the University of Toronto, discusses the roles that various individuals play in the socialization of the "Barbary apes" that are found in Gibraltar.

• • •

The adult female is the first environment for the neonate, and thus the context for the most important biological maturation. The infant first sees and makes its first oriented movements within her arms. However, since the leader male takes the neonate from as early as the first day of life, it is the leader male who is the preeminent influence in socialization of the infant until it is approximately two weeks of age. The leader male encourages biological maturation by reinforcing the infant's mouth sucking movements until they become the social chatter gesture, and by encouraging the infant in locomotor skills and to take the dorsal trans-port position [on the back of an adult]. These basic body movements are of prime importance for all future social contact. The leader male reorients the infant away from its mother, himself, and other adults, by permitting subadult males to first snatch the infant, and later to take it away for greater and greater distances. It is while the infant is under the care of the subadult males that it increasingly contacts juveniles and later age-mates. From contact with subadult males and juveniles, as well as from this social context, the infant learns, as he develops motor skills, all the necessary information to become a functioning member of the troop. The information ranges from what to eat to available routes from one area to another.

Source: Frances D. Burton, "The Integration of Biology and Behavior in the Socialization of *Macaca Sylvana* of Gibraltar," in Frank E. Poirier, ed. *Primate Socialization* (New York: Random House, Inc., 1972), pp. 55–56.

# APES

Apes are more closely related to humans than are monkeys. As a rule, they are larger than monkeys, more intelligent, and better equipped and able to manipulate objects with their hands and their feet. In other words, apes are remarkably like us, and this is reflected in the way that apes and humans are classified. Together, we are the only living representatives of the superfamily **Hominoidea**, a term that refers to primates that possess ape-like traits. However, whereas taxonomists assign humans to the family **Hominidae**, they arrange apes into two families: **Hylobatidae**, which includes gibbons and siamangs, and **Pongidae**, which includes orangutans, gorillas, and chimpanzees. See Table 2.4.

## GIBBONS AND SIAMANGS
## (SUPERFAMILY HOMINOIDEA, FAMILY HYLOBATIDAE)

The gibbon and the siamang, which are also called small or lesser apes, live in the tropical rain forests of Southeast Asia — gibbons in various locations but siamangs only on the island of Sumatra in Indonesia (see Figure 2.10). Befitting their designation they are the smallest of the apes; all adult lesser apes stand between 45 centimetres and 65 centimetres tall, weigh between 5.5 kilo-

**TABLE 2.4** A taxonomy of living apes and humans that distinguishes apes from humans on a familial level

| SUPERFAMILY | FAMILIES | SUBFAMILIES | GENERA | COMMON NAME | NUMBER OF SPECIES |
|---|---|---|---|---|---|
| Hominoidea | Hylobatidae | Hylobatinae | *Hylobates* | Gibbon | 8 |
| | | | *Symphalangus* | Siamang | 1 |
| | Pongidae | | *Pongo* | Orangutan | 1 |
| | | | *Gorilla* | Gorilla | 1 |
| | | | *Pan* | Chimpanzee | 2 |
| | Hominidae | | *Homo* | Humankind | 1 |

grams and 10.5 kilograms, and have a brain size of about 100 cm³ (cubic centimetres). Like the majority of apes they eat mainly vegetable foods, especially ripe fruit that they pick in clusters from the branches of trees. They do not, however, live together in well-defined social groups. Instead, when gibbons and siamangs mature at about seven years of age, they tend either to live alone or to form small family units composed of a single adult male, a single adult female, and up to four dependent offspring. All small apes are also completely arboreal and are extremely good **brachiators**, which means that they travel by using their hands to swing underneath branches from one branch to another in the trees (see QQ 2.11). Gibbons and siamangs can and do move rapidly in this way because of their relatively small size, their exceptionally long arms, and their remarkably powerful and precise grip.

**FIGURE 2.10** The present geographic range of Hylobatidae

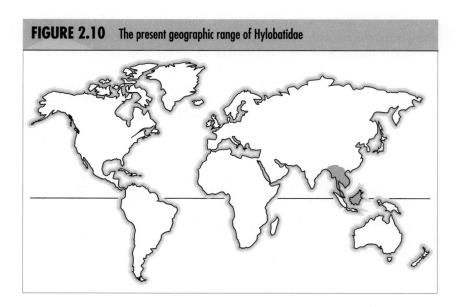

Gibbon
brachiating

# 2.11 GIBBON BRACHIATION

**Stephen Rosen discusses the gibbon's ability to brachiate.**

. . .

Of all the primates, the gibbon is the most fully adapted to brachiation. Except for 10 to 15 percent of locomotor time, the gibbon is a *full-time brachiator* . . . . The skill and beauty of the gibbon's brachiation cannot be adequately described, even still pictures do not do it justice . . . . Its grace and agility rival that of gazelles and gliding birds. Almost its entire anatomy is designed for brachiation . . . . The *forelimbs* are *extremely long*, the arm length being approximately two and one-half times its trunk length. Contrary to popular belief, [however,] the gibbon actually has long hindlimbs, which cannot be fully extended at the knee joint . . . [and that] are tucked up to the trunk in brachiation . . . .

[Gibbons] also walk bipedally atop tree limbs; in fact, the genus name, *Hylobates* means "tree-walker." If placed upon the ground, the gibbon will generally prefer semierect bipedal locomotion. When walking bipedally, the gibbon's arms are raised outwardly above the head with the hands bent at the wrist, the upper limbs acting as balancing organs. [However, i]t should be noted that in its natural environment the gibbon is wholly arboreal.

Source: Stephen I. Rosen, *Introduction to the Primates: Living and Fossil*, © 1974, pp. 112–115, 167–170. Reprinted by permission of Prentice Hall, Englewood Cliffs, New Jersey.

# Orangutans
## (Superfamily Hominoidea, Family Pongidae, Genus *Pongo*)

The great apes, which are considerably larger than the small apes, include the orangutan, the gorilla, and the chimpanzee. Among these, the orangutan (*Pongo pygmaeus*), whose long shaggy red coat makes it easy to identify, is the only great ape that is currently found in Asia (see Figure 2.11). The orangutan lives exclusively in the lowland and hilly tropical rain forests of the Indonesian archipelago, on the islands of Borneo and Sumatra, where the aboriginal inhabitants call it "the old man of the forest" in recognition of its resemblance to humankind.

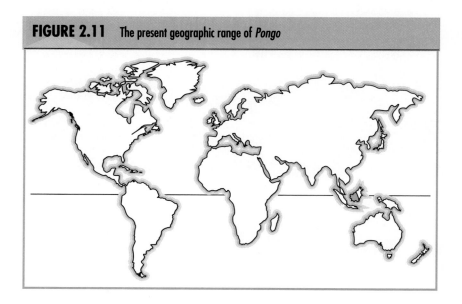

**FIGURE 2.11**    The present geographic range of *Pongo*

As is the case with gibbons and siamangs, orangutans are arboreal frugivores. However, in sharp contrast to the small apes, orangutans are **sexually dimorphic**, which means they are characterized by distinctive non-reproductive anatomical differences between males and females. The overall size of the orangutan's body is an example: whereas adult males are about 1.4 metres tall and weigh between 60 kilograms and 90 kilograms, adult females are about 1.2 metres tall and weigh between 40 kilograms and 50 kilograms. Also unlike the small apes, orangutans do not rely strictly on their long arms and long thin hands to move through their dense forest habitat. Instead, they are **quadrumanual arborealists**, using both their hands and their feet to grasp and suspend themselves from branches and limbs. Orangutans also have a larger cranial capacity than the small apes — about 400 cm$^3$ — and are much more intelligent.

## SOCIAL LIFE

Thanks in large part to the research efforts of Canadian primatologist Biruté Galdikas, it has also been discovered that orang-utans are the least social of the apes (see QQ 2.12). In fact, aside from occasional sexual liaisons, which last for several days at most, males and females spend the majority of their time apart. Males travel and forage alone, as do females except when they are rearing their young, which leave the mother when they are between seven and ten years of age.

Orangutan

© 1993 Frans Lanting MA 112/ Minden Pictures

## QQ 2.12 AMONG THE ORANGUTANS

Reporter Daniel Wood introduces Biruté Galdikas, a Canadian primatologist, and one of the world's foremost experts on orangutans.

• • •

Biruté Galdikas pours herself another cup of coffee. "It's easy in the Borneo rainforest," she says. "It's harder to be happy in Canada. The pressure there: to compete, achieve, accumulate goods. I've learned to be at peace with myself . . . here with the orang-utans" . . . .

But then, Galdikas is not a typical scientist. A greying and wilful forty-one-year-old Canadian, she is the youngest member of a trio of female primatologists that [Louis] Leakey selected . . . [who would] have the patience and fortitude for what might well be decades of study [among the great apes] . . . .

His first recruit was Jane Goodall — slim, blond, English, and the daughter of an old friend . . . . [The second was] a tall, thirty-four-year-old American occupational therapist, Dian Fossey.

Courtesy of Daniel Wood

Biruté Galdikas in the field

And Galdikas? She was the daughter of parents who fled Europe after the Soviet occupation of Lithuania and brought her as a small child to Toronto . . . .

[Since 1971] Galdikas [has] spent thousands of hours observing orang-utans and filled scores of note-books. She chronicled male territorial conflicts and found that orang-utans, unlike chimpanzees and gorillas, very seldom fought. Instead, they threatened each other with grunts and the furious shaking of branches. She watched female seductions of males and occasional adolescent male rape of females. She studied the animal's mating habits — how much like humans they were in the variety of positions they employed, in the frequent tenderness of their contact, in the nuzzling and the leisureliness of their move-ments. In documenting their breeding habits, she was the first person to be able to record the birth of an orang-utan in the wild . . . .

"When you live in the forest a long time," says Galdikas as she checks her trousers for leeches, "you come to realize how little humans need to be happy. All we need is shelter, food, friendship, and a sense of family. That's *all* we need. Look at orang-utans. All *they* need is a fruit tree and some branches to make a nest. But humans aren't like orang-utans. Orang-utans are loners. Humans are more like chimpanzees or gorillas. They're a gregarious species."

Source: Daniel Wood, "Persons of the Forest," *Saturday Night*, Vol. 103, No. 1 (1988), pp. 47–53.

## GORILLAS AND CHIMPANZEES

Towards the end of the nineteenth century, R. L. Garner, a zoologist, went into the tropical forests of Gabon, West Africa, to study gorillas and chimpanzees in the wild. He built a big cage, not to capture apes, but to lock himself inside for protection against what he believed were wild and ferocious beasts. Although Garner's study marked the beginning of observational studies of apes in their natural habitat, as we now know his tactics were flawed; in fact, having met with little success after 112 days, Garner abandoned his quest. However, since Garner's day, with the advent of increasingly sophisticated observational research techniques and the inauguration of new genetic and biochemical studies, our understanding of gorillas and chimpanzees has grown at an enormous rate.

R.L. Garner standing outside his cage

From R. L. Garner, *Gorillas and Chimpanzees*, 1896 – Plate 45

Among other things, these studies have demonstrated that the genetic structures of gorillas and chimpanzees (which determine their anatomical and physiological features) are remarkably similar to our own. In the case of the gorilla, for instance, there is a more than 95 percent structural overlap with humans, and in the case of the chimpanzee a more than 99 percent overlap. Based on facts such as these, some taxonomists have suggested that instead of grouping orangutans, gorillas, and chimpanzees in the family Pongidae and humans in the family Hominidae, it makes better sense to position orangutans, gorillas, chimpanzees, and humans in the family Hominidae, and then to divide that family into two subfamilies: **Ponginae**, represented exclusively by orangutans; and **Homininae**, which includes gorillas, chimpanzees, and ourselves. Since the traditional classification is still quite popular, however, we will follow the older arrangement here.

## GORILLAS (SUPERFAMILY HOMINOIDEA, FAMILY PONGIDAE, GENUS *GORILLA*)

The gorilla (*Gorilla gorilla*) lives in the tropical secondary forests of equatorial Africa (see Figure 2.12). While the genus includes three subspecies — the western lowland gorilla (*Gorilla gorilla gorilla*), the eastern lowland gorilla (*Gorilla gorilla graueri*), and the mountain gorilla (*Gorilla gorilla beringei*) — there are only minor anatomical differences between these. Far more impressive is the gorilla's overall size and its sexual dimorphism. Whereas adult male gorillas are between 1.7 metres and 1.8 metres tall and weigh between 140 kilograms and 180 kilograms, adult females are about 1.5 metres tall and weigh about 90 kilograms. Male gorillas also have large

**FIGURE 2.12**   The present geographic range of *Gorilla*

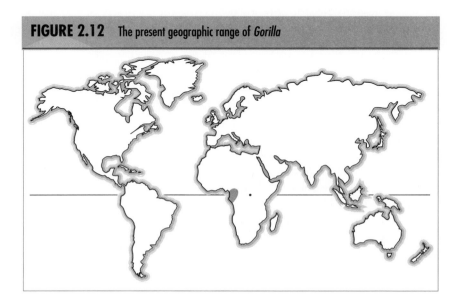

canine teeth and a **sagittal crest** — a ridge of bone that runs along the midline of the top of the skull from the front to the back, which helps to support the muscles that are used in chewing. By way of contrast, many female gorillas lack these features.

Like orangutans, gorillas have a relatively large cranial capacity — about 500 cm$^3$ — and they are very intelligent. Unlike orangutans, however, gorillas are predominantly folivores, rather than frugivores. They are also mostly terrestrial; although young gorillas are excellent brachiators, the adults, because of their size, spend most of their time on the ground. There they generally use all four limbs for walking, resting the weight of the lower body on the soles of the feet while they bend their fingers under their hands so that they can rest the weight of the upper body on their knuckles when they move. This method of locomotion is called **knuckle-walking**. Adults also usually sleep on the ground in nests that they fashion from large-leaved plants and from high standing grasses.

## SOCIAL LIFE — GORILLAS

The social organization of gorillas is as intriguing as their size. Although gorillas live together in small cohesive troops containing about a dozen individuals that occupy a home range of about 160 hectares, these groups are formed in an unusual way. In sharp contrast to the majority of non-human primate societies, gorilla troops are not based on a core of related females to which males become attached after they reach adolescence. Instead, the mainstay of the troop is the **silverback,** an adult male at least 11 years old to which unrelated adolescent females and mature females become attached. His position in the troop is pre-eminent. Field studies indicate that the silverback is responsible not only for protecting the members of the troop from predators, but also for leading them as they move from place to place. The silverback also has access to more sexual partners and is **groomed** the most often — an interactive behaviour, it will be recalled, in which one animal picks dirt and parasites from another's hair with its hands and teeth to initiate social contact. Those under the silverback's care, however, are not always under his direct control. This is especially true of females that frequently leave one troop in favour of another either at adolescence or later in life, sometimes because of competition over them by males. When this happens, although gorillas are generally peaceful, the male competitors may become aggressive. As a result, the females are occasionally injured and their offspring sometimes accidentally killed.

Unfortunately, like many other species of non-human primates, gorillas are declining in number. However, as Dian Fossey and others have pointed out, this is not because of internal strife. Instead, the greatest danger that gorillas face is from humans: from poachers and hunters who kill them, and from traders who purchase gorilla hands and heads to sell to collectors. If these practices continue, gorillas may soon become extinct. See QQ 2.13.

# 2.13 THE VANISHING MOUNTAIN GORILLA

Terry Maple and Michael Hoff discuss the plight of the mountain gorilla (*Gorilla gorilla beringei*).

• • •

In 1978 the Fauna Preservation Society launched a campaign to raise funds for the protection of mountain gorillas in the *Parc National des Volcans in Rwanda*. This effort followed publicity concerning the killing of a wild gorilla named *Digit* which had gained fame by appearing in a television documentary about the research of Dian Fossey . . . .

The Death of *Digit*, and later *Uncle Bert* and *Macho*, were the result of a growing trade in gorilla heads and hands which began in 1976. Although little is known about the mechanics of this trade, it appears to be dependent on the interaction of poachers, middleman entrepreneurs, and white residents and touring trophy hunters. This is a serious problem in view of the fact that the use of high velocity weapons has led to greater efficiency in killing these animals. Since the Virunga Volcanoes may be

Courtesy of The Dian Fossey Gorilla Fund, Englewood, Colorado

▼

Digit, decapitated and handless

▲

the last refuge of the mountain gorilla, the danger of rapid extermination is very real.

Source: Terry L. Maple and Michael P. Hoff, *Gorilla Behaviour* (New York: Van Nostrand Reinhold Company, 1982), pp. 257–258.

## CHIMPANZEES (SUPERFAMILY HOMINOIDEA, FAMILY PONGIDAE, GENUS *Pan*)

Like gorillas, chimpanzees are found in equatorial Africa, in habitats ranging from dense tropical forests and deciduous woodlands to sparsely treed savanna or grassland (see Figure 2.13). Although it was once thought that there were as many as 14 living species, careful research has reduced that number to two: the common chimpanzee (*Pan troglodytes*), which is distributed across central Africa from Senegal to Tanzania north of the Zaïre River, and the pygmy chimpanzee or Bonobo (*Pan paniscus*), which lives exclusively in Zaïre between the Zaïre and Kasai rivers.

Despite its name, the pygmy chimpanzee is about the same size as its common counterpart. Adult males of both species stand between 73 centimetres and 92 centimetres tall and weigh about 40 kilograms, and adult females stand between 70 centimetres and 85 centimetres tall and weigh about 30 kilograms. Both species also have a cranial capacity that ranges between about 300 cm$^3$ and 400 cm$^3$ among adults. In addition, while the members of both species are knuckle-walkers, they generally spend about half of their waking hours in trees and bed down in tree nests at night.

**FIGURE 2.13** The present geographic range of *Pan*

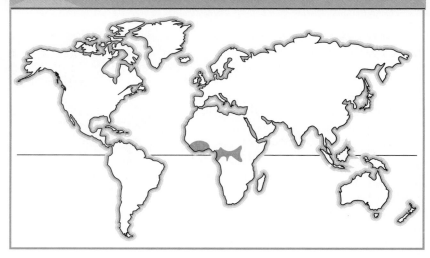

## SOCIAL LIFE — CHIMPANZEES

Careful research has also revealed several fascinating things about chimpanzee troops, which contain between 15 and 120 individuals that occupy home ranges that vary in size from about 800 hectares to about 2200 hectares. One is that while chimpanzee troops are formed in much the same way as gorilla troops, instead of becoming attached to a single male, adolescent female chimpanzees become attached to a group of related males. And instead of moving from one group to another from time to time, females tend to remain in the group to which they become attached throughout their lives. Adult females also frequently travel exclusively in the company of their dependent offspring, if any, or alone if they have no dependent offspring. Adult

Chimpanzees eating a bushbuck (an African antelope)

David Bygott/Anthro-Photo, #2393

males, on the other hand, typically travel in small groups that contain between three and six individuals. Chimpanzee troops are consequently less cohesive than gorilla troops with respect to feeding and travelling.

Another important discovery about chimpanzee troops is that although it was once thought that their members were **carnivores** or meat eaters, and later that they were strictly **herbivores** or vegetarians, there is now evidence that, like humans, chimpanzees are **omnivores** that rely on both meat and vegetable foods. While the bulk of their diet consists of ripe fruit, chimpanzees occasionally "hunt" for meat cooperatively and share the products of their labour — traits that are likewise reminiscent of humankind. See QQ 2.14.

## 2.14 THE OMNIVOROUS CHIMPANZEE

Geza Teleki looks at the social significance of the omnivorous diet of the chimpanzees that live in the Gombe National Park.

• • •

It . . . [was once] widely believed that apes and monkeys are vegetarians, and that man is alone among the primates in preying on other animals . . . .

Today, after some 40 years of field observations of ape and monkey behavior, it is quite clear that man is not the only primate that hunts and eats meat. Many other primates are omnivorous. One in particular — the chimpanzee — not only cooperates in the work of the chase but also engages in a remarkably socialized distribution of prey after the kill. The chimpanzees whose predatory behavior has been most closely observed are semi-isolated residents of the Gombe National Park in western Tanzania. The area, formerly known as the Gombe Stream Chimpanzee Reserve, is where Jane . . . Goodall began her notable long-term field study of chimpanzees in 1960. I myself spent 12 months watching the predatory behavior of these apes in 1968–1969 . . . .

The Gombe chimpanzees can be described in summary as omnivorous forager-predators that supplement a basically vegetarian diet in various ways, including the optional practice of hunting other mammals, with fellow primates being their favorite prey. The fact that the chimpanzees' meat-eating is optional means that in terms of meat consumption they do better than purely opportunistic omnivores that "collect" and eat an immobile prey animal now and then. At the same time the Gombe chimpanzees are free of the exclusive dependence on a meat diet that is characteristic of most carnivores.

The primary significance of the Gombe chimpanzees' predatory behavior is not, however, dietary. What is far more important is the behavior that accompanies the predatory episodes: cooperation in the chase and the sharing of the prey. The more we learn about primate behavior, the smaller the differences between human and nonhuman primates.

Chimpanzees can also be aggressive and, thanks to the pioneering research efforts of primatologist Jane Goodall and her colleagues, this side of their nature has recently been revealed. In 1970, after observing common chimpanzees in the Gombe National Park for a decade, Goodall and her co-workers witnessed a remarkable event. The troop of 21 chimpanzees they

**FIGURE 2.14** Chimpanzee territorial conflict

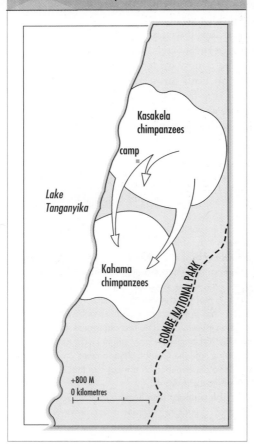

were studying began to form two separate groups. Although the troop had once shared a home range whose borders were regularly patrolled by small groups of adult males, by 1972, two regionally distinct communities had been established: the Kasakela chimpanzees in the north and the Kahama chimpanzees in the south. See Figure 2.14.

During the next two years the relationship between the members of the two troops was peaceful, and this seemed to confirm Goodall's original impression that chimpanzees are not normally aggressive, which in fact is the case. However, in 1974 the Kasakela males began to attack their male and female counterparts in the south, and by 1977 the Kahama community had all but been destroyed. Why this rare display of aggressive behaviour occurred has yet to be determined. Perhaps, as Goodall and her colleagues suggested, the Kasakela males were bent on occupying new territory in the south, or perhaps, as others have argued, the attacks were caused by competition over food given to them by researchers. In either event, one thing is clear: although our closest living relatives are generally peaceful, under certain conditions, they can be extremely aggressive. See QQ 2.15.

# QQ 2.15 CHIMPANZEE AGGRESSION

Jane Goodall and her colleagues describe an attack on a Kahama male by a group of Kasakela chimpanzees.

• • •

On the morning of January 7, 1974, a large mixed group of Kasakela individuals travelled slowly southward, with periods of feeding. At 1415 a party of six adult males (Hugo, Humphrey, Faben, Figan, Jomeo, and Sherry), an adolescent male (Goblin), and the ster-

ile female (Gigi, in oestrus) detached itself from the others and began to travel more purposefully southward. During this journey calls were heard to the south. The party began to travel quickly and silently in that direction. Finally they came upon Godi [a Kasakela male] who was feeding in a tree and clearly caught unaware. He leapt down and fled, screaming. Humphrey, Jomeo, and Figan were close on his heels, running three abreast with the rest of the party following. Humphrey grabbed Godi's leg and pulled him to the ground, then jumped on Godi as he tried to escape. The other chimpanzees began hitting Godi as he sat, pinned to the ground by Humphrey who was leaning forward and holding on to the victim's legs with both hands. During the ensuing attack, Humphrey remained in this position: Godi had no chance to escape or to defend himself . . . .

Godi remained motionless on the ground when Humphrey got off [after about 10 minutes] . . . but, as the attackers moved off, he got up and looked after them, screaming. It was clear that he was badly wounded: he had a great gash, from his lower lip down the left side of his chin, and this was bleeding profusely. Blood was also coming from his nose, his upper lip was swollen, and there were bleeding cuts in the right corner of his mouth. There were puncture marks on his right leg and between his ribs on the right side. He had a few small wounds on his left forearm. Since that moment Godi has not been seen, despite the fact that field assistants and students have been working with the Kahama community since April 1974.

Source: Jane Goodall, *et al.*, "Intercommunity Interactions in the Chimpanzee Population of the Gombe National Park," in David A. Hamburg, and Elizabeth R. McCown, eds. *The Great Apes: Perspectives on Human Evolution, Vol. V* (Menlo Park, California: The Benjamin/Cummings Publishing Company, Inc., 1979), pp. 35–36.

# COMMUNICATIVE SKILLS

The communicative skills of the African apes are also very like our own. In 1966, Beatrice and Allen Gardner began to teach their chimpanzee Washoe to use **American Sign Language (ASL)** — the sign language that many deaf persons use in Canada and the United States. By 1969, Washoe had learned about 130 signs and was putting them together in order to construct simple phrases such as "go in," "tickle Washoe," and "Washoe sorry." The Gardners concluded not only that Washoe was capable of symbolic communication, but also that future research would substantiate and build upon their findings. See QQ 2.16.

## 2.16 WASHOE'S COMMUNICATIVE SKILLS

The Gardners discuss the results of their work with Washoe.

. . .

At an earlier time we would have been more cautious about suggesting that a chimpanzee might be able to produce extended utterances to communicate information. We believe now that it is the writers — who would predict just what it is that no chimpanzee will ever do — who must proceed with caution. Washoe's achievements will probably be exceeded by another chimpanzee, because it is unlikely that the conditions of training have been optimal in this first attempt. Theories of language that depend upon the identification of aspects of language that are exclusively human must remain tentative until a considerably larger body of intensive research with other species becomes available.

Source: R. Allen Gardner and Beatrice T. Gardner, "Teaching Sign Language to a Chimpanzee," *Science*, Vol. 165, No. 3894 (1969), pp. 671–672.

Koko, a female gorilla, made it clear that the Gardners were correct, at least with respect to signing. By the mid-1980s, Koko, who was trained by Francine Patterson, had mastered the signs for 600 ASL words. Nor is ASL the only symbolic communication system that the great apes have mastered in captivity. A chimpanzee named Sara was taught to associate concepts such as "prefer," "different," and "same as" with plastic tokens of various sizes and shapes. She was even able to "put the banana in the pail and the apple in the dish" when she was "asked" to do so by her trainers. Meanwhile, another chimpanzee named Lana had been taught to communicate via a specially designed computer language called Yerkish. By using a keyboard, Lana was able not only to issue and follow complex commands, but also to distinguish between grammatically correct and incorrect Yerkish statements. More recently, a male bonobo named Kanzi has matched their achievements; on the basis of the way in which he has learned to manipulate 256 geometric symbols attached to a board, Kanzi's trainers estimate that he possesses the grammatical skills of a two-and-a-half-year-old child.

▼

Kanzi
and human
observers

▲

© Michael Nichols/Magnum Photos, Inc.

The performance of Washoe and the other great apes that have been taught to communicate with their trainers without uttering spoken words is remarkable; it demonstrates that primates other than ourselves are capable of symbolic communication. What remains to be determined is whether apes and possibly other non-human primates communicate symbolically in the wild. Preliminary research indicates that this is almost certainly the case. As early as 1940 it was reported that gibbons employ nine different **calls** or characteristic sounds that are equivalent to statements such as "I am here," "follow me," "here is food," "danger," and "I am hurt." Since then it has been discovered that chimpanzees have an even larger repertoire of calls, each of which seems to be associated with a specific environmental configuration. Although there is still considerable controversy as to whether these calls can be regarded as language, that is to say, words and rules for their use, it is clear that the sounds that apes utter in the wild are much more than simply hoots and grunts.

# HUMANKIND (SUPERFAMILY HOMINOIDEA, FAMILY HOMINIDAE, GENUS *Homo*)

Although gorillas and chimpanzees are our closest relatives there are obvious differences between us, and for this reason living humans are accorded their own family — Hominidae — their own genus — *Homo* — and their own species — *sapiens*.

## BIPEDALISM

One of the most striking differences between humankind and the African apes is that we are **bipedal** — we walk upright on our feet (see Figure 2.15). True, gorillas and chimpanzees can also walk erect, but only for short distances and even then with difficulty. This is because their heads tilt forward when they stand, their spines are straight, their arms are longer than their legs, and they have opposable big toes that tend to curl under their feet. Nor can they straighten their legs at the knee, which makes them crouch when they stand. Upright posture is much easier for humans because our head is well-balanced on the neck, our spine is S-shaped, our pelvis is short and wide, our legs are longer than our arms, and we do not have opposable big toes. Taken together, this combination of features makes it possible for us to walk with our feet firmly on the ground. See QQ 2.17.

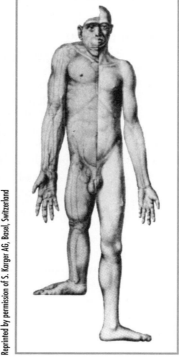

**FIGURE 2.15** A comparative look at the proportions of chimpanzees and humans

From "Post Embryonic Age Changes" by A.H. Schultz in PRIMATOLOGIA, I.S. Karger, 1956. Reprinted by permission of S. Karger AG, Basel, Switzerland

## ℚ  2.17 THE HUMAN GAIT

John Napier discusses human walking.

• • •

Anyone who has watched other people walking and reflected a little on the process has noticed that the human stride demands both an up-and-down and a side-to-side displacement of the body . . . . General observations of this kind were reduced to precise measurements during World War II when a group at the University of California at Berkeley led by H.D. Eberhart conducted a fundamental investigation of human walking in connection with requirements for the design of artificial legs. Eberhart and his colleagues found that a number of functional determinants interacted to move the human body's center of gravity through space with a minimum expenditure of energy. In all they isolated six major elements related to hip, knee and foot movement that, working together . . . [made it possible for people to walk upright]. If any one of these six elements was disturbed, an irregularity was injected into the normally smooth, undulating flow of walking, thereby producing a limp. What is more important, the irregularity brought about a measurable increase in the body's energy output during each step.

Source: John Napier, "The Antiquity of Human Walking," *Scientific American,* Vol. 216, No. 4, p. 57. Copyright © 1967 by Scientific American Inc. All rights reserved.

## THE SKULL AND THE BRAIN

Humans and apes also have different skulls and brains. For one thing, the human face is much flatter. Nor do we have a thick, ape-like bone **brow ridge** that protrudes outward over our eyes, or possess a sagittal crest and large canine teeth. The size of our brain is also much larger — ranging from 900 cm$^3$ to 2300 cm$^3$ — and while cranial capacity is certainly not the only factor that determines intelligence, we are much more intelligent than apes, perhaps, as some have suggested, because the structure of our brain makes it possible for us, more than any other animal, to act in accord with something that we may suddenly remember. This is because, in addition to receiving sensory input from various parts of the body and coordinating bodily responses to that input, the brain can initiate bodily responses by virtue of its own internal activity. See Figure 2.16.

## SEXUAL DIMORPHISM

It is also interesting to note that while we are much less sexually dimorphic than many other primate species, there are some recognizable anatomical differences between contemporary males and females. Males, for instance, tend to be larger than females from the same population. They also tend to have larger muscles than females from the same population, and the areas where those muscles are attached are more ruggedly constructed. Moreover, although humankind is not characterized by protruding brow ridges, what brow ridges modern humans do possess are more fully developed in males

**FIGURE 2.16** Cross-section of a modern human brain

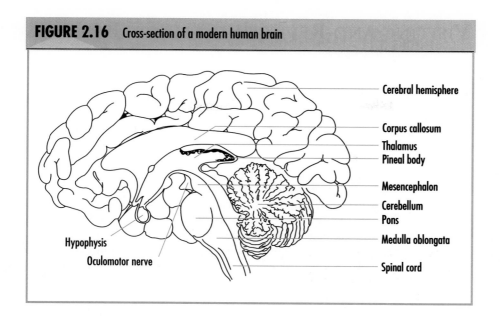

Cerebral hemisphere

Corpus callosum
Thalamus
Pineal body

Mesencephalon
Cerebellum
Pons
Medulla oblongata

Spinal cord

Hypophysis

Oculomotor nerve

than in females. In addition, whereas males tend to have somewhat slanted foreheads, those of females are more rounded. The greatest difference between males and females, however, is in the shape of the pelvis, which is broader and has wider openings in females than in males. A pelvis of this sort makes it possible for females to give birth to offspring whose heads are exceedingly large in relation to the overall size of the body at birth, and is part of our biological heritage. See QQ 2.18.

## 2.18 THE HUMAN PELVIS

Robert Tague looks at pelvic dimorphism in modern humans.

• • •

Anatomists and anthropologists have long known that in humans the pelvis is more spacious in females than males, whereas for other measures of the skeleton males are larger than females. In both sexes the pelvis functions in the transfer of weight between the upper body and lower limbs, as the site of attachment of major blocks of muscles associated with posture and locomotion, and in the support and protection of viscera. Only in females, however, does the pelvis serve as a birth canal. As form relates to function, the inference is that evolution of pelvic dimorphism is associated with obstetrics.

Source: Robert G. Tague, "Sexual Dimorphism in the Human Bony Pelvis, With a Consideration of the Neandertal Pelvis from Kebara Cave, Israel," *American Journal of Physical Anthropology*, Vol. 88, No. 1 (Copyright © 1992), p. 1. Reprinted by permission of Wiley Liss, a division of John Wiley and Sons, Inc.

# VISITING AND REVISITING YELLOW BABOONS

Although there are differences between humankind and the other living primates, it can be informative to draw analogies, especially between ourselves and the higher primates whose biosocial attributes closely resemble our own. Such analogies, however, must be constructed with care. What has been learned about yellow baboons (*Papio cynocephalus*) illustrates the point.

## HABITAT, SIZE, AND DOMINANCE

Yellow baboons occupy the forest edge and grasslands of eastern and central sub-Saharan Africa, where they feed on a wide variety of plant foods and some meat that they acquire by capturing small animals such as hares, young ungulates, and vervet monkeys. Compared to other Old World monkeys, yellow baboons are quite large. This is especially true of adult males, which have a combined head and body length that ranges between 60 centimetres and 80 centimetres, and weigh about 23 kilograms. Adult females, on the other hand, have a combined head and body length that ranges between 40 centimetres and 60 centimetres and weigh about 12 kilograms. Both live together with their infants in troops containing between 40 and 80 individuals that occupy home ranges of about 4000 hectares.

When these troops were first studied in the field in the late 1950s and early 1960s it was the way in which their social organization was described that made yellow baboons stand out. The adult males, it was reported, were organized in a **dominance hierarchy** — a social arrangement in which those with the highest rank possess special rights and responsibilities. Observers

A yellow baboon troop on the move

Irven DeVore/Anthro-Photo, #4416

noted that the most dominant males had access to more sexual partners, were groomed the most often, and occupied the best feeding and resting areas. It was also reported that these males were responsible for preventing serious conflicts from breaking out within the troop and for protecting females and infants from predators. The implication was that dominant adult males controlled virtually every aspect of the troop's social affairs.

## MISREPRESENTING THE FACTS

There are two problems with this sociological characterization. The first is that it misrepresents the facts. Although more recent and more detailed research has confirmed that there is a male dominance hierarchy in yellow baboon troops, it has also been demonstrated that this hierarchy is not the foundation on which the social solidarity of yellow baboon troops is based. Like almost all monkeys, yellow baboons are organized around a core of related females that remain in the same troop throughout their lives, that have their own dominance hierarchy, and to which males become attached when they leave their natal troops at adolescence. Also like most monkeys, yellow baboons possess a complex social system that is characterized by an intricate network of social relationships that exist among and between males, females, and infants. No one group, in other words, controls the social affairs of the troop.

## THE MAN-THE-HUNTER SCENARIO

The second problem with the characterization is more subtle. As already mentioned, one of the reasons that primatologists study our close living relatives is to help us to better understand ourselves. The dominance hierarchy among male baboons is a case in point: it was once thought to be an appropriate model to help us to better understand the origin of our own social institutions. In fact, assuming that yellow baboon societies and early human societies were organized along similar lines, some of those who believed that dominant male baboons controlled the social affairs of the troop also alleged that dominant males played an analogous role in the social affairs of our early savanna-dwelling human ancestors. If this belief were correct, then it would lend considerable support to what is known as the **man-the-hunter scenario** — the idea that the origin of human society can be traced back to a group of dominant and aggressive male hunters who attracted females to themselves on account of their superior subsistence skills. Since the females did not possess these skills, they remained subordinate and subservient to the males.

However, if the baboon society-early human society analogy is accurate, then it makes far better sense, given what has been learned about yellow baboon society, to argue that both sexes played critical roles in shaping our original social institutions. Considering what ethnologists have learned about modern hunter-gatherer societies and those of the recent past, the idea is all the more likely. In those societies, both sexes contribute to the social solidarity and economic well-being of the group. Under the circumstances, it is not surprising that support for the man-the-hunter scenario has collapsed. Instead, almost all anthropologists now maintain that the flexibility that was required for our ancestors to be able to form and live in social groups depended on the individual and collective skills of *all* of the members of the group. See QQ 2.19.

## 2.19 THE MAN-THE-HUNTER SCENARIO

Adrienne Zihlman comments on women's and men's roles in shaping early human society.

• • •

The flexibility of the human adaptation over the long haul allowed exploitation of a wide variety of plant, animal, and fish species, through the invention of tools and new behavioral patterns. Ultimately, the success of human reproduction means producing and socializing offspring that not only survive but also in turn have offspring of their own. Women's critical contribution to shaping the human adaptation must be integrated into

an evolutionary picture in order to explore their interrelation with men's roles. If we are to advance our understanding of sex roles in prehistory, which in turn may further understanding of the sexes today, we must ask questions in ways not previously asked, and most importantly we must break away from the traditional "man the hunter" formulation. Only then can we begin to redress the imbalance of history, and embark on new avenues of research to broaden our insights into human behavior.

Source: Adrienne L. Zihlman, "Women as Shapers of the Human Adaptation," in Frances Dahlberg, ed., *Woman the Gatherer* (New Haven: Yale University Press), p. 111. Copyright © 1981 Yale University Press. All rights reserved.

Given this reformulation, it is clear that analogies between ourselves and our close animal relatives must be based on a thorough examination of the facts. Keeping this principle in mind, as more research is done in the future, both in laboratories and in the field, we will learn more about the similarities and the differences between ourselves and our primate relations. In the meantime, it is important to remember that although the differences reflect the fact that each primate species has its own evolutionary history, the similarities between us point to the same conclusion that Linnaeus reached more than two centuries ago, namely, that we are not independent of nature, but are part of the natural world.

# KEY TERMS AND CONCEPTS

ANATOMY

PHYSIOLOGY

TAXONOMY

KINGDOM ANIMALIA

CHORDATA (*kor'-da-ta*)

NOTOCHORD

Vertebrata

Mammalia

Postorbital bar

Stereoscopic vision

Digits

Opposable thumb

Opposable big toe

Primate pattern

Arboreal habitats
(*ahr-bor'-ee-uul + habitats*)

Primary rain forests

Secondary rain forests

Woodland forests

Gallery forests

Home range

Savanna

Prosimii (*proh-sim'-ee-eye*)

Anthropoidea (*an'-throh-poy'-dee-a*)

Tupaiidae (*too-py'-i-day*)

Lemuriformes (*lee-mur'-i-formz*)

Arboreal quadrupeds
(*arboreal + kwah'-droo-pedz*)

Vertical clingers and leapers

Diurnal (*dy-ur'-nal*)

Nocturnal

Crepuscular (*kree-pus'-kyoo-lur*)

Tarsiiformes (*tar'-si-formz*)

Tarsal bones

Platyrrhini (*pla-ti-ree'-nee*)

Platyrrhine nose (*pla'-ti-reen + nose*)

Callitrichidae (*kal-i-trik'-i-day*)

Cebidae (*seh'-bi-day*)

Prehensile tail

Troops

Insectivores (*in-sek'-ti-vorz*)

Folivores (*foh'-li-vorz*)

Frugivores (*froo'-gi-vorz*)

Catarrhini (*ka-ti-ree'-nee*)

Catarrhine nose (*ka'-ti-reen + nose*)

Cercopithecoidea (*sur-koh-pith-i-koy'-dee-a*)

Cercopithecinae (*sur-koh-pith'-i-sin-ay*)

Colobinae (*koh-loh'-bin-ay*)

Terrestrial quadrupeds

Xenografts (*zee'-noh + grafts*)

Hominoidea (*haw-min-oy'-dee-a*)

Hominidae (*haw-min'-i-day*)

Hylobatidae (*hy-loh-bat'-i-day*)

Pongidae (*pon'-ji-day*)

Brachiators (*bray'-kee-ay-torz*)

Sexually dimorphic

Quadrumanual arborealists

Ponginae (*pon'-ji-nay*)

Homininae (*haw-min'-i-nay*)

Sagittal crest (*saj'-i-tal + crest*)

Knuckle-walking

Silverback

Grooming

Carnivores

Herbivores

Omnivores

American Sign Language (ASL)

Calls

Bipedal

Brow ridge

Dominance hierarchy

Man-the-hunter scenario

# SELECTED READINGS

## INTRODUCTION

Willey, B. *The Eighteenth Century Background: Studies on the Idea of Nature in the Thought of the Period.* New York: Columbia University Press, 1957.

## EDWARD TYSON

Montagu, A. *Edward Tyson, M.D. F.R.S. 1650–1708, and the Rise of Human and Comparative Anatomy in England.* Philadelphia: American Philosophical Society, 1943.

Tyson, E. *Orang-Outang, sive Homo Sylvestris,* or *The Anatomy of a Pygmie Compared with that of a Monkey, an Ape, and a Man.* London: Bennett and Brown, 1699.

## LINNAEUS

Frangsmyr, T., ed. *Linnaeus: The Man and His Work.* Berkeley: University of California Press, 1983.

Larson, J. *Reason and Experience: The Representation of Natural Order in the Work of Carl von Linné.* Berkeley: University of California Press, 1971.

## HUMANS AND OTHER ANIMALS

Goto, H. *Animal Taxonomy.* London: Edward Arnold, 1982.

## THE PRIMATE PATTERN

Ankel-Simons, F. *A Survey of Living Primates and their Anatomy.* New York: Macmillan, 1983.

Jolly, A. *The Evolution of Primate Behavior,* (2nd ed.). New York: Macmillan, 1985.

Kinsey, W., ed. *The Evolution of Human Behavior: Primate Models.* Albany: State University of New York, 1987.

Napier, J. and Napier, P. *A Handbook of Living Primates.* New York: Academic Press, 1976.

Swindler, D. and Erwin, J., eds. *Comparative Primate Biology,* (4 vols.). New York: Alan Liss, 1986.

Szalay, F., Rosenberger, A., and Dagosto, M. "Diagnosis and Differentiation of the Order Primates." *Yearbook of Physical Anthropology,* Vol. 30 (1987), pp. 75–105.

## PRIMATE HABITATS

Smuts, B., *et. al.,* eds. *Primate Societies.* Chicago: University of Chicago Press, 1987.

Wolfheim, J. *Primates of the World: Distribution, Abundance, and Conservation.* Seattle, Washington: University of Washington Press, 1983.

## PROSIMII

Doyle, G. and Martin, R. *The Study of Prosimian Behavior.* New York: Academic Press, 1979.

Jolly, A. *Lemur Behavior: A Madagascan Field Study.* Chicago: University of Chicago Press, 1967.

Napier, J. and Walker, A. "Vertical Clinging and Leaping, A Newly

Recognized Category of Locomotor Behaviour Among Primates." *Folia Primatologica*, Vol. 6 (1967), pp. 180–203.

Richard, A. *Behavioral Variation: Case Study of a Malagasy Lemur.* Lewisburg, Pennsylvannia: Bucknell University Press, 1978.

Schwartz, J. "What is a Tarsier?" In Eldredge, N., and Stanley, S., eds. *Living Fossils.* New York: Springer-Verlag, 1984, pp. 38–49.

Wright, P. "Lemurs Lost and Found." *Natural History*, Vol. 97 (1988), pp. 56–60.

## MONKEYS

Anderson, J. "Study of Primate Organ Transplants Set for New Year." *Western News*, Vol. 28 (1992), pp. 1–2.

Hershkovitz, P. *Living New World Monkeys (Platyrrhine), with an Introduction to the Primates* (volume 1). Chicago: University of Chicago Press, 1977.

Lindberg, D., ed. *Macaques: Studies in Ecology, Behavior and Evolution.* New York: Van Nostrand Reinhold, 1980.

Napier, J. and Napier, P., eds. *Old World Monkeys.* New York: Academic Press, 1970.

Strasser, E. and Delson, E. "Cladistic analysis of cercopithecid relationships." *Journal of Human Evolution*, Vol. 16 (1987), pp. 81–100.

Taub, D. "Geographic Distribution and Habitat Diversity of the Barbary Macaque, *Macaca sylvanus* L." *Folia Primatologica*, Vol. 27 (1977), pp. 108–133.

Terborgh, J. *Five New World Primates: A Study in Comparative Ecology.* Princeton, New Jersey: Princeton University Press, 1983.

## APES

Galdikas, B. "Orangutans, Indonesia's 'People of the Forest'." *National Geographic*, Vol. 148 (1975), pp. 444–473.

Galdikas, B. "Living with Orangutans." *National Geographic*, Vol. 157 (1980), pp. 830–853.

Preuschoft, H., Chivers, D., Brockelman, W., and Creel, N., eds. *The Lesser Apes: Evolutionary and Behavioral Biology.* Edinburgh: Edinburgh University Press, 1984.

Schwartz, J. *Aspects of the Biology of the Orangutan.* Oxford: Oxford University Press, 1988.

## GORILLAS AND CHIMPANZEES

Fossey, D. *Gorillas in the Mist.* Boston: Houghton-Mifflin, 1983

Fossey, D. "His name was Digit." *International Primate Protection League Newsletter*, Vol. 13 (1986), pp. 10–15.

Goodall, J. *The Chimpanzees of Gombe: Patterns of Behavior.* Cambridge, Massachusetts: Harvard University Press, 1986.

Mowat, F. *Virunga: The Passion of Dian Fossey.* Toronto: McClelland and Stewart, 1987.

Reynolds, V. "How wild are the Gombe chimpanzees?" *Man*, n.s., Vol. 10 (1975), pp. 123–125.

Schaller, G. *The Mountain Gorilla: Ecology and Behavior.* Chicago: University of Chicago Press, 1963.

Teleki, G. *The Predatory Behaviour of Wild Chimpanzees.* Lewisburg, Pennsylvannia: Bucknell University Press, 1973.

Wrangham, R. "Artificial Feeding of Chimpanzees and Baboons in Their Natural Habitats." *Animal Behaviour*, Vol. 22 (1974), pp. 83–93.

## COMMUNICATIVE SKILLS

De Luce, J. and Wilder, H., eds. *Language in Primates.* New York: Springer-Verlag, 1983.

Linden, E. *Silent Partners: The Legacy of the Ape Language Experiments.* New York: Times Books, 1986.

Patterson, F. and Linden, E. *The Education of Koko.* New York: Holt, Rinehart and Winston, 1981.

Rumbaugh, D., ed. *Language Learning by a Chimpanzee: The Lana Project.* New York: Academic Press, 1977.

## HUMANKIND

Passingham, R. *The Human Primate.* San Francisco: Freeman, 1992.

## VISITING AND REVISITING YELLOW BABOONS

DeVore, I. "Male Dominance and Mating Behavior in Baboons." In Beach, F., ed. *Sex and Behavior.* New York: Krieger, 1965, pp. 266–289.

Fedigan, L. "The Changing Role of Women in Models of Human Evolution." *Yearbook of Physical Anthropology*, Vol. 15 (1986), pp. 25–66.

Haraway, D. *Primate Visions: Gender, Race, and Nature in the World of Modern Science.* New York: Routledge, Chapman & Hall, 1989.

Lee, R. *The !Kung San: Men, Women and Work in a Foraging Society.* Cambridge, Massachusetts: Cambridge University Press, 1979.

Smuts, B. *Sex and Friendship in Baboons.* Chicago: Aldine, 1985.

Strum, S. *Almost Human: A Journey into the World of Baboons.* New York: Random House, 1987.

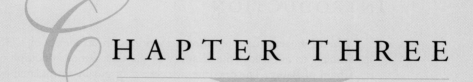

# CHAPTER THREE

# THE MEANING OF
# EVOLUTION

*contents at a glance*

# INTRODUCTION

When Linnaeus first published his writings on taxonomy in 1735, he forced scholars to reconsider the accuracy of the then famous doctrine of the **Great Chain of Being**. The doctrine, which was originally formulated by the Greek philosopher Plato (427–347 B.C.) and refined by his pupil Aristotle (384–322 B.C.), was intended to reveal the permanent natural order of all things in the universe, from the simplest forms of matter through humankind to God — the creator of the universe and all of its parts. Incorporated into subsequent Western European philosophical and religious ideas about the orderliness of the universe, the doctrine held that nature was arranged like a ladder, with inanimate matter, plants, and non-human animals located on the bottommost rungs, angels and God on the uppermost, and humankind midway between — part body and part spirit. See Figure 3.1.

Linnaeus contradicted this view, for his taxonomy suggested that humankind should be regarded not only as an animal, but also one that strongly resembled the apes. In fact, although Linnaeus maintained that humankind was created in God's image and possessed what he referred to as a "portion of intellectual divinity," he also wrote: "I demand of you, and of the whole world, that you show me a generic character … by which to distinguish between Man and Ape. I myself most assuredly know of none." It was because he could not find such a trait that, in the tenth edition of *Systema naturae*, Linnaeus grouped humans together with the other primates of which he was aware into a taxonomic category that he labelled **Anthropomorpha**. See Figure 3.2.

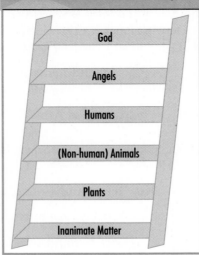

**FIGURE 3.1** A diagrammatic representation of The Great Chain of Being as it would have been conceived of in Linnaeus' day

God

Angels

Humans

(Non-human) Animals

Plants

Inanimate Matter

**FIGURE 3.2** Linnaeus' arrangement of the genus *Homo*.

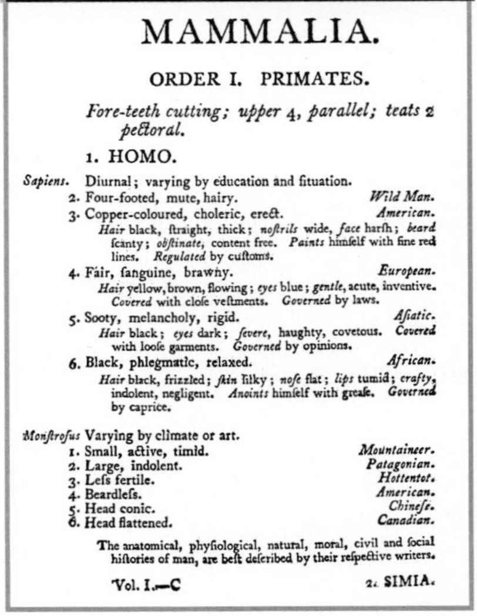

## MAMMALIA.

### ORDER I. PRIMATES.

*Fore-teeth cutting; upper 4, parallel; teats 2 pectoral.*

1. HOMO.

*Sapiens.* Diurnal; varying by education and situation.
2. Four-footed, mute, hairy.     *Wild Man.*
3. Copper-coloured, choleric, erect.     *American.*
    *Hair* black, straight, thick; *nostrils* wide, *face* harsh; *beard* scanty; obstinate, content free. *Paints* himself with fine red lines. *Regulated* by customs.
4. Fair, sanguine, brawny.     *European.*
    *Hair* yellow, brown, flowing; *eyes* blue; *gentle*, acute, inventive. *Covered* with close vestments. *Governed* by laws.
5. Sooty, melancholy, rigid.     *Asiatic.*
    *Hair* black; *eyes* dark; *severe*, haughty, covetous. *Covered* with loose garments. *Governed* by opinions.
6. Black, phlegmatic, relaxed.     *African.*
    *Hair* black, frizzled; *skin* silky; *nose* flat; *lips* tumid; *crafty*, indolent, negligent. *Anoints* himself with grease. *Governed* by caprice.

*Monstrosus* Varying by climate or art.
1. Small, active, timid.     *Mountaineer.*
2. Large, indolent.     *Patagonian.*
3. Less fertile.     *Hottentot.*
4. Beardless.     *American.*
5. Head conic.     *Chinese.*
6. Head flattened.     *Canadian.*

    The anatomical, physiological, natural, moral, civil and social histories of man, are best described by their respective writers.

**Vol. I.—C**           2. SIMIA.

But this was as far as Linnaeus was willing to go. Fearful that Church officials would place him under an ecclesiastical ban, which would have terminated his scientific career, Linnaeus did not propose a scientific explanation to account for the resemblances between apes and humans. Instead, in the introduction to *Systema naturae*, Linnaeus wrote that the similarities reflected what God had intended when He created the world, and that, as a taxonomist, it was his task not to question God's intentions, but to use his powers of observation to classify the plants and animals that God had seen fit to create. See QQ 3.1.

 **3.1  LINNAEUS ON TAXONOMY**

Linnaeus describes his view of the role of a taxonomist.

• • •

Man, the last and best of created works, formed after the image of his Maker, endowed with a portion of intellectual divinity, the governor and subjugator of all other beings, is, by his wisdom alone, able to form just conclusions from such things as present themselves to his senses, which can only consist of bodies merely natural. Hence, the first step of wisdom is to know these bodies; and to be able, by those marks imprinted on them by nature, to distinguish them from each other, and to affix to every object its proper name.

Source: Linnaeus quoted in Carl P. Swanson, *The Natural History of Man* (Englewood Cliffs, New Jersey: Prentice-Hall, Inc., 1963), p. 7.

# DARWIN'S VIEW

## AN IMPORTANT PROPOSAL

For many years after Linnaeus' death, the assumption that the resemblances between species were the result of God's handiwork prevailed. However, in the second half of the nineteenth century Charles Darwin (1809–1882) proposed "that the innumerable species, genera, and families with which the world is peopled are descended from common parents and have all been modified in the course of descent." This statement has two important implications. The first is that, whatever their differences, all forms of life are related. Given this proposition, it is understandable why primates, and in fact all animals, have similar physical traits; this is exactly what one would expect if they had evolved from "common parents." The second implication of Darwin's statement is that, notwithstanding their common parentage, all forms of life are subject to change. Thus, it stands to reason that each life form is characterized by its own evolutionary history, during which it has acquired the traits that set it apart. Darwin, however, did more than simply comment on the difference and similarities between species; he also proposed a theory to account for their natural history. The theory was intended to explain the evolution of species — how species change over time, and how these changes inevitably result in the emergence of new species. See Figure 3.3.

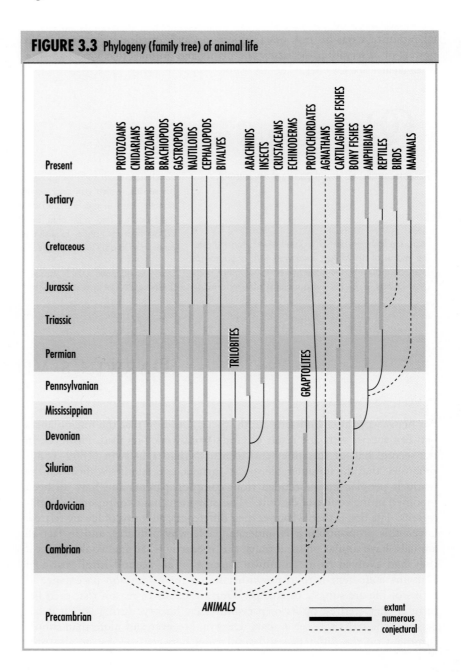

**FIGURE 3.3** Phylogeny (family tree) of animal life

# DARWIN'S GENIUS

The contents of the theory bear witness to Darwin's genius, which was based on his ability to use his exceptionally keen powers of observation and reasoning to deduce how the world's species had evolved. For instance, one of the facts on which Darwin's theory is based is that although all species have an enormously high reproductive potential, that potential can never be fully realized because the earth's limited resources cannot support continuous growth. It was this limiting factor that led Darwin to conclude that all creatures must be involved in an ongoing **struggle for existence**, sometimes with their own kind and sometimes with members of other species, in which many more perish than survive.

In addition, Darwin knew from his own observations that the individuals within a species exhibited inherited variation in their features. In other words, even though they resembled one another, the members of a species differed because they had inherited different modified traits. From this observation he deduced that those individuals whose inherited modifications made them better suited to the environment than others stood the best chance to reach maturity, acquire mates, and pass on their modified traits to their

The Bettmann Archive

A portrait of Charles Darwin (1809–1882) as a young man

descendants. It was as a result of this process of descent with modification, Darwin argued, that species composed of newly constructed organisms eventually appeared. Thus, one part of Darwin's genius was that he identified the environmental conditions that promoted evolution, and the other was that he developed a scientific explanation to account for the process. See QQ 3.2.

## QQ 3.2 DARWIN'S GENIUS

David Hull comments on the distinction between Darwin's argument in favour of evolution and the theory that bears his name.

• • •

One must . . . distinguish between Darwin's argument for the existence of evolution and a full blown theory of evolution. The outline of Darwin's argument for the existence of evolution is easy to follow. Organisms tend to produce young in such quantities that if they all lived to reproduce themselves in the same plentitude and their offspring did the same indefinitely, the earth would rapidly be overpopulated even by a single species. Clearly this does not happen. Some species increase in numbers seasonally, only to decrease later. Others increase during certain geological periods only to become extinct later. Fluctuations do occur in the number of organisms that exist at any one time in the history of the earth, but there is nothing like the rapid expansion that one would expect on the basis of their reproductive potentiality. Hence, Darwin reached the rather obvious conclusion that a high percentage of organisms produced die before they in turn are able to reproduce.

Two possibilities present themselves at this point. Either the destruction of all these organisms could occur entirely at random or it could result at least in part from the relation between each individual organism and its environment — the environment including, of course, other organisms of its own kind and other species. Certainly many organisms die in a haphazard manner unrelated to any differences between them and other organisms. For example, few if any organisms could survive immersion in lava, regardless of any specific characteristics they might or might not have. In some cases, however, it seems likely that there is a differential survival of organisms that is a function of the relation between them and their environment; they survive because in the main they are better adapted to their environment than other organisms are to theirs. In the struggle for existence, the fitter organisms tend to survive more frequently than the less fit [and it is these that give rise to new species].

Source: David Hull, *Philosophy of Biological Science* (Englewood Cliffs, New Jersey: Prentice-Hall, Inc., 1974), p. 50.

# THE PEPPERED MOTH

The evolutionary history of the **peppered moth** (*Biston betularia*) in nineteenth-century England illustrates the accuracy of Darwin's conclusions. The facts of the case are straightforward. There are two varieties of peppered moth: one relatively light in colour and the other comparatively dark. During the first half of the nineteenth century the light-coloured moths dominated the population. However, during the second half of the century the dark-coloured variety became the commoner of the two. Why? Considering what Darwin had to say about the origin of species, the change in colouration can be easily explained.

▼

Dark (left) and light (right) coloured forms of *Biston betularia*

▲

© M.W.F. Tweedie, National Audubon Society/Photo Researchers, Inc.

As will be recalled, based on the fact that the members of a species display inherited variation in their features, Darwin concluded that those best suited to the environment by virtue of their modified features are the most likely to survive, and this is exactly what happened in the case of the peppered moth. When the peppered moth is not in flight it alights on rocks and on the trunks of trees where it is preyed upon by insect-eating birds. Before the industrial revolution had significantly altered the character of the English countryside, the places where the peppered moth set down were light or covered in lichen, and while this camouflaged the light-coloured variety it left the dark ones vulnerable to the birds. Hence, more light than dark. However, during the second half of the nineteenth century, because of the large quantities of soot and other airborne pollutants that were pouring out of the chimneys of English houses and factories, the places where the moths habitually alighted became increasingly dark. It was then that the dark-coloured variety became the more numerous form. The statistics from heavily industrialized areas of England indicate the magnitude of the change. Whereas in some of these areas dark-coloured moths comprised less than one percent of the peppered moth population in 1848, only fifty years later they comprised more than 95 percent of the population. See QQ 3.3.

 **3.3 A TALE OF TWO COLOURS**

H. D. B. Kettlewell, who was one of the first to call attention to the theoretical significance of moths such as *Biston betularia*, explains the impact of the industrial revolution on the English countryside and describes the habitat of dark-coloured moths.

• • •

Ever since the Industrial Revolution commenced in the latter half of the 18th century, large areas of the earth's surface have been contaminated by an insidious and largely unrecognized fallout of smoke particles. In and around industrial areas the fallout is measured in tons per square mile per month; in places like Sheffield in England it may reach 50 tons or more. It is only recently that we have begun to realize how widely the lighter smoke particles are dispersed, and to what extent they affect the flora and fauna of the countryside.

In the case of the flora the smoke particles not only pollute foliage but also kill vegetative lichens on the trunks and boughs of trees. Rain washes the pollutants down the boughs and trunks until they are bare and black. In heavily polluted districts rocks and the very ground itself are darkened.

Now in England there are some 760 species of larger moths. Of these more than 70 [including *Biston betularia*] have exchanged their light color and pattern for dark or even all-black coloration.

Source: H.D.B. Kettlewell, "Darwin's Missing Evidence," *Scientific American* Vol. 200, No. 3, p. 48. Copyright © 1959 by Scientific American Inc. All rights reserved.

It is also interesting to note that in areas of the country where strict pollution abatement measures have been implemented, the light-coloured form of *Biston betularia* has made a numerical comeback, and this too is consistent with Darwin's conclusions.

Although Darwin was brilliant, he did not develop his ideas in an intellectual vacuum. Nor was he the first to attempt to come to grips with the phenomenon of biological change. While their arguments were often far-fetched, great thinkers before him had also speculated about the origin of species. The first to do so were Greek and Roman philosophers.

## THE EARLY GREEKS AND ROMANS

The Greek philosopher Thales (c. 640–546 B.C.) was one of the earliest to comment on the origin of the species. Foreshadowing what would later be said about the **prebiotic soup** — the conglomeration of organic molecules in the atmosphere that were washed into oceans and rivers where they may have been transformed into the first living organisms — Thales came to the conclusion that all living things were derived from the water. His countryman, the renowned Greek physician Hippocrates of Cos (460–357 B.C.), also held evolutionary views. He maintained that the physical features of humankind were influenced by the environment — an idea that is still central to evolutionary theory. It was the Roman poet-philosopher Titus Carus Lucretius (95–55 B. C.), however, who came closest to modern evolutionary thought. Like Darwin, Lucretius believed that the physical features that were preserved in a species were those that increased the members' chance of survival, either in the wild where they relied on their own hardiness to survive, or else under human supervision where their attributes were selected by their keepers. Lions, foxes, and stags, Lucretius wrote, were an example of the former; dogs, sheep, and cattle of the latter. See QQ 3.4.

## 3.4 EVOLUTION ACCORDING TO LUCRETIUS

In his book, *The Nature of the Universe*, the Roman poet-philosopher Lucretius proposed evolutionary ideas that set him ahead of his time by eighteen centuries.

• • •

Every species that you now see drawing the breath of life has been protected and preserved from the beginning of the world either by cunning or by prowess or by speed. In addition, there are many that survive under human protection because their usefulness has commended them to our care. The surly breed of lions, for instance, in their native ferocity have been preserved by prowess, the fox by cunning and the stag by flight. The dog, whose loyal heart is alert even in sleep, all beasts of burden of whatever breed, fleecy sheep and horned cattle, over all these . . . man has established his protectorate. They have gladly escaped from predatory beasts and sought peace and the lavish meals, procured by no effort of theirs, with which we recompense their service. But those that were gifted with none of these natural assets, unable either to live on their own resources or to make any contribution to human welfare, in return for which we might let their race breed in safety under our guardianship — all

© Zoological Society of San Diego 1993

Siberian tiger, which relies on its own hardiness to survive

Courtesy of The Smits Family

Domestic cat, which survives under human protection

# THE CREATION STORY

Although the early Greeks and Romans made some interesting observations about evolution, during the Middle Ages (*c.* 500–1300 A.D.) their ideas were either forgotten or lay dormant. At the time, Judeo-Christian principles dominated scientific thinking regarding the origin of species. Scientists of the period accepted the Bible as the sole authority on the subject, and its contents were interpreted in a way that stifled intellectual speculation about biological change. Theologians insisted that the **creation story** in the Book of Genesis proved that God had created the plants and the animals in the Garden of Eden, and it was assumed that, once created, a species never changed its form (see QQ 3.5). This denies the possibility of evolution since it implies that species can only reproduce exact replicas of their own kind, a principle that is known as the **immutability of species**. Humankind was also thought to be immutable for another reason; the Bible says that humans were created in God's image, and since God is perfect and unchanging, it was assumed that the same must be true of humankind. Thus, given the intellectual climate during the Middle Ages, there was no impetus to think about biological change.

# 3.5 THE CREATION STORY

In the Book of Genesis, the first book of the Bible, the origin of plants and animals is explained in divine terms.

. . .

In the beginning God created heaven and earth . . . .

God said, "Let the earth produce vegetation: seed-bearing plants, and fruit trees on earth, bearing fruit with their seed inside, each corresponding to its own species." And so it was . . . .

God said, "Let the waters be alive with a swarm of living creatures, and let birds wing their way above the earth across the vault of heaven." And so it was . . .

God said, "Let the earth produce every kind of living creature in its own species: cattle, creeping things and wild animals of all kinds." And so it was . . . .

God said, "Let us make man in our own image, in the likeness of ourselves, and let them be masters of the fish of the sea, the birds of heaven, the cattle, all the wild animals and all the creatures that creep along the ground." And so it was . . . .

God created man in the image of himself,
in the image of God he created him,
male and female he created them . . . .

Such was the story of heaven and earth as they were created.

Source: Henry Wansbrough, ed. *The New Jerusalem Bible* (New York: Doubleday & Company, Inc., 1985), pp. 17–18. Copyright © 1985 by Darton, Longman and Todd Ltd. and Doubleday, a division of Bantam Doubleday Dell Publishing Group, Inc. Reprinted by permission.

Michelangelo's depiction of The Creation Story

The Bettmann Archive

# GEORGES LOUIS DE BUFFON (1707–1778)

During the eighteenth and nineteenth centuries, despite the pervasive influence of the Bible, scholars once again began to speculate about biological change. One of the first to challenge the biblical view was the great French natural philosopher Georges Louis de Buffon. Among his many other accomplishments Buffon wrote *Histoire Naturelle*, a massive 15-volume study published between 1749 and 1767 that included a number of essays on the origin and evolution of the earth. It was in these essays that Buffon rejected the idea that all of the world's species were created in their present form; in fact, in Buffon's opinion plants and animals that were quite unlike their ancestors were capable of being produced (see QQ 3.6). He even went so far as to coin the term **degradation** to call attention to the process, by which he meant that the natural history of modern life forms could be traced back to several pure original ancestors from which they had "degenerated" when the original ancestors spread into new territory and encountered new environmental conditions. In the case of mammals, for instance, Buffon claimed that the 200 or so mammalian species of which he was aware had "degenerated" from 38 original types.

▼

George Louis de Buffon

(1707–1778)

▲

The Hulton Deutsch Collection, London

## 3.6 BUFFON'S VIEW

Buffon was convinced that the natural world was not constant but subject to incremental change.

• • •

Nature being contemporaneous with matter, space, and time, her history is that of all substances, all places, all ages; and although it appears at first view that her great works never alter or change, and that in her productions, even the most fragile and transitory, she always shows herself to be constantly the same, since her primary models regularly reappear before our eyes in new representations; however, in observing her closer, it will be seen that her course is not absolutely uniform; it will be recognized that she admits sensible variations, that she receives successive alterations, that she even lends herself to new combinations, to mutations of matter and of form; that, finally, much as she seems fixed as a whole, she is variable in each of her parts; and if we encompass her in all her extent, we can no longer doubt that she is today very different from what she was at the beginning.

Source: Buffon quoted in Carl P. Swanson, *The Natural History of Man* (Englewood Cliffs, New Jersey: Prentice-Hall, Inc., 1973), p. 94.

But while Buffon anticipated some of what Darwin would have to say, he failed to identify the mechanism that was responsible for evolutionary change. In addition, he argued that if a modern life form were placed in its original habitat it would revert to the ancestral type, and this conflicts with modern evolutionary thought. The contemporary view is that species are incapable of reverting to the ancestral forms from which they have evolved. In any event, Buffon's arguments so angered ecclesiastical officials that they warned him that he would be excommunicated from the church if he did not change his views, which he did shortly before he died.

## JEAN BAPTISTE DE LAMARCK (1744–1829)

Another French scholar, Jean Baptiste de Lamarck, also believed that plants and animals had evolved, although in a curious way. While Lamarck was willing to concede that species were created by God, he was adamant that, once created, species were then subject to biological modifications that were neither directed nor restricted by God. He also proposed an interesting but inaccurate theory to explain the nature of evolutionary change.

▼ Jean Baptiste de Lamarck (1744–1829) ▲

The Hulton Deutsch Collection, London

Lamarck's theory, which is known as **Lamarckism**, is based on the idea that species are capable of changing their physical features in order to become better suited to the environment. If a plant or animal were poorly suited to its environment, Lamarck said, then it could compensate by forcing itself to use certain parts of its body to perform new tasks. This led him to the rather obvious conclusion that the parts that were used in this way would become more fully developed during the organism's life. Thereafter, Lamarck claimed, the newly developed features were passed on to the organism's descendants, making it easier for them to survive because of the **inheritance of acquired characters**. Evolutionary change, he concluded, was the inevitable outcome of the process.

The giraffe was one of his examples. Lamarck reasoned that, in the beginning, giraffes had short legs and necks. This made it hard for them to survive since they had to compete with many other, similarly constructed animals for food. To overcome the difficulty, giraffes began to stretch their legs and necks to feed on the leaves at the tops of trees. Eventually, after many generations, the process worked — giraffes had long legs and necks and a much more reliable source of food.

Of course, Lamarck's reasoning was incorrect; giraffes can no more transmit the physical features they acquire through exercise to their offspring

## 3.7 THE MOLE'S EYES

Lamarck relied on the idea of disuse to explain why the mole's vision is poor.

• • •

▼
Mole
▲

(c) Eric Hosking, National Audubon Society/ Photo Researchers, Inc.

Eyes in the head are characteristic of a great number of different animals, and essentially constitute a part of the plan of organization of the vertebrates.

Yet the mole, whose habits require a very small use of sight, has only minute and hardly visible eyes, because it uses its organ so little . . . .

Light does not penetrate everywhere; consequently animals which habitually live in places where it does not penetrate, have no opportunity of exercising their organs of sight . . . . [I]t becomes clear that the shrinkage and even disappearance of the organ in question are the results of a permanent disuse of that organ.

Source: Jean Baptiste Lamarck, *Zoological Philosophy: An Exposition with Regard to the Natural History of Animals*, trans. Hugh Elliot (New York: Macmillan Publishing Company, Inc., 1963), p. 116.

than human body-builders can pass on their well-developed muscles to their children. Nor does the disuse of an organ, as Lamarck claimed, inevitably lead to its loss (see QQ 3.7). The fact that blind parents can produce sighted offspring illustrates the point. Still, Lamarck did regard evolution as a fact, a point of view that contemporary evolutionists share.

## BARON GEORGES CUVIER (1769–1832)

By the time Lamarck came forward with his theory, Baron Georges Cuvier, another French scholar, had already established a reputation as the leading anatomist of the day. He had gained the reputation by perfecting the techniques of comparative anatomy, meticulously comparing and contrasting the structural arrangements of various animals in order to determine the specific anatomical similarities and differences between them. So great was his reputation that it was alleged that he could identify an animal by looking at only one of its bones.

▼
Baron Georges Cuvier
(1769–1832)
▲

The Hulton Deutsch Collection, London

THE MEANING OF EVOLUTION *91*

Cuvier relied on his anatomical expertise when he began to study fossilized bones and teeth that had been discovered by construction workers in the limestone quarries of Paris. After comparing the fossils with the bones and teeth of living animals, Cuvier concluded that the materials represented animals that had become extinct. But Cuvier was not an evolutionist. Instead, he attempted to explain the presence of these ancient remains by a theory that is known as **catastrophism**. According to the theory, the remains of extinct organisms are simply examples of species originally created by God, but then destroyed by miraculous catastrophic events such as the flood described in Genesis. After each catastrophe — and the flood was but one of many — God recreated life but in a slightly different form. It was this novel idea that allowed Cuvier to maintain his faith in the Bible, and, at the same time, promote and encourage the scientific study of fossils. Despite his commitment to science, however, Cuvier was not above using his considerable reputation to discredit Lamarck, whom he apparently detested. See QQ 3.8.

## QQ 3.8 CUVIER AND LAMARCK

Richard Burkhardt, Jr. comments on the role that Cuvier played with respect to the cold reception that was accorded to Lamarck's evolutionary ideas by the French scientific establishment.

. . .

With the exception of a few brief and scattered comments Lamarck's evolutionary ideas were publicly received in silence . . . [in large part because of] the posture towards Lamarck's ideas [that were] adopted by the dominant figure of French natural science at the time: Georges Cuvier . . . .

It is difficult to estimate just how much the posture of Cuvier toward Lamarck's evolutionary ideas may have influenced contemporaries who might otherwise have been well disposed to give Lamarck's ideas some serious attention. Presumably Cuvier's influence in this regard was considerable.

Source: Richard W. Burkhardt, Jr. "Lamarck, Evolution, and the Politics of Science." *Journal of the History of Biology* Vol. 3, No. 2 (1970), pp. 291–296. Reprinted by permission of Kluwer Academic Publishers.

## CHARLES LYELL (1797–1885)

Although catastrophism was popular with the Church and among Cuvier's followers, it failed to win broad-based scientific support. Charles Lyell, a Scottish geologist, was one of the most outspoken critics of the theory. In his book, *Principles of Geology*, published in three volumes between 1830 and 1833, Lyell argued that species did not become extinct because of sudden, divinely inspired cataclysmic events. Instead, like James Hutton (1726–1797), another Scottish geologist, Lyell maintained that age-old forces such as earthquakes and erosion were constantly changing the world — an idea that is known as **uniformitarianism**. He even went to Sicily to test out the accuracy of the idea. There he studied Mt. Etna, a volcano whose massive cone, Lyell concluded, had been formed gradually over the years by ongoing natural forces rather than suddenly in historic times by the hand of God. And while Lyell did not say so himself, he clearly implied that it was these ongoing natural forces that regularly caused species to arise and eventually to become extinct. He also said that the earth was very old, certainly old enough for species to have changed gradually

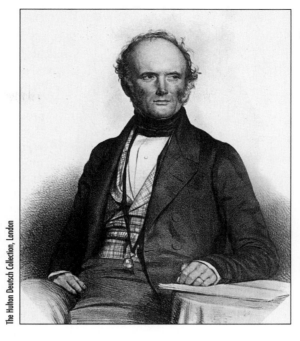

The Hulton Deutsch Collection, London

▼

Charles Lyell
(1797–1885)

▲

rather than as a result of sudden, miraculous catastrophes (see QQ 3.9). The idea was in direct conflict with the Church — according to James Ussher (1581–1656), Archbishop of Armagh, Ireland, the successive "begats" in the Bible proved that earth had been created in 4004 B.C.

## QQ 3.9 CATASTROPHISM *VS.* UNIFORMITARIANISM

Francis Hitching describes the differences between catastrophism and uniformitarianism.

• • •

Catastrophism, in the mind of the general public at the time, meant simply a belief in Noah's Flood, the single great event which had changed the face of the Earth. To . . . creationists of the time — Baron Cuvier . . . and others — catastrophism was rather different. Examining the tortured geology of the rocks, with its folds and upheavals, and evidence of past extinc-

tions, they concluded that catastrophes had struck Earth repeatedly . . . .

Lyell sought an alternative that avoided the miraculous, and said that geological history could perfectly well be understood by observing processes still going on . . . .

But Lyell went on to infer . . . that the processes on Earth had always been much the same as now, and that rates of geological change had been uniform through time. It was a concept eagerly adopted by Darwin . . . .

Source: Francis Hitching, *The Neck of the Giraffe: Darwin, Evolution, and the New Biology* (New York: New American Library, Inc., 1982), p. 129.

# Thomas Malthus (1766–1834)

Thomas Malthus
(1766–1834)

The Hulton Deutsch Collection, London

There is little doubt that Darwin was indebted to all of his intellectual forerunners — to Buffon who contemplated the possibility of biological evolution, to Lamarck who theorized about the process, to Cuvier who recognized that fossils represented extinct forms of life, and to Lyell who implied that modern life forms had been produced over an extended period of time by natural causes. Darwin, however, might never have formulated his own theory of evolution were it not for another outstanding scholar, Thomas Malthus (1766–1834), an English clergyman and political economist.

In 1798, Malthus published *An Essay on the Principle of Population* in which he argued that, while the number of individuals in a population increased at a geometric rate, their food supply increased at an arithmetic rate. The result, according to Malthus, was that populations would inevitably outstrip their food supply. When this happened what he referred to as a "struggle for existence" took place, in which those least fit to compete suffered dire consequences. Scrawny plants and non-human animals paid the ultimate price — they perished. Meanwhile, among humans, those least fit to compete were condemned to poverty-stricken lives. It is because of this so-called **great restrictive law** that Darwin is indebted to Malthus (see QQ 3.10). It was this law that encouraged Darwin to think that species were involved in an ongoing struggle for existence, and in that struggle only the fittest survived to reproduce.

## QQ 3.10 THE GREAT RESTRICTIVE LAW

Thomas Malthus explains why populations outstrip their food supply and what happens when they do.

• • •

I say that the power of the population is indefinitely greater than the power of the earth to produce subsistence for man . . . since [p]opulation, when unchecked, increases in a geometric ratio, [but] subsistence increases only in an arithmetical ratio. A slight acquaintance with numbers will shew the immensity of the first power in comparison with the second.

Throughout the animal and vegetable kingdoms, nature has scattered the seeds of life abroad with the most profuse and liberal hand. She has been comparatively sparing in the room and nourishment necessary to rear them . . . . The race of plants and the race of animals shrink under this great restrictive law. And the race of man[kind] cannot by any effort of reason

# CHARLES DARWIN (1809–1882)

## EARLY YEARS

Indebted though he was to others, Charles Darwin's contribution to our understanding of evolution was nonetheless profound; it ranks as one of the greatest scientific achievements of all time. Darwin, however, was not always the serious scholar that many no doubt imagine. Born into a well-known English family — his mother Susannah was the daughter of Josiah Wedgwood, the famous Staffordshire potter; his father Robert a prosperous country physician; and his paternal grandfather Erasmus an evolutionary thinker in his own right — young Darwin was a mediocre student at best. Although he showed an early interest in natural history and read widely, when his father sent him to Edinburgh University at the age of sixteen to study medicine, Charles quickly discovered that he was cold to the idea of following in his father's footsteps. When he rejected medicine as a career

The Hulton Deutsch Collection, London

Christ's College, Cambridge, in Darwin's day

his father sent him to Christ's College, Cambridge, to study theology, but there too his lack-lustre performance failed to set him apart; instead of devoting himself to his studies, Darwin preferred to hunt and gamble and spend his time with other students who were similarly inclined. Darwin may, in fact, have learned more about natural history outside of class than in, particularly while on excursions to the countryside that were organized by John Stevens Henslow (1796–1861), a botany professor at Cambridge who became young Darwin's mentor and friend.

## THE BEAGLE

In 1831 Darwin's life began to change. Recommended for the position by Henslow, and over the strenuous objections of his father, twenty-two year old Charles accepted an invitation to serve as a naturalist on the British scientific research vessel *H.M.S. Beagle*, which the Admiralty had commissioned to chart the coast of South America and more accurately fix longitude by undertaking chronological reckonings around the world (see Figure 3.4). In his five years of service, in which he visited South America and Australia while the *Beagle* sailed around the world, Darwin spent thousands of hours studying the relationships between organisms and their environments. He later claimed that it was this voyage, more than anything else, that shaped his subsequent scientific career. See QQ 3.11.

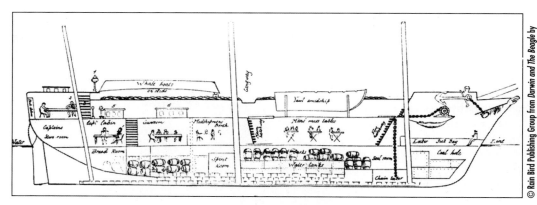

Side-elevation of *H.M.S. Beagle*

## QQ 3.11 THE VOYAGE OF THE *BEAGLE*

Darwin recalls the importance of the voyage of the *Beagle* to his subsequent scientific career.

. . .

The voyage of the Beagle has been by far the most important event in my life and has determined my whole career . . . . I have always felt that I owe to the voyage the first real training or education of my mind.

I was led to attend closely to several branches of natural history, and thus my powers of observation were improved . . . .

Therefore, my success as a man of science, whatever this may have amounted to, has been determined, as far as I can judge, by complex and diversified mental qualities and conditions. Of these the most important have been — the love of science — unbounded patience in long reflecting over any subject

**FIGURE 3.4** The *Beagle's* ports of call

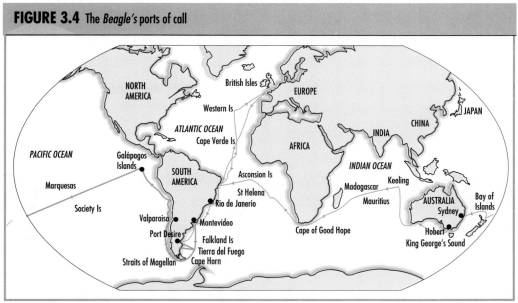

Redrawn and adapted from ORIGINS by Richard E. Leakey and Roger Lewin. Copyright © 1977 by Richard E. Leakey and Roger Lewin. Used by permission of the publisher, Dutton, an imprint of New American Library, a division of Penguin Books USA Inc.

## NATURAL SELECTION

When Darwin returned to England in 1836 he continued to pursue his research, adding to what he had already learned about the relationships between organisms and their environments while travelling around the world. He also began to formulate his own theory of evolution, a theory that, above all else, called attention to the process of **natural selection**. In his monumental work, *On the Origin of Species*, which was published in 1859, Darwin explained in detail what he meant by natural selection.

Drawing on what Malthus had said about the "struggle for existence," and taking into account the fact that species exhibit variation in their features, Darwin reasoned that some individuals in a species must be better suited to the environment than others. When all of these organisms are forced to compete amongst themselves or with the members of other species, he said, and not only for food as Malthus had argued but for *anything* that contributes to their survival, then the ones that are best **adapted** or suited to their environment because of their inherit-

ON

THE ORIGIN OF SPECIES

BY MEANS OF NATURAL SELECTION,

OR THE

PRESERVATION OF FAVOURED RACES IN THE STRUGGLE
FOR LIFE.

By CHARLES DARWIN, M.A.,

FELLOW OF THE ROYAL, GEOLOGICAL, LINNÆAN, ETC., SOCIETIES;
AUTHOR OF "JOURNAL OF RESEARCHES DURING H. M. S. BEAGLE'S VOYAGE
ROUND THE WORLD."

LONDON:
JOHN MURRAY, ALBEMARLE STREET.
1859.

*The right of Translation is reserved.*

The Bettmann Archive

ed features will survive; the others will perish. And since it is generally only the survivors that are able to reproduce, it is mostly their modified biological characteristics that will be passed on in the course of descent. Darwin referred to the process as natural selection to emphasize that, in the struggle for existence, it is strictly nature or the environment that selects the survivors, and that new species are a natural and a predictable outcome of the process. See QQ 3.12.

## QQ 3.12 NATURAL SELECTION

In the last paragraph of *The Origin of Species*, Darwin summarized his view of natural selection.

• • •

It is interesting to contemplate an entangled bank, clothed with many plants of many kinds, with birds singing on the bushes, with various insects flitting about, and with worms crawling through the damp earth, and to reflect that these elaborately constructed forms, so different from each other, and dependent on each other in so complex a manner, have all been produced by . . . Natural Selection, entailing Divergence of Character and the Extinction of less-improved forms. Thus, from the war of nature, from famine and death, the most exalted object which we are capable of conceiving, namely, the production of the higher animals, directly follows. There is grandeur in this view of life, with its several powers, having been originally breathed into a few forms or into one; and that, whilst this planet has gone cycling on according to the fixed law of gravity, from so simple a beginning endless forms most beautiful and most wonderful have been, and are being, evolved.

Source: Charles Darwin, *On the Origin of Species By Means of Natural Selection, or the Preservation of Favoured Races in the Struggle for Life* (London: John Murray, 1859), pp. 489–90.

## DARWIN'S FINCHES

Darwin developed the concept of natural selection in stages. During the *Beagle's* voyage he had the opportunity to study a wide variety of plants and animals. Among these were a number of finches that lived on the Galápagos Islands, about 1000 kilometres off the northwest coast of South America. The *Beagle* cruised for just over a month among the islands, from mid-September through mid-October, 1835.

While he was there Darwin identified thirteen species of finches, which have since become known as **Darwin's finches**. Although all had short tails, laid white eggs spotted with pink, and built roofed nests, their beaks were slightly different. These slight differences were important, for they allowed the small birds to occupy different ecological niches and rely on different sources of food.

There are two main ecological zones on the Galápagos Islands — an arid coastal region and a humid forested region in the interior — and each was inhabited by finches whose beaks were adapted to particular sources of food within each zone. For example, whereas six species of finches lived on the ground near the sea, three of these had relatively blunt, powerful beaks that were used to break open hard seeds, one had a comparatively long, pointed beak that was best suited to feeding on the prickly pear, and two possessed beaks that were suitable for feeding on cactus as well as on seeds. And while seven species lived in the interior of the islands, one sported a parrot-like beak that allowed it to feed on buds and fruits, four had beaks that were best suited to capturing insects in trees, one had a woodpecker-like beak that it used to root out wood-boring insects, and another possessed a warbler-like beak that allowed it to capture insects in bushes. Equally important, none of the species was in direct competition with any of the others for food. The three seed-eating varieties, for instance, each fed on different sizes of seeds, and the four species that pursued insects in trees captured different prey. See Figure 3.5.

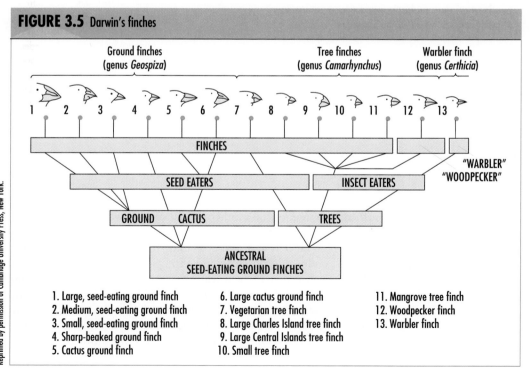

**FIGURE 3.5** Darwin's finches

Ground finches (genus *Geospiza*)    Tree finches (genus *Camarhynchus*)    Warbler finch (genus *Certhicia*)

1 2 3 4 5 6 7 8 9 10 11 12 13

FINCHES

SEED EATERS    INSECT EATERS

"WARBLER"
"WOODPECKER"

GROUND   CACTUS    TREES

ANCESTRAL
SEED-EATING GROUND FINCHES

1. Large, seed-eating ground finch
2. Medium, seed-eating ground finch
3. Small, seed-eating ground finch
4. Sharp-beaked ground finch
5. Cactus ground finch

6. Large cactus ground finch
7. Vegetarian tree finch
8. Large Charles Island tree finch
9. Large Central Islands tree finch
10. Small tree finch

11. Mangrove tree finch
12. Woodpecker finch
13. Warbler finch

The fact that the finches never had any other direct competitors or natural enemies on the islands is also important. Those who have since studied Darwin's finches reason that this must have helped to promote the **adaptive radiation** of the birds — the diversity that results when the members of a rapidly evolving population become adapted to a wide variety of ecological niches that are available for them to exploit. They also maintain that if it were not for the absence of competitors and predators, then it would have been far more difficult for so many species of Darwin's finches to have evolved. See QQ 3.13.

 **QQ**

## 3.13 DARWIN'S FINCHES

David Lack comments on the adaptive radiation of Darwin's finches.

• • •

That Darwin's finches are so highly differentiated suggests that they colonized the Galapagos Islands considerably ahead of the other land birds [that live there today]. Therefore, for a period, perhaps a very long period, they were probably without food competitors of other species, while at the same time a variety of foods and habitats was available to them. Even at the present time there are few other passerine birds [or those whose feet are adapted for perching] in the Galapagos, and these are predominantly insectivorous species, some of which may perhaps compete with the warbler-finch *Certhidea*, and possibly with the insectivorous tree-finches, but they compete scarcely, if at all, with the other forms of Darwin's finches. In particular, the later arrivals include no seed-eating, cactus-feeding, or wood-boring birds.

The absence of other land birds has had a most important influence on the evolution of Darwin's finches, since it has allowed them to evolve in directions which otherwise would have been closed to them . . . .

The absence of predators [likewise had an important influence, for it] probably means that Darwin's finches have been limited in their numbers primarily by their food supply. When this is the case, adaptations in feeding methods are likely to be of special importance in determining the survival of species, so that the absence of predators may well have accelerated the adaptive radiation of the finches.

Source: David Lack, *Darwin's Finches — an Essay on the General Biological Theory of Evolution* (Cambridge, England: Cambridge University Press, 1947), pp. 113–114. Reprinted with the permission of Cambridge University Press.

## ARTIFICIAL SELECTION

Darwin turned his attention to another species of bird when he returned to England. There he studied domesticated pigeons; in fact, he not only corresponded with pigeon fanciers and read everything he could about pigeons, but also joined two London pigeon clubs and kept pigeons himself. It was as a result of this experience that Darwin learned that the natural history of the domesticated pigeon could be traced back to the wild rock-dove, and that its tame descendants had been kept by fanciers since the heyday of the Roman Empire (between 500 B.C. and 500 A.D.). But what impressed Darwin most was that fanciers could develop new physical features in a flock by allowing only certain birds to mate, that is, through a procedure that is known as **artifi-**

cial selection. He concluded that it was this procedure that had given rise to so many breeds, including, among others, the carrier, tumbler, pouter, runt, barb, turbit, Jacobin, trumpeter, laugher, and fantail, each with its own habits, voice, and characteristic appearance. Darwin wrote about these in the first chapter of the *Origin of Species*, in which he examined the effects of artificial selection. See QQ 3.14.

 ## 3.14 THE DOMESTICATED PIGEON

**Darwin describes some breeds of the domesticated pigeon.**

• • •

The diversity of the breeds is something astonishing. Compare the English carrier and the short-faced tumbler, and see the wonderful difference in their beaks, entailing corresponding differences in their skulls. The carrier, more especially the male bird, is also remarkable from the wonderful development of the carunculated [or small fleshy growth of] skin about the head; and this is accompanied by greatly elongated eyelids, very large external orifices to the nostrils, and a wide gape of mouth. The short-faced tumbler has a beak in outline almost like that of a finch; and the common tumbler has the singular and strictly inherited habit of flying at a great height in a compact flock, and tumbling in the air head over heels. The runt is a bird of great size, with long massive beak and large feet; some of the sub-breeds of runts have very long necks, others very long wings and tails, others singularly short tails. The barb is allied to the carrier, but, instead of a very long beak,

has a very short and broad one. The pouter has a much elongated body, wings, and legs; and its enormously developed crop, which it glories in inflating, may well excite astonishment and even laughter. The turbit has a short and conical beak, with a line of reversed feathers down the breast; and it has the habit of continually expanding, slightly the upper part of the oesophagus. The Jacobin has the feathers so much reversed along the back of the neck that they form a hood; and it has, proportionally to its size, much elongated wing and tail feathers. The trumpeter and laugher, as their names express, utter a very different coo from the other breeds. The fantail has thirty or even forty tail-feathers, instead of twelve or fourteen, the normal number in all the members of the great pigeon family, and these feathers are kept expanded, and are carried so erect that in good birds the head and tail touch . . . . Several other less distinct breeds might be specified.

*Source: Charles Darwin, On the Origin of Species By Means of Natural Selection, or the Preservation of Favoured Races in the Struggle for Life (London: John Murray, 1859), pp. 21–22.*

▼

Pouter

pigeon

▲

## NATURAL SELECTION AND DARWIN'S FINCHES

It was while he was learning about pigeons that Darwin's mind went back to the Galápagos Islands. There were many types of finches on the islands. How had they become different? Obviously no animal breeder had produced the variation. Nor did Darwin believe that the Creator had made them that way in accord with a predetermined plan. Instead, he reasoned that the thirteen species had evolved from an extinct common ancestor that had come to the islands from South America in the remote past. With no natural enemies and plenty of food the original finch population began to increase, and as it grew it became more physically varied because of the acquisition of modified traits. Eventually, however, the population became so large that the various types of finches that had evolved from the original ancestor were forced to compete with each other for food. When this happened the ones whose modified beaks enabled them to get food survived and reproduced; the others perished. The thirteen species that Darwin observed were the result of the process.

## DARWIN AND WALLACE

Today, biological anthropologists maintain that the living primates have evolved from a common ancestor much like Darwin's finches. Darwin did not say this in his book; he also carefully avoided the topic of human evolution since he knew that natural selection contradicted the idea that humans were a special creation of God. In fact, Darwin made it clear that he would have postponed writing about natural selection if that were possible, preferring instead to collect ever more evidence in support of his theory before he brought it forward. But, in the meantime, another English naturalist, Alfred Russel Wallace (1823–1913), who earned his livelihood by collecting biological specimens from remote locations and selling these in Europe, had come to the same conclusions as Darwin. He wrote to Darwin from Indonesia in 1858, appending a short paper on natural selection to his letter. The paper was

Alfred Russel Wallace
(1823–1913)

The Hulton Deutsch Collection, London

titled "On the Tendency of Varieties to Depart Indefinitely from the Original Type," and when Darwin read the contents, which Wallace had conceived of while suffering from a bout of malaria, he was struck by the remarkable similarities between Wallace's conclusions and his own.

After consulting with his scientific colleagues as to how best to solve the thorny problem of priority, Darwin concluded that it was appropriate that Wallace be given equal credit for the idea of natural selection as long as his own priority was acknowledged (see QQ 3.15). As a result, in 1858, Wallace's paper along with an excerpt from Darwin's work on natural selection, which was written before he had heard from Wallace, were read before the Linnaean Society of London. However, the next year, when Darwin published his book, it was he rather than Wallace who was given credit for the theory.

## 3.15 DARWIN AND WALLACE

**Stephen Jay Gould tells the story of Darwin and Wallace.**

• • •

Darwin, to recount the famous tale briefly, developed his theory of natural selection in 1838 and set it forth in two unpublished sketches of 1842 and 1844. Then, never doubting his theory for a moment, but afraid to expose its revolutionary implications, he proceeded to stew, dither, wait, ponder, and collect data for another fifteen years. Finally, at the virtual insistence of his closest friends, he began to work over his notes, intending to publish a massive tome that would have been four times as long as the *Origin of Species*. But, in 1858, Darwin received a letter and manuscript from a young naturalist, Alfred Russel Wallace, who had independently constructed the theory of natural selection while lying ill with malaria on an island in the Malay Archipelago. Darwin was stunned by the detailed similarity. Wallace even claimed inspiration from the same nonbiological source — Malthus' *Essay on Population.*

Darwin, in great anxiety, made the expected gesture of magnanimity, but devoutly hoped that some way might be found to preserve his legitimate priority. He wrote to Lyell: "I would far rather burn my whole book than that he or any other man should think that I have behaved in a paltry spirit." But he added a suggestion: "If I could honorably publish, I would state that I was induced now to publish a sketch . . . from Wallace having sent me an outline of my general conclusions." Lyell and [and another colleague named] Hooker took the bait and came to Darwin's rescue. While Darwin stayed home, mourning the death of his young child from scarlet fever, they presented a joint paper to the Linnaean Society containing an excerpt from Darwin's 1844 essay together with Wallace's manuscript. A year later, Darwin published his feverishly compiled "abstract" of the longer work — the *Origin of Species.* Wallace had been eclipsed.

Source: Stephen Jay Gould, *The Panda's Thumb: More Reflections in Natural History* (New York: W. W. Norton & Company, Inc., 1980), p. 48.

## AS DARWIN HAD FEARED

The *Origin of Species* was a sensation; the 1250 copies that were originally printed sold out the day they appeared in bookstores, and, since then, the book has become one of the most popular scientific treatises of all time. However, as Darwin had feared, despite the fact that he scarcely mentioned humankind in the book, writing only that he hoped that his work would throw "light … on the origin of man and his history," his conclusion that the evolution of species was due to natural selection was interpreted by most clerics of the day as an attack on Christian dogma. Nevertheless, Darwin did not debate his critics in public — that was done by his colleagues, including the famous British zoologist, Sir Thomas Henry Huxley (1825–1895), who was nicknamed "Darwin's Bulldog." One of the most famous debates of this sort took place in 1860 between Huxley and Samuel Wilberforce, Bishop of Oxford (1805–1873), who was also known as "Soapy Sam" because of the eloquence of his sermons. See QQ 3.16.

 **3.16 HUXLEY DEFENDS DARWIN**

In 1860, at the annual meeting of the British Association for the Advancement of Science, there was a dramatic exchange between Sir Thomas Henry Huxley and Bishop Samuel Wilberforce.

• • •

Bishop Wilberforce, throwing a glance at Huxley, ended a suave and superficial speech by asking him "as to his belief in being descended from an ape. Is it on his grandfather's or his grandmother's side [,he asked,] that the ape ancestry comes in?" Huxley did not rise till the meeting called for him; then he let himself go . . . .

"I asserted, and I repeat, [he replied,] that a man has no reason to be ashamed of having an ape for his grandfather. If there were an ancestor whom I should feel shame in recalling it would rather be a *man* — a man of restless and versatile intellect — who, not content with success in his own sphere of activity, plunges into scientific questions with which he has no real acquaintance, only to obscure them by aimless rhetoric, and distract attention of his hearers from the real point at issue by eloquent digressions and skilled appeals to religious prejudice."

Source: Edward Clodd, *Thomas Henry Huxley* (New York: Dodd, Mead and Company, 1902), pp. 21–22.

## THE DESCENT OF MAN

Darwin did finally turn his attention to human evolution in *The Descent of Man*, published in 1871, in which he argued, in one passage, that "the facts … declare, in the plainest manner, that man is descended from some lower form, notwithstanding that connecting links have not hitherto been discovered." As for the connecting links themselves, Darwin wrote something that biological anthropologists would confirm in the twentieth century, namely, that "[t]he Simidae then branched off into two great stems, the New World and Old World monkeys; and from the latter at a remote period, Man, the wonder and the glory of the universe, proceeded." Nine years after the book was published, at the age of seventy-three, Darwin died. He was laid to rest next to the famous physicist Sir Isaac Newton in Westminster Abbey, London.

The Hulton Deutsch Collection, London

▼

Thomas Henry Huxley (1825–1895)

▲

The Hulton Deutsch Collection, London

▼

Bishop Samuel Wilberforce (1805–1875)

▲

# AMENDMENTS

## POLYGENESIS

Although Darwin's conclusions were based on solid factual evidence, much of which he himself amassed, some of that evidence was unavailable in Darwin's day. Under the circumstances, it is not surprising that certain of his proposals have been amended over the years. The contemporary view that all of the world's species have evolved from but one common ancestor is a case in point. In 1859 Darwin wrote that it was possible that life had originated not once, but perhaps on several occasions, and, although Darwin never said so himself, this encouraged some of his colleagues to conclude that there was more than one ancestral form of life. The belief that the world's species evolved from several original ancestors, rather than one, is called **polygenesis**.

Since Darwin's day, however, scientists have discovered that all forms of life are composed of the same chemical compounds. More specifically, scientists now know that the bodies of *all* life forms are composed of proteins; that proteins are composed of amino acids; and that amino acids are composed of various combinations of four chemical bases (adenine, cytosine, guanine, and thymine). These bases provide the information that is contained in deoxyribonucleic acid or **DNA** — the hereditary material that governs how each and every living thing is constructed. This has led scientists to reject polygenesis in favour of the view that all species owe their existence to a common ultimate ancestor that originally possessed this universal chemical code, and whose descendants possess the code because it makes life possible. See QQ 3.17.

## 3.17 POLYGENESIS

Mark Ridley explains why modern evolutionists support the view that all species evolved from a single common ancestor rather than from multiple ancestors.

• • •

Bodies are built from the hereditary material, DNA, by the translation of a sequence made up of four [chemical] bases, which are symbolized by the letters **A**, **C**, **G**, and **T** [**A** for adenine, **C** for cytocine, **G** for guanine, and **T** for thymine]. A triplet of these bases specifies an amino acid; a sequence of triplets specifies a sequence of amino acids; a sequence of amino acids

makes up a protein; and (roughly speaking) bodies are built of many different proteins. What matters here is that the code . . . is known to be universal . . . . The universality of the code is easy to understand if every species is descended from a common ancestor. Whatever code was used by the common ancestor would, through evolution, be retained. It would be retained because any change in it would be disastrous. A single change would cause all the proteins of the body, perfected over millions of years, to be built wrongly; no such body could live.

Source: Mark Ridley, *The Problems of Evolution* (Toronto: Oxford University Press, 1985), pp. 10–11. Reprinted by permission.

## PUNCTUATED EQUILIBRIUM

Another amendment to one of Darwin's proposals is currently being debated, and it too was prompted by the accumulation of new evidence. When Darwin first proposed his theory scientists assumed that evolution was a slow, gradual, and continuous process in which one species shaded almost imperceptibly into the next as a result of incremental biological change — an idea that is known as **gradualism**. Darwin himself believed this was true. In fact, along with the idea that evolution was not directed toward the attainment of any particular goal, gradualism was one of Darwin's basic tenets.

However, in 1972, palaeontologists Niles Eldredge and Stephen Jay Gould pointed out that the steady accumulation of fossils during the twentieth century indicated that the evolution of species was characterized by sudden rather than gradual changes, with species remaining virtually unchanged for perhaps millions of years and then suddenly being transformed into new species — *suddenly* in this case meaning thousands of years. In other words, based on the fossil record, Eldredge and Gould concluded that evolutionary change was discontinuous rather than continuous. In concert with this view, many evolutionists have now rejected gradualism in favour of **punctuated equilibrium** —Eldredge and Gould's idea that the evolution of species is characterized by **saltations**, that is, by relatively long periods of biological stability interrupted by the comparatively sudden appearance of new forms. See Figure 3.6 and QQ 3.18.

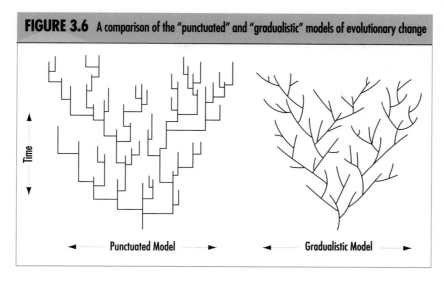

**FIGURE 3.6** A comparison of the "punctuated" and "gradualistic" models of evolutionary change

Time

◄— **Punctuated Model** —►      ◄— **Gradualistic Model** —►

 **3.18 GRADUALISM *VS.* PUNCTUATED EQUILIBRIUM**

Niles Eldredge and Michelle Eldredge comment on the nature of evolutionary change.

• • •

Traditionally, paleontologists have emphasized time over geographic distribution as the more important element of the evolutionary process. Strong evidence, suggests, however, that evolution is not simply slow, steady change of an entire species through long periods of time . . . .

The fossil record is full of apparently sudden evolutionary jumps, where a parent species is followed by its daughter species without intermediate fossil links connecting the two. The traditional explanation for such jumps is an incomplete fossil record, but our findings contradict tradition . . . .

Perhaps the most amazing feature of . . . [the evolutionary process] is stasis — a persistence against change — through vast amounts of time. Contrary to popular belief, evolutionary change seems to occur infrequently, and usually in small, isolated populations in a short span of time. The bulk of a species' history is stasis, and there is no inexorable, progressive evolutionary march through time.

Source: Niles Eldredge and Michelle J. Eldredge, "A Trilobite Odyssey," *Natural History* Vol. 81, No. 10 (1972), pp. 53–59.

Whether punctuated equilibrium can account for the evolutionary history of all the world's species, or even of a majority, remains to be seen — this is a problem that only can be settled by research. In any case, the punctuated equilibrium model does not conflict with Darwin's concept of natural selection; it is best regarded as a way of characterizing the process. Thus, with or without the Eldredge-Gould amendment, Darwin's contention that natural selection governs the evolution of species remains an extremely powerful intellectual tool; it is not only the basis of sciences such as biology and zoology, but is also the conceptual foundation on which biological anthropology is based. Biological anthropologists rely on the principle of natural selection to explain the biological history of the primates — how we and our closest living relatives evolved.

# MISGUIDED ATTACKS

Scientists do not accept theories on faith. In order for the scientific community to embrace a theory it must, above all else, account for the facts that it attempts to explain in a more efficient and a more effective manner than competing theories. Darwin's contention that natural selection governs the evolution of species is no exception. Although it does not explain everything about the process, it has stood up to more than a century of careful scrutiny by countless scholars from a wide variety of disciplines. It is on account of this fact that Darwin's ideas currently enjoy broad-based scientific support and appeal. There have, however, been some misguided, religiously-inspired attacks on Darwin, and these are worth reporting.

## THE MONKEY TRIAL

One of the most celebrated of these attacks was launched in Tennessee, in 1925, at what has since become known as the **monkey trial**. In that trial, John T. Scopes, a school teacher in Dayton, Tennessee, was charged with teaching the theory of evolution to his high school biology class contrary to Section One of Tennessee's 1925 *Anti-Evolution Law*. The section stated "that it shall be unlawful for any teacher in any of the universities, normal and all other public schools of the state, which are supported in whole or in part by the public-school funds of the state, to teach any theory that denies the story of the divine creation of man as taught in the Bible and to teach instead that man has descended from a lower order of animals."

Not surprisingly, the eleven-day trial attracted international media coverage, for it pitted William Jennings Bryan for the prosecution against Clarence Darrow for the defence — two legendary legal figures of the day. Bryan, who was a three-time presidential candidate and a former Secretary of State under Woodrow Wilson, happened to be attending a meeting of the World Fundamentalist Society in Memphis as the trial date approached, and when he was asked to serve as special prosecutor he gladly accepted the invitation free of charge in order to defend his Christian views. "We cannot afford to have a system of education that destroys the religious faith of our children," he said. Meanwhile, Darrow had agreed to act for Scopes free of charge after having being invited to do so by the American Civil Liberties Union.

Clarence Darrow
(1857–1938) with
hands folded and
William Jennings Bryan
(1860-1925) at the
monkey trial

Huge crowds jammed the small courthouse and overflowed onto the courthouse lawn in anticipation of a fiery debate, and the spectators were not disappointed. In an unorthodox move Darrow called Bryan to the stand to testify as an expert on the Bible, and, as the transcript of the trial shows, he all but demolished the credibility of his opponent who, incidentally, died within a week. Although Scopes was found guilty and fined $100.00, the decision was overturned on appeal. See QQ 3.19.

 **3.19 THE *ANTI-EVOLUTION LAW***

In his remarks to the court on the third day of the monkey trial, Clarence Darrow, the lawyer who defended Scopes, discussed the implications of Tennessee's *Anti-Evolution Law.*

• • •

If today you can take a thing like evolution and make it a crime to teach it in the public schools, tomorrow you can make it a crime to teach it in the private schools, and next year you can make it a crime to teach it in the hustings or in the church. At the next session [of the legislature] you may ban books and the newspapers. Soon you may set Catholic against Protestant, and try to foist your own religion upon the minds of men.

After a while, your Honor, it is the setting of man and creed against creed, until with flying banners and beating drums we are marching backward to the glorious ages of the sixteenth century when

## CHRISTIAN FUNDAMENTALISTS AND SCIENTIFIC CREATIONISTS

The monkey trial foreshadowed things to come. Since 1925 other attempts have been made to prohibit or restrict the teaching of evolutionary theory, particularly by **Christian fundamentalists** who insist on a literal interpretation of the Bible. This flows from their belief that the Bible contains the unadulterated word of God, and that its contents must be understood without resorting to mysticism or allegory or metaphor. More specifically, in accord with the creation story in the Book of Genesis, Christian fundamentalists maintain that God created plants and animals in their present form in the Garden of Eden. They also believe, in concert with the number of generations mentioned in the Bible since Adam and Eve, that the earth is at best a few thousand years old. Under the circumstances, it is hardly surprising that some Christian fundamentalists claim that Darwinian theory should not be taught in public schools.

Others among them have adopted a different approach. They call themselves **scientific creationists**, and rather than oppose the teaching of Darwinian theory they insist that the creation story should be incorporated into the science curriculum of public schools and taught in science classes. This demand is reasonable, they say, because the Bible is much more than just a great theological work. From the standpoint of scientific creationists the Bible is also a reliable source of scientific knowledge, far more reliable, in fact, than any other source, including Darwin's *Origin of Species*. Aggressive lobbying on their part has yielded them political dividends: in 1981 they encouraged the Arkansas legislature to pass a law that required public school teachers to teach the creation story to their science classes. Although that law and others like it in Mississippi and Louisiana were subsequently deemed to be unconstitutional, by 1982 similar laws were pending in 18 other states.

## THE PROBLEM WITH SCIENTIFIC CREATIONISM

Anthropologists reject the agenda of scientific creationists and not without reason. However enthusiastic scientific creationists may be, there is overwhelming scientific evidence that the earth is billions of years old, that the physical features of plants and animals have changed through time, and that natural selection is the most efficient and the most effective way to explain how species evolve. Under the circumstances, scientific creationism is contradicted by the facts, and this makes it an empty doctrine. However, the most telling criticism of scientific creationism is that it is not really science at all. Science is based on observation and testing, not on faith which is the cornerstone of belief in the Bible, and to confuse them is a monumental disservice to both. Misrepresenting religion as science is an affront to both intellectual traditions. See QQ 3.20.

## 3.20 SCIENTIFIC CREATIONISM

Michael Ruse, Professor of History and Philosophy at the University of Guelph, minces no words in his criticism of scientific creationism.

• • •

. . . I believe Creationism is wrong: totally, utterly, and absolutely wrong. I would go further. There are degrees of being wrong. The Creationists are at the bottom of the scale. They pull every trick in the book to justify their position. Indeed, at times, they verge right over into the downright dishonest. Scientific Creationism is not just wrong: it is ludicrously implausible. It is a grotesque parody of human thought, and a downright misuse of human intelligence. In short, to the Believer, it is an insult to God . . . .

Let me conclude with one last reflection. Obviously, I love and cherish Darwinian evolutionary theory, as one of the great intellectual achievements of all time. But my pleading is not just for Darwinism,

or any kind of evolutionism. It is for all human inquiry, particularly all scientific inquiry. If Darwinism is beaten down by the Creationists, who falls next? Remember that the Bible speaks of the sun stopping for Joshua. Both Luther and Calvin took this as textual evidence against Copernicus. Will we have to make room for religion in physics, also? And if religion, why not astrology, and all the other world systems? There is no shortage of believers prepared to fight for their causes. And . . . if Scientific Creationism is taught as a viable alternative, there cannot fail to be a deadening of the critical faculties. What is known to be fallacious will then be judged valid, and what is seen to be inadequate will be taken as proven. Hence, my fight is not just a fight for one scientific theory. It is a fight for all knowledge.

Source: Michael Ruse, *Darwinism Defended: A Guide to the Evolution Controversies* (Toronto: Addison-Wesley Publishing Company, 1982), pp. 303 and 329.

Finally, it is worthwhile to point out that not all Christians object to Darwin's views. Some Christian theologians hold that while humankind was certainly created in God's *spiritual* image, this does not deny the possibility that humankind's *biological* features have evolved. The Roman Catholic Church has adopted exactly this position. On 26 April, 1986, Pope John Paul II told the participants in a symposium on faith and evolution that the Roman Catholic faith does not oblige its adherents to reject evolutionary views, as long as evolutionists focus their attention on biological phenomena and not on the evolution of humankind's spiritual and moral attributes.

# KEY TERMS AND CONCEPTS

GREAT CHAIN OF BEING

ANTHROPOMORPHA
(*an-throh-poh-mor'-fah*)

EVOLUTION OF SPECIES

STRUGGLE FOR EXISTENCE

PEPPERED MOTH

PREBIOTIC SOUP

CREATION STORY

IMMUTABILITY OF SPECIES

DEGRADATION (*deg-reh-day'-shun*)

LAMARCKISM (*la-mahr'-kiz-em*)

INHERITANCE OF ACQUIRED CHARACTERS

CATASTROPHISM (*ka-tas'-troh-fiz-em*)

UNIFORMITARIANISM
(*yoo'-ni-for-mi-ter'-ee-an-iz-em*)

GREAT RESTRICTIVE LAW

NATURAL SELECTION

ADAPTED

DARWIN'S FINCHES

ADAPTIVE RADIATION

ARTIFICIAL SELECTION

POLYGENESIS

DNA (DEOXYRIBONUCLEIC ACID)
(*dee-ox'-ee-ry-bow-noo'-klay-ik + acid*)

GRADUALISM

PUNCTUATED EQUILIBRIUM

SALTATIONS (*sal-tay'-shunz*)

MONKEY TRIAL

CHRISTIAN FUNDAMENTALISTS

SCIENTIFIC CREATIONISTS

# SELECTED READINGS

## INTRODUCTION

Bethell, T. "Agnostic Evolutionists."
*Harper's*, Vol. 220 (1985), pp. 49–61.

Bynum, W. "The Great Chain of
Being After Forty Years: an Appraisal."
*History of Science*, Vol. 13 (1975),
pp. 1–28.

Lovejoy, A. *The Great Chain of Being:
A Study of the History of An Idea.*
New York: Harper, 1936.

## DARWIN'S VIEW

Hoskin, M. "A Chat with Charles
Darwin." In Hunter, D. and Whitten,
P., eds. *Readings in Physical
Anthropology and Archaeology.* New
York: Harper & Row, 1978, pp. 20–25.

Miller, J. *Darwin for Beginners.*
New York: Pantheon, 1982.

Simpson, G. *Principles of Animal
Taxonomy.* New York: Columbia
University Press, 1953.

Stone, I. *The Origin: A Biographical Novel of Charles Darwin.* New York: New American Library, 1980.

## THE PEPPERED MOTH

Kettlewell, H. *The Evolution of Melanism.* Oxford: Oxford University Press, 1973.

Mayr, E. *Animal Species and Evolution.* Cambridge, Massachusetts: Harvard University Press, 1963.

## EVOLUTION BEFORE DARWIN

Bowler, P. *Evolution: The History of An Idea* (revised edition). Berkeley, California: University of California Press, 1989.

Buffon, G. *Natural History, General and Particular* (new edition translated by W. Smellie). London: Cadell and Davies, 1812.

Chase, A. *The Legacy of Malthus.* New York: Knopf, 1977.

Clodd, E. *Pioneers of Evolution: From Thales to Huxley.* Freeport, New York: Books for Libraries, 1897.

Coleman, W. *Georges Cuvier, Zoologist: A Study in the History of Evolutionary Theory.* Cambridge, Massachusetts: Harvard University Press, 1964.

Lyell, C. *Principles of Geology: Being an Attempt to Explain the Former Changes of the Earth's Surface by Reference to Causes now in Operation* (3 volumes). London: John Murray, 1830–1833.

Peterson, W. *Malthus.* Cambridge, Massachusetts: Cambridge University Press, 1979.

## CHARLES DARWIN

Appleman, P., ed. *Darwin.* New York: Norton, 1970.

Barrett, P., ed. *The Collected Papers of Charles Darwin.* Chicago: University of Chicago Press, 1977.

Bock, W. "The Definition and Recognition of Biological Adaptation." *American Zoologist,* Vol. 20 (1980), pp. 217–227.

Brackman, A. *A Delicate Arrangement: The Strange Case of Charles Darwin and Alfred Russel Wallace.* New York: Times Books, 1980.

Darwin, C. *The Descent of Man, and Selection in Relation to Sex* (2 volumes). London: John Murray, 1871.

Ehrlich, P. and Ehrlich, A. *The Causes and Consequences of the Disappearance of Species.* New York: Random House, 1981.

Eiseley, L. *Darwin's Century: Evolution and the Men Who Discovered It.* New York: Doubleday, 1958.

Grant, P. "Speciation and Adaptive Radiation of Darwin's Finches." *American Scientist,* Vol. 69 (1981), pp. 653–663.

Grant, P. *Ecology and Evolution of Darwin's Finches*. Princeton, New Jersey: Princeton University Press, 1986.

Huxley, T. *Evidence as to Man's Place in Nature*. London: Williams & Norgate, 1863.

Keynes, R. *The Beagle Record*. New York: Cambridge University Press, 1979.

Lack, D. "Darwin's Finches." *Scientific American*, Vol. 188 (1953), pp. 66–72.

Lewontin, R. "Adaptation." *Scientific American*, Vol. 239 (1978), pp. 212–230.

McKinney, H. *Wallace and Natural Selection*. New Haven: Yale University Press, 1972.

Moorehead, A. *Darwin and the Beagle*. New York: Harper & Row, 1969.

Sober, E. *The Nature of Selection*. Cambridge, Massachusetts: MIT Press, 1984.

Stanley, S. *Extinction*. New York: Scientific American Books, 1987.

Stern, J., Jr. "The Meaning of 'Adaptation' and its Relation to the Phenomenon of Natural Selection." In Dobzhansky, T., Hecht, M., and Steere, W., eds. *Evolutionary Biology*, Vol. 4 (1970), pp. 38–66.

Vorzimmer, P. *Charles Darwin: The Years of Controversy: The Origin of Species and its Critics*, 1859–1882. Philadelphia, Pennsylvania: Temple University Press, 1970.

Young, R. *Darwin's Metaphor: Nature's Place in Victorian Culture*. London: Cambridge University Press, 1985.

## AMENDMENTS

Eldredge, N. *Time Frames: The Rethinking of Darwinian Evolution and the Theory of Punctuated Equilibria*. New York: Simon and Schuster, 1985.

Eldredge, N. and Gould S. "Punctuated Equilibria: an Alternative to Phyletic Gradualism." In Schopf, T., ed. *Models in Paleobiology*. San Francisco: Freeman, Cooper, 1972, pp. 82–115.

Gould, S. "Is a new and general theory of evolution emerging?" *Paleobiology*, Vol. 6 (1980), pp. 119–130.

Gould, S. and Eldredge, N. "Punctuated Equilibrium at the Third Stage." *Systematic Zoology*, Vol. 35 (1986), pp. 143–148.

Levinton, J. and Simon C. "A Critique of the Punctuated Equilibria Model and Implications for the Detection of Speciation in the Fossil Record." *Systematic Zoology*, Vol. 29 (1980), pp. 130–142.

Vrba, E. "Evolution, Species and Fossils: How does Life Evolve?" *South African Journal of Science*, Vol. 76 (1980), pp. 61–84.

## MISGUIDED ATTACKS

Futuyma, D. *Science on Trial: The Case for Evolution*. New York: Pantheon, 1983.

Gilkey, L. *Creationism on Trial: Evolution and God at Little Rock*. Minneapolis, Minnesota, Winston, 1985.

Godfrey, L., ed. *Scientists Confront Creationism*. New York: Norton, 1983.

Greene, J. *The Death of Adam: Evolution and its Impact on Western Thought*. Ames, Iowa: Iowa State University Press, 1959.

Morris, H. *Scientific Creationism*. San Diego, California: Creation-Life Publishers, 1974.

Newell, N. "Evolution Under Attack." *Natural History*, Vol. 83 (1974), pp. 32–39.

Ruse, M. *Taking Darwin Seriously: A Naturalistic Approach to Philosophy*. New York: Basil Blackwell , 1986.

Scopes, J. *Center of the Storm*. New York: Holt, Rinehart and Winston, 1967.

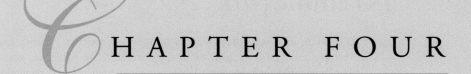

# CHAPTER FOUR

# THE IMPORTANCE OF HEREDITY

*contents at a glance*

# INTRODUCTION

As brilliant as it was, there were two aspects of Darwin's theory that puzzled even his most ardent and vocal supporters. One was that while Darwin maintained that species with large populations displayed more variation in their physical traits than species with small populations, he could not explain why this was the case. As he himself put it: "Our ignorance of the laws of variation is profound." Nor could Darwin explain how some individuals in a population inherited the modified traits that made them and their kind reproductively distinct. Although he toyed with the idea that such modified traits were produced when parental traits were blended, in the end he candidly admitted that he was unable to solve the problem. See QQ 4.1.

## 4.1 DARWIN ON INHERITANCE

Darwin admitted he did not understand the principles of inheritance in the *Origin of Species*.

• • •

Every one must have heard of cases of albinism, prickly skin, hairy bodies, *etc.*, appearing in several members of the same family. If strange and rare deviations of structure are really inherited, less strange and commoner deviations may be freely admitted to be inheritable. Perhaps the correct way of viewing the whole subject would be to look at the inheritance of every character whatever as the rule, and non-inheritance as the anomaly.

[Unfortunately, however, t]he laws governing inheritance are quite unknown. No one can say why the same peculiarity in different individuals of the same species, and in individuals of different species, is sometimes inherited and sometimes not so; why the child often reverts in certain characters to its grandfather or grandmother or other much more remote ancestor; why a peculiarity is often transmitted from one sex to both sexes, or to one sex alone . . . .

Source: Charles Darwin, *On The Origin of Species By Means of Natural Selection, or the Preservation of Favoured Races in the Struggle for Life* (London: John Murray, 1859), p. 13.

While Darwin's supporters regarded these frank admissions as a reflection of the limits of scientific knowledge of the day, his opponents were enraged. They insisted that if Darwin could not explain how modified traits arose and were inherited, then he could not explain how old species evolved into new ones — the implication being that his theory did not deal with the "origin" of species at all. At best, his detractors said, natural selection explained only how existing species became better suited to the environment by producing new breeds of the same species, not how one species gave rise to another. Although the same refrain is sometimes heard today — tele-evangelist David Mainse, for instance, still insists that "the frog does not become a prince" — scientists have long since solved the problems that baffled Darwin.

▼

Charles Darwin
(1809–1882),
age sixty-six

▲

The source of variation within species and the inheritance of traits that make it possible for the species barrier to be crossed are both due to the factors that govern **heredity** — the process by which biological characteristics originate and are transmitted from one generation to the next.

# MENDEL'S VIEW

## AN EXPERIMENTAL GARDEN

Although the significance of his work did not become clear until the turn of the present century, the modern study of heredity began with the research of Gregor Johann Mendel (1822–1884), an Augustinian monk. In his monastery garden, in what was then the Austrian town of Brünn and is now the city of Brno in the Czech Republic, Mendel performed meticulous breeding experiments that demonstrated that heredity is not a matter of accident but follows distinct and predictable patterns.

Mendel's most outstanding research was on common garden peas (*Pisum sativum*). These familiar plants are normally **self-fertilizing** — the male pollen, produced by the anthers of the flower, fertilizes the female ovules of the same flower and the peas (seeds) form in the pod. But in order to control his experiments Mendel used a procedure called **cross-fertilization**, in which he removed the pollen-producing anthers from some plants and fertilized their ovules with pollen from other plants. He used this technique to study seven attributes of the pea plant, including the shape of the peas, the colour of the first leaves, the colour of the peas, the shape of the pods, the colour of pods, the location of the flowers, and the length of the stem, each of which featured a pair of contrasting traits. The shape of the peas, for instance, was either round or wrinkled, the colour of the first leaves either yellow or green, the colour of the seed coat either grey or white, the form of the pod either inflated or constricted, the colour of the pod either green or yellow, the position of the flower either axial (along the stem) or terminal (at the top of the stem), and the length of the stem either long (about 2000 cm) or short (about 500 cm). See QQ 4.2.

An artist's rendering of Gregor Johann Mendel (1822–1884) in his garden

The Bettmann Archive

# QQ  4.2 MENDEL AND HIS GARDEN

Hugo Iltis describes Mendel and his famous experimental garden.

• • •

[The] special strip of garden . . . adjoining the monastery wall, is only one hundred and twenty feet [37 metres] long and a little over twenty feet [six metres] wide, but, small though it is, it has become of historical significance. During the 'fifties and 'sixties of the nineteenth century, any one passing this way might have seen, on fine spring days, a vigorous, short, rather sturdily built and somewhat corpulent man engaged in a laborious occupation which would have been puzzling to any uninstructed observer. Here there were to be seen, clinging to staves, the branches of trees, and stretched strings, hundreds of pea-plants of the most various kinds . . . . The gardener would move from one flower to another.

Source: Hugo Iltis, *Life of Mendel,* trans. Eden and Cedar Paul (New York: Hafner Publishing Company, 1966), p. 107.

## A FAMOUS EXPERIMENT

Mendel began one of his most famous experiments with two "purebred" varieties of plants, so-called because their seeds consistently bred true to form: one variety whose peas always produced plants with long-stemmed vines, and another whose peas always produced plants with short-stemmed vines. He then cross-pollinated the two varieties in order to create a new "hybrid" generation of peas, so-called because they were the offspring of two different purebred parents. Much to Mendel's surprise, when these hybrid peas were planted they yielded nothing but long-stemmed plants. Similar results were obtained when Mendel experimented with the other six pairs of contrasting traits that he examined; in all cases the distinguishing feature of only one of the purebred parents was displayed among the hybrids: round for the shape of the peas, yellow for the colour of the first leaves, grey for the colour of the seed coat, inflated for the shape of the pod, green for the colour of the pod, and axial for the position of the flower. See Table 4.1.

**TABLE 4.1**    Mendel's seven experiments with garden peas

| TRAIT | PARENTS | OFFSPRING |
|---|:---:|:---:|
| Seed form | round × wrinkled | round |
| Colour of first leaves | yellow × green | yellow |
| Colour of seed coat | grey × white | grey |
| Form of pod | inflated × constricted | inflated |
| Colour of pod | green × yellow | green |
| Position of flower | axial × terminal | axial |
| Length of stem | long × short | long |

Given these results, the question that came immediately to Mendel's mind was: what happened to the other traits? For eight years he searched for the answer. But Mendel was patient; in the end he had cross-bred his 72 original plants 287 times. Nor did he rush into print with his results. Only when he was satisfied with the accuracy of his findings did he prepare a report titled *Experiments in Plant Hybridization*. In that report, which was published in 1865, Mendel made it clear that the only way in which the principles of heredity would be revealed was by making precise and repeated measurements and keeping detailed records. See QQ 4.3.

## 4.3 *PISUM SATIVUM*

Mendel explains why he decided to experiment on common garden peas.

• • •

The value and utility of any experiment are determined by the fitness of the material to the purpose for which it is used, and thus in the case before us it cannot be immaterial what plants are subjected to experiment . . . .

The selection of the plant group which shall serve for experiments of this kind must be made with all possible care if it be desired to avoid from the outset every risk of questionable results.

The experimental plants must necessarily —

1. Possess constant differentiating characters.
2. The hybrids of such plants must, during the flowering period, be protected from the influence of all foreign pollen, or be easily capable of such protection.
3. The hybrids and their offspring should suffer no marked disturbance in their fertility in the successive generations.

Accidental impregnation by foreign pollen, if it occurred during the experiments and were not recognized, would lead to entirely erroneous conclusions. Reduced fertility or entire sterility of certain forms, such as occurs in the offspring of many hybrids, would render the experiments very difficult or entirely frustrate them. In order to discover the relations in which the hybrid forms stand towards each other and also towards their progenitors it appears to be necessary that all members of the series developed in each successive generation should be, *without exception*, subjected to observation.

At the very outset special attention was devoted to *Leguminosae* [legumes] on account of their peculiar floral structure. Experiments which were made with several members of this family led to the result that the genus *Pisum* was found to possess the necessary qualifications.

Source: Gregor Mendel, "Experiments in Plant Hybridization." In Peters, James A., ed. *Classic Papers in Genetics* (Englewood Cliffs, New Jersey: Prentice-Hall, Inc., 1959), pp. 2–3.

## ALLELES AND GENES

The theory that Mendel developed to explain the heredity of the pea plant was ingenious. Although they could not be seen, he suspected that a single pair of invisible "factors" inside the plant was responsible for each trait that he studied. By convention, these factors are represented by letters. For example, in the purebred tall plant the factors controlling the length of the stem are represented by the uppercase letters (T) and (T), and in the purebred short one by the lowercase letters (t) and (t). Today these paired factors are called **alleles** — a term that refers to alternate forms of a particular gene. **Genes**, in turn, can be defined as chemical blueprints that determine the structure and function of the various parts of an organism's body. Thus, in the case of the gene that controls the length of the stem in peas, (T) is best understood as the allele that is responsible for long-stemmed plants, and (t) as the allele that is responsible for short-stemmed ones. If an organism possesses identical alleles of the same gene, say, for instance, either (T) and (T) or (t) and (t), then it said to be **homozygous** for the gene. If, on the other hand, it possesses alternate alleles of the same gene, say, for instance (T) and (t), then it is said to be **heterozygous** for the gene.

# DEOXYRIBONUCLEIC ACID (DNA)

Although Mendel recognized genes, he did not identify the molecular units they contain that account for their ability to transmit hereditary information. That material is DNA, which, as suggested earlier, is a complex organic molecule that determines how each and every living thing is constructed. DNA molecules are found in **cells** — the units that make up the bodies of all living things, from viruses to organisms such as ourselves. The DNA molecules located in the nucleus of the cell are referred to as **nuclear DNA**; and those situated in the minuscule, granular, mitochondrial bodies inside the cell but outside the nucleus (which we will discuss in a later chapter) are known as **mitochondrial DNA**.

In the 1950s the structure of nuclear DNA was brought to light by James Watson and Francis Crick. Watson and Crick demonstrated that the DNA molecule is a double spiral or helix made up of two, long, intertwined thread-like strands. They also showed that each strand was composed of a long chain of molecular units, each unit containing a phosphate, a sugar, and one of four chemical bases, either adenine, cytosine, guanine, or thymine. In addition, they demonstrated that the strands that formed the double helix were connected via a series of simple hydrogen bonds between complementary chemical bases on adjoining strands — adenine invariably bonding with thymine, and cytosine with guanine. See QQ 4.4.

 **4.4 DNA**

James Watson and Francis Crick describe the structure of DNA.

. . .

We . . . [believe that the] structure . . . of deoxyribose nucleic acid . . . has two helical chains each coiled round the same axis . . . .

The structure is an open one, and its water content is rather high. At lower water contents we would expect the bases to tilt so that the structure could become more compact.

The novel feature of the structure is the manner in which the two chains are held together by . . . [complementary] bases . . . . They are joined together in pairs, a single base from one chain being hydrogen-bonded to a single base from the other chain so that the two lie side by side . . . .

[We have] found that only specific pairs of bases can bond together. These pairs are adenine . . . with thymine . . . and guanine . . . with cytosine . . . .

In other words, if an adenine forms one member of a pair, on either chain, then . . . the other member must be thymine; similarly for guanine and cytosine. The sequence of bases on a single chain does not appear to be restricted in any way. However, if only specific pairs of bases can be formed, it follows that if the sequence of bases on one chain is given, then the sequence on the other chain is automatically determined.

Source: J.D. Watson and F.H.C. Crick, "Molecular Structure of Nucleic Acids" *Nature* Vol. 171, p. 738. Reprinted with permission from *Nature*. Copyright 1953, Macmillan Magazines Limited.

Looking much like a spiral staircase (see Figure 4.1), the intertwined strands are capable of reproducing themselves. This is accomplished via a process called **replication**, in which the intertwined strands of the DNA molecule unwind as the hydrogen bonds between the bases dissolve, and the unpaired bases on both strands attract new complementary bases that are present as raw materials in the cell. It is by producing four strands where only two existed before that hereditary information is generated, the nature of which depends upon the specific arrangement of the chemical bases an allele contains. Alternate alleles of the same gene — such as (**T**) and (**t**) in the case of Mendel's plants — contain different sequences of chemical bases, and these provide different hereditary instructions with respect to the trait the gene controls. The reason is that different sequences of chemical bases produce different amino acids; different sequences of amino acids produce different proteins; and different proteins produce alternate types of cells, including, in the case of Mendel's plants, cells that form long and short-stemmed vines, and, in the case of human beings, cells that are responsible for the colour of the skin, hair, and eyes, the configuration of the face, the overall size of the body, and so on. Alleles consequently contain the molecules that govern the biological properties of each and every cell in an organism's body.

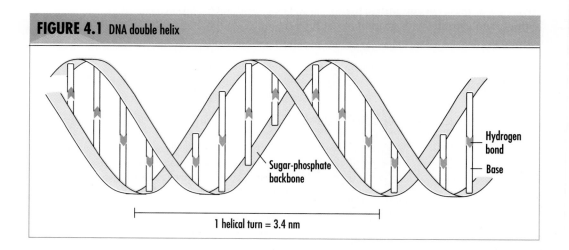

**FIGURE 4.1** DNA double helix

Sugar-phosphate backbone

Hydrogen bond

Base

1 helical turn = 3.4 nm

# MITOSIS AND MEIOSIS

Returning to Mendel, although he knew nothing about nucleic acids and was unaware of the chemistry of alleles, he understood how they functioned, and he acquired this knowledge by anticipating the principles behind **mitosis** — the way that **somatic** or body cells divide — and **meiosis** — the way that **gametes** or sex cells divide. Today these processes are well understood.

## MITOSIS

Since every multicellular organism that reproduces sexually begins its life as a **zygote**, a single cell that is formed by the union of a female gamete and its male counterpart, the zygote must be capable of generating all of the cells that make up the various parts of the body. This is where mitosis comes into play; through this type of cell division, the nuclear DNA in a cell causes that cell to produce twin cells that are genetic replicas of the parent. For example, among humankind, the process begins when the nuclear DNA in the zygote instructs that cell to twin. Within the first week of life, about 100 such undifferentiated cells are formed in the same way. Thereafter, by responding to only some of the instructions contained in the DNA code, these generalized cells give rise to all of the specialized cells that are necessary for a functioning human being to survive. Although it is a continuous, carefully regulated process, for the sake of convenience mitosis can be divided into a number of stages in which the **chromosomes** — the thread-like structures on which alleles and hence DNA are located — are first replicated and are then evenly distributed into the twin cells that arise from the parent (see Figure 4.2). It is via this process of asexual (somatic cell) reproduction that an organism grows to maturity and is able to maintain its cellular integrity until it dies.

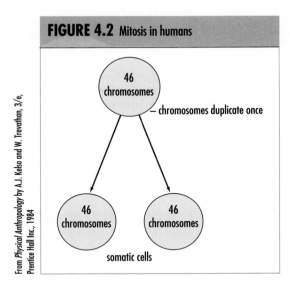

FIGURE 4.2 Mitosis in humans

From *Physical Anthropology* by A.J. Kelso and W. Trevathan, 3/e, Prentice Hall Inc., 1984

## MEIOSIS

But because sexually reproducing organisms such as ourselves must have offspring with roughly the same number and types of genes as their parents, the sex cells of the parents must also be able to divide in a way that helps to ensure that they contain only one half of each organism's total number of alleles. Otherwise the offspring would possess twice as many alleles as its parents and this would be fatal. It is here that meiosis or reduction division comes into play. Meiosis occurs only in the sex cells, and it takes place when the nuclear DNA in those cells instructs them to divide in a way that helps to ensure that, if and when an ovum is fertilized by sperm, each gamete will contain only one-half of both parent's complete complement of alleles. Like mitosis, meiosis is also a continuous and a carefully regulated process, and it too is represented in stages for the sake of convenience. See Figure 4.3 and QQ 4.5.

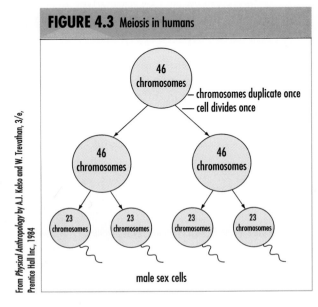

FIGURE 4.3 Meiosis in humans

From *Physical Anthropology* by A.J. Kelso and W. Trevathan, 3/e, Prentice Hall Inc., 1984

## 4.5 MITOSIS AND MEIOSIS

Anthony Smith describes the functions of mitosis and meiosis.

• • •

In a sexual system all offspring acquire their inheritance from both parents. The sperm and ovum fuse to form a single cell, with all the inherited material being within this single cell. Therefore, because of this fusing, each parent must have a system for presenting only half a human's cell's vital needs. The two halves then unite in fertilization to form a whole cell. That united cell must have a system for dividing to form more cells, each similar, each carrying the same quotient of inheritance, and each dividing equally . . . .

Now for some names. The system for forming half a cell is called meiosis. After fertilization the system of normal and equal division is called mitosis.

Source: Anthony Smith, *The Body* (New York: Walker and Company, 1968), p. 205.

# SEGREGATION AND RECOMBINATION

Mendel's contemporaries did not understand meiosis. Although scientists before Mendel knew that offspring were produced by the union of two gametes — one male and the other female — while some mistakenly assumed that the male gamete contained all of the hereditary information and that the female gamete provided the nourishment, others mistakenly believed the opposite. Mendel disagreed with both groups; he maintained that both gametes possessed the same amount of hereditary information.

Equally important, because Mendel anticipated at least some of what would later be learned about meiosis, he reasoned that the alternate alleles that offspring inherited from their parents were transmitted singly rather than in tandem. If this were the case, then the alternate alleles in the sex cells must separate before reproduction occurs. It was this conclusion that led Mendel to formulate the **Principle of Segregation**, which states that, during meiosis, alternate alleles of the same gene separate and become housed in different gametes. Mendel also maintained that the single set of alleles contributed by the male gamete and the female gamete recombined when the offspring was formed — a law he named the **Principle of Recombination**. See QQ 4.6.

## 4.6 HALDANE'S ANALOGY

J.B.S. Haldane explains Mendel's principles of segregation and recombination by way of a clever analogy.

• • •

Nonbiologists often find a certain difficulty in following Mendel's laws, so I am going to use an analogy rather freely. In Spain every person has two surnames, one of which is derived from the father and one from the

mother. For example, if you are called Ortega y Lopez you derive the name Ortega from your father, it was his father's name; and Lopez from your mother, as it was her father's name. Now I want you to imagine a . . . nation in which everybody has two surnames; and when a child is born there is a . . . ceremony by which he receives one of the surnames from one parent and one from the other, these being drawn at random by a priest, whose business it is. For example, if Mr. Smith-Jones marries Miss Brown-Robinson the children may get the name Smith or Jones from the father and Brown or Robinson from the mother; and will be called Smith-Brown, Smith-Robinson, Jones-Brown, or Jones-Robinson. If the priest draws at random those four types occur with equal frequency . . . . If you try to remember that simple scheme you will find no great difficulty in understanding the laws which govern heredity. The things which are analogous to the surnames are called genes.

▼

J.B.S. Haldane (1892–1964)

▲

Source: J.B.S. Haldane, *Heredity and Politics* (New York: W. W. Norton & Company, 1938), pp. 46-47.

Thus, to return again to our original example of Mendel's hybrid peas, what had happened was that the hybrids had inherited one allele controlling the length of the stem from each of its homozygous parents — (T) from the long-stemmed parent since it possessed a pair of (T) alleles, and (t) from the short-stemmed one since it possessed a pair of (t) alleles. The result was a plant that contained a new pair of alleles of the gene controlling the length of the stem — (Tt), a heterozygote. Using (T) to represent tallness and (t) to indicate shortness, the cross-breeding of the parents and the results can be expressed in the following algebraic equation: (TT) × (tt) = (Tt) or (Tt) or (Tt) or (Tt). See QQ 4.7.

## 4.7 THE PUNNETT SQUARE

Named after their inventor, geneticist Reginald C. Punnett, Punnett squares can be used to determine the kinds and frequencies of the various types of offspring that can be expected to result from a mating.

• • •

An alternative method for combining the . . . [alleles of the parents] is to place the female['s] . . . [alleles]

# DOMINANCE AND RECESSIVENESS

Yet in Mendel's hybrids only the trait of tallness was visible, not shortness. This, geneticists now know, is because when two alleles that control the expression of a specific trait recombine, in some cases, one of them is **dominant** and the other is **recessive**. Although these designations may conjure up images of the dominant allele being more powerful or intrinsically better than the recessive, this is not what experts have in mind when they employ these terms. The labels are simply used to indicate that when a dominant and a recessive allele controlling the same trait are both present in an organism's underlying genetic make-up or **genotype**, then it is only the dominant allele that reveals itself in the organism's outward appearance or **phenotype**. The recessive factor, although still present in the organism's genotype, is masked by its dominant counterpart. Mendel confirmed this in the case of his experimental plants by allowing the hybrid generation to reproduce, that is, by crossing $(Tt) \times (Tt)$, which, in algebraic terms, yields $(TT)$ or $(Tt)$ or $(tT)$ or $(tt)$. See Tables 4.2 and 4.3.

**TABLE 4.2**  The genotypic results of Mendel's seven experiments with hybrid garden peas

| TRAIT | HYBRID PARENTS' GENOTYPES | OFFSPRING'S GENOTYPES | GENOTYPIC RATIO |
|---|---|---|---|
| Seed form | $(Rr) \times (Rr)$ | $(RR) + (Rr) + (rR) + (rr)$ | 1:2:1 |
| Colour of first leaves | $(Yy) \times (Yy)$ | $(YY) + (Yy) + (yY) + (yy)$ | 1:2:1 |
| Colour of Seed coat | $(Pp) \times (Pp)$ | $(PP) + (Pp) + (pP) + (pp)$ | 1:2:1 |
| Form of pod | $(Ii) \times (Ii)$ | $(II) + (Ii) + (iI) + (ii)$ | 1:2:1 |
| Colour of pod | $(Gg) \times (Gg)$ | $(GG) + (Gg) + (gG) + (gg)$ | 1:2:1 |
| Position of flower | $(Aa) \times (Aa)$ | $(AA) + (Aa) + (aA) + (aa)$ | 1:2:1 |
| Length of stem | $(Tt) \times (Tt)$ | $(TT) + (Tt) + (tT) + (tt)$ | 1:2:1 |

## TABLE 4.3   The phenotypic results of Mendel's seven experiments with hybrid garden peas

| TRAIT | HYBRID PARENTS' PHENOTYPES | OFFSPRING'S PHENOTYPES (ACTUAL RESULTS) | OFFSPRING'S PHENOTYPIC RATIO |
|---|---|---|---|
| Seed form | round | 5474 round: 1850 wrinkled | 2.96:1 |
| Colour of first leaves | yellow | 6022 yellow: 2001 green | 3.01:1 |
| Colour of seed coat | grey | 705 grey: 224 white | 3.15:1 |
| Form of pod | inflated | 882 inflated: 299 constricted | 2.95:1 |
| Colour of pod | green | 428 green: 152 yellow | 2.85:1 |
| Position of flower | axial | 651 axial: 207 terminal | 3.14:1 |
| Length of stem | long | 787 long: 277 short | 2.84:1 |

The results can be interpreted in two ways: Genotypically, one-quarter of the new generation possessed a pair of (T) alleles like its original tall grandparents (TT); one-half possessed a combination of the (T) allele and (t) allele like its hybrid parents (Tt or tT); and one-quarter possessed a pair of (t) alleles like its short grandparents (tt). The genotypic ratio, therefore, is 1:2:1. Phenotypically, however, the observable feature of tallness — which is manifested when the allele combinations (TT) or (Tt) or (tT) is present — occurs in three-quarters of the plants, and the observable feature of short-ness — which is manifested when a pair of (t) alleles is present (tt) — shows up in one-quarter. The phenotypic ratio of tall plants to short ones is thus 3:1. As the contents of Tables 4.2 and 4.3 indicate, Mendel obtained similar genotypic and phenotypic ratios when he crossed the hybrids that showed only the dominant observable feature of the other six pairs of characteristics that he studied; in all cases the recessive allele that was present in the geno-type, but whose expression was masked in the phenotype, reappeared when the hybrids were crossed.

# HUMAN HEREDITY

## EARLOBES

Although humans are obviously much more complex organisms than common garden peas, what Mendel discovered about his experimental plants is nonethe-less instructive in helping us to better understand our own heredity. The alter-nate alleles of the gene that determines whether human beings have "free" earlobes or "attached" earlobes provide a good example; they operate in much the same way as the alleles that determine the length of the stem in pea plants. What geneticists have discovered is that the allele for free earlobes (F) is domi-nant over its counterpart for attached earlobes (f). This means that while indi-viduals who exhibit free earlobes may possess the allele combinations (FF), (Ff), or (fF), those with attached earlobes always possess a pair of (f) alleles (ff).

Consequently, if two people with attached earlobes mate, then their children will have attached earlobes since: (ff) x (ff) = (ff) or (ff) or (ff) or (ff).

▼
Free earlobe
(left) and
attached earlobe
(right)
▲

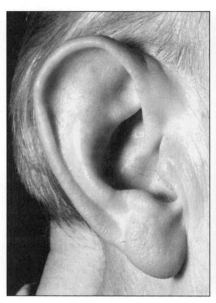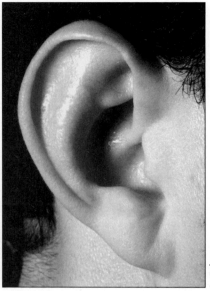

Roy Taylor

On the other hand, if two people with free earlobes produce offspring, then a number of possibilities exist. If both parents are homozygous, then the children will have free earlobes since: (FF) x (FF) = (FF) or (FF) or (FF) or (FF). If one parent is homozygous and the other a hybrid, then the children once again will have free earlobes since: (FF) x (Ff) = (FF) or (Ff) or (Ff) or (Ff). However, if both parents are hybrids, then there is a 3 in 4 chance that the children will exhibit free earlobes, and a 1 in 4 chance that their earlobes will be attached. The reason: (Ff) x (Ff) = (FF) or (Ff) or (fF) or (ff). See QQ 4.8.

## QQ 4.8 DOES SHE OR DOESN'T SHE?

**M. W. Farnsworth asks a question about earlobes.**

• • •

A man with attached earlobes marries a woman with free earlobes. Their seven children have free earlobes. One son marries and of his children, half have free and half have attached earlobes . . . .

[Question:] What is the phenotype of the son's wife?

Source: M. W. Farnsworth, *Genetics* (New York: Harper & Row, Publishers, Inc., 1978), p. 17. Copyright © 1978 by M. W. Farnsworth. Reprinted by permission of HarperCollins Publishers, Inc.

(*Ibid.*, p. 577).

[Answer: The s]on's wife has attached ear lobes.

## ALBINISM

The shape of the earlobes is not the only human trait that is determined by a single pair of alleles, either one or both of which may be dominant or recessive. Another is **albinism**, a hereditary disorder that is manifested by a lack of pigmentation in the skin, eyes, and hair. Albinism, which occurs occasionally in many mammals, is caused by the presence of a pair of recessive alleles — usually represented as (**aa**) — that prevent the cells that are responsible for normal pigmentation from producing the enzyme tyrosinase. Without this enzyme, although the skin appears to be pink because of the underlying blood vessels, and the eyes seem to have pink irises and red pupils, the cells that are responsible for the pigmentation of the skin and the eyes actually remain colourless and the hair is snow white. Although heterozygous individuals (**Aa** or **Aa**) may pass on the disorder to their offspring, only those who are homozygous recessive show the trait, and they represent about one in 20 000 persons in the general population.

Neg. No. 118/Field Museum of Natural History, Chicago

▼

Albino Hopi girl in the company of two normally pigmented Hopi girls

▲

# THE COMPLEXITY OF HUMAN HEREDITY

The way that the shape of the earlobes and albinism are inherited are rare examples of human traits that are determined by well-recognized, simple genetic patterns. Since Mendel's day scientists have discovered that most of our physical features are due to complex genetic patterns that involve more than one pair of alleles, that is to say, multiple genes. Such traits are called **polygenic**.

## EYE COLOUR

Consider the alleles that are responsible for the colour of our eyes. For many years it was mistakenly believed that our eye colour was controlled by a single pair of alternate alleles: (**BB**) and (**Bb**) or (**bB**) accounting for brown eyes, and (**bb**) for blue. At the same time, it was also widely believed that whenever the allele for brown (**B**) recombined with the one for blue (**b**), then the brown would mask the blue. More recent research has shown that these propositions are only partly correct. While brown is dominant with respect to blue, it is also clear that our eye colour is not limited to these categories — there are different shades and mixtures of brown and blue not to mention grey, green, and hazel and variations among these. Under the circumstances, it is not surprising that the idea that the colour of our eyes is controlled by a single pair of alternate alleles has been rejected; like almost all of our physical features, eye colour is a polygenic trait that is produced by multiple genes. See QQ 4.9.

## 4.9 BROWN EYES AND BLUE

**Alice Brues looks at eye colour.**

• • •

Eye color in man is sometimes classified into blue and brown. Close inspection of a few eyes will show that this classification is very inadequate. Even a rather impressionistic evaluation of eye color includes ambiguous shades such as green and hazel. A careful examination of the iris shows that many individuals in populations in which light shades of iris color are present, have intermediate colors — actually mosaics combining light and dark areas . . . . The total amount of variation in mixed-color eyes is enormous because of differences in component colors, in amount of various colors, and in the particular pattern found . . . .

The inheritance of eye color is certainly not as simple as it is sometimes declared to be in genetics texts, which like to describe "blue" and "brown" as an example of recessive and dominant traits in man. The mere fact that there are so many varieties of iris color other than simple blue and brown shoots down this system. More careful study indicates that the pure, uniform brown shades — the typical eye color of most human populations — are probably dominant over

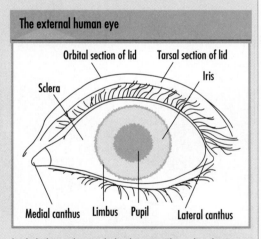

**The external human eye**

both light and mixed shades . . . [but also that] . . . there must be a number of modifying factors affecting shades of blue and gray, the amount and pattern of dark markings in mixed eyes, and so forth, that have yet to be identified.

Source: Alice M. Brues, *People and Races* (New York: Macmillan Publishing Co., Inc., 1977), pp. 102–103. Copyright © 1977. Reissued 1990 by Waveland Press, Inc., Prospect Heights, Illinois. Reprinted with permission from the publisher.

## SEX DETERMINATION

Whether a newborn human infant is male or female is determined in yet another way. The sex of the infant depends upon its complement of chromosomes — the thread-like structures on which alleles are located. By studying photographic representations of chromosomes, known as **karyotypes**, geneticists have determined that humans possess 46 chromosomes arranged into 23 pairs. During meiosis each pair normally separates, so that when a human zygote is formed through the union of sperm and egg, 23 chromosomes are inherited from each parent. Among those 23 pairs is one pair of **sex chromosomes** (one-half of each pair contributed by each parent), which determine whether an individual is male or female. These chromosomes are represented by the letters **X** and **Y**, and the presence or absence of maleness is now believed to be determined by a single gene located on the **Y** chromosome. Unless this gene is activated, femaleness — the basic developmental pattern — will emerge.

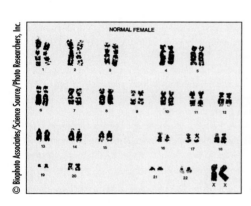
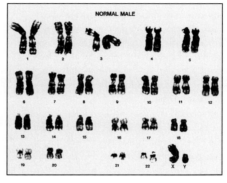

A karyotype of a normal human female (left) and male (right)

Another important difference between the sex chromosomes is their parental origin. Whereas females produce only X-bearing eggs, males produce both X-bearing and Y-bearing sperm. For this reason, when the sex chromosomes of the parents recombine, only two combinations are normally possible: **XX**, which results in a female and occurs when both parents contribute an **X**; and **XY**, which results in a male and occurs when the mother contributes an **X** and the father a **Y**. Since there is an equal chance that the father will contribute an X chromosome or a Y, the probability that the child will be male or female is the same — .50 in each case.

## SEX PRE-SELECTION

Despite claims to the contrary, no reliable method has yet been developed to alter the ratio of male to female births — 106 males to 100 females worldwide. But there is little doubt that new reproductive technologies will soon make sex pre-selection possible, and will force those who are capable of administering the technologies and those who wish to choose the sex of their offspring to consider the social implications of their actions. In October 1989 the Government of Canada attempted to come to grips with the issue by establishing the Royal Commission on New Reproductive Technologies. Among other things, the Commission's mandate included gathering information on the techniques that might be used to pre-select the sex of a child, and assessing the implications of allowing Canadians to make such a choice. See QQ 4.10.

## 4.10 SEX PRE-SELECTION

The Royal Commission on New Reproductive Technologies explains how it is dealing with the issue of sex pre-selection.

• • •

The direction and magnitude of the social consequences of sex selection will depend on the extent to which Canadians prefer one sex over another and are willing to use the techniques, the extent to which techniques are available, and how effective the techniques are. To assess the implications fully, the Commission is carrying out several studies:

The Commission is surveying Canadians about their attitudes regarding the preferred sex of their children and the selection and pre-selection of sex, in order to assess public attitudes, values, and knowledge concerning sex pre-selection. In addition, the Commission is critically appraising what data are available in the literature on preferred sex of one's

children, public attitudes, values, and knowledge about the use of sex pre-selection methods, and public knowledge of such methods and their availability.

The Commission is evaluating methods of sex pre-selection in terms of their reliability rates and the prospects for changes in these rates, and to estimate the frequency with which these techniques are currently used.

The Commission is surveying genetic counsellors in Canada to determine their attitudes to pre-natal diagnosis of sex for non-medical reasons. This study is building on a previous study in this area, and will therefore be able to document changes in these attitudes over time, as well as to assess the willingness of counsellors to recommend pre-natal diagnosis for this reason.

Source: Royal Commission on New Reproductive Technologies, *Update* (Ottawa: Royal Commission on New Reproductive Technologies, 1992), pp. 16–17.

# POPULATION GENETICS

## GENE POOLS

Given what has already been said about human heredity, one may well wonder how many genes our chromosomes house. Scientists currently estimate the number to be about 100 000. They have also discovered that the alleles that make up our genes recombine in constantly changing combinations, to the extent that, aside from identical twins who arise from a single egg and thus have identical genotypes, not even individuals who have the same biological parents would ever be exactly alike. The same is true of almost all species, and it is for this reason, just as Darwin observed, that large populations are more physically varied than small ones. The greater the number of individuals in a population the larger will be its **gene pool**, or the total number of alleles in the population, and the larger its gene pool the greater the number of possible allele combinations. See Table 4.4.

| **TABLE 4.4** Number of types of progeny that can be produced when their parents differ in selected numbers of dominant genes | |
| --- | --- |
| GENE DIFFERENCES | TYPES AMONG THE PROGENY |
| 1 | 2 |
| 2 | 4 |
| 4 | 16 |
| 5 | 32 |
| … | … |
| 10 | 1 024 |
| … | … |
| 20 | 1 048 576 |
| … | … |
| 30 | 1 073 741 824 |
| … | … |
| n | $2^n$ |

Source: Adapted from L. C. Dun and Theodosius Dobzhansky, *Heredity, Race and Society*, revised and enlarged edition (New York: New American Library, Inc., 1952), p. 53. Used by permission of the publisher, Dutton, an imprint of New American Library, a division of Penguin Books USA Inc.

## THE HARDY-WEINBERG THEOREM

The fact that genes can be thought of as existing in pools is fundamental to **population genetics**, which is the study of the way that genetic information is exchanged among individuals who can and do interbreed. One of the most important building blocks of this branch of science is the **Hardy-Weinberg theorem.** Developed independently in 1908 by a British mathematician, G.H. Hardy, and a German physician, W. Weinberg, the theorem holds that, under certain conditions, the genotypic frequencies that prevail in a population will remain inherently stable through time. The provisos are that there must be random mating or **panmixis** in the population; that the population must be relatively large and stable; that the genotypes that characterize the population must stand an equal chance of survival; and that the alleles that comprise the genotypes must not be subject to change.

Consider the following example. Imagine that the abovementioned conditions are met in a population with a single trait that is controlled by one pair of alleles, say, for instance, (**T**) and (**t**), where (**T**) represents the dominant allele, and (**t**) represents the recessive allele. According to the Hardy-Weinberg theorem, the proportion of those individuals in the population who are homozygous dominant (**TT**) for the trait, homozygous recessive (**tt**) for the trait, and heterozygous (**Tt** or **tT**) for the trait will, in algebraic terms, remain constant from generation to generation. As Hardy and Weinberg demonstrated, the relationship between the three genotypes is defined by the equation $p^2 + 2pq + q^2 = 1$, where p represents the frequency of (**T**), q represents the frequency of (**t**), and $p + q = 1$, which means that there are no other alleles in the population. See QQ 4.11.

## 4.11 FOR THE MATHEMATICALLY INCLINED

G.H. Hardy discusses the algebraic underpinnings of the Hardy-Weinberg theorem.

• • •

In the *Proceedings of the Royal Society of Medicine* . . . Mr. Yule is reported to have suggested, as a criticism of the Mendelian position, that if brachydactyly [a genetic disorder among humans that features abnormally short fingers and toes] is dominant "in the course of time one would expect, in the absence of counteracting factors, to get three brachydactylous persons to one normal."

It is not difficult to prove, however, that such an expectation would be quite groundless. Suppose that *Aa* is a pair of Mendelian characters, *A* being dominant, and that in any given generation the numbers of pure dominants (*AA*), heterozygotes (*Aa*), and pure recessives (*aa*) are as $p:2q:r$. Finally, suppose that the numbers are fairly large, so that the mating may be regarded as random, that the sexes are evenly distributed among the three varieties, and that all are equally fertile. A little mathematics of the multiplication-table type is enough to show that in the next generation the numbers will be as $(p+q)^2 : 2 (p+q)(q+r) : (q+r)^2$, or as $p1:^2q1:r1$, say.

The interesting question is — in what circumstances will this distribution be the same as that in the generation before? It is easy to see that the condition for this is $q^2 = pr$. And since $q1^2 = p1r1$, whatever the values of *p*, *q* and *r* may be, the distribution will in any case continue unchanged.

Suppose to take a definite instance, that *A* is brachydactyly, and that we start from a population of pure brachydactylous and pure normal persons, say in the ratio of 1:10 000. Then $p = 1$, $q = 0$, $r = 10 000$ and $p1 = 1$, $q1 = 10 000$, $r1 = 100 000 000$. If brachydactyly is dominant, the proportion of brachydactylous persons in the second generation is 20 001:100 020 001, or practically 2:10 000, twice that in the first generation; and this proportion will afterwards have no tendency whatever to increase. If, on the other hand, brachydactyly were recessive, the proportion in the second generation would be 1:100 020 001, or practically 1:100 000 000, and this proportion would afterwards have no tendency to decrease.

In a word, there is not the slightest foundation for the idea that a dominant character should show a tendency to spread over a whole population, or that a recessive should die out.

Source: G.H. Hardy, "Mendelian Proportions in a Mixed Population" *Science*, Vol. 28 (1908), pp. 49–50. Copyright © 1908 by the AAAS.

## GENETIC DRIFT

The Hardy-Weinberg theorem is important to biological anthropologists because it helps them to identify the factors that are responsible for genetic change in a population. Such change provides the raw materials upon which natural selection operates, and it occurs when one of the provisos on which the theorem is based has not been satisfied. For example, sometimes the gene pool of a population changes on account of genetic drift. **Genetic drift** refers to fluctuations in the allele frequencies in the gene pool of a population that are caused by chance events. Such fluctuations almost always occur in relatively small populations in which factors such as accidental death and inbreeding due to small population size increase or decrease the frequency of the alleles that are predicted by the Hardy-Weinberg theorem.

The Dunkers are an interesting example. The Dunkers are a small religious sect of Old German Baptist Brethren whose strict marriage customs have kept their members reproductively isolated since their ancestors migrated from Germany to the farmlands of eastern Pennsylvania in the eighteenth century. As such, the Dunkers represent an ideal population in which the impact of genetic drift can be studied, and in the early 1950s Bentley Glass and his colleagues undertook such a study among the almost 300 Dunkers then living in Franklin County, Pennsylvania. What they discovered was that genetic drift had caused certain alleles in the Dunker population to vary significantly from what the Hardy-Weinberg theorem predicts. The alleles that produced their blood types, for instance, set the Dunkers apart from both their German ancestors and their modern American neighbours, blood type A occurring far more frequently among the Dunkers than expected, and blood type O somewhat less frequently. The alleles that controlled the presence of mid-digital hair on the fingers, "hitch-hiker's thumb" [or hyperextensibility of the thumb], and attached earlobes had also been affected by genetic drift; the Dunkers exhibited lower frequencies of these traits than did their neighbours. See QQ 4.12.

 **4.12 GENETIC DRIFT AMONG THE DUNKERS**

Bentley Glass explains what he and his colleagues discovered about genetic drift among the Dunkers.

• • •

It occurred to Glass, *et al*, . . . that conclusive evidence of the operation of genetic drift might be obtained from a study of "a genetic isolate of known size, age, and origin and which in particular shares an environment indistinguishable from that of the major population with which it is to be compared. The situation ideal for study would be that of a genetic isolate interspersed within a larger population, so intermingled that the individuals of the isolate do not differ from those of the general population in any aspect of life except their assortative mating restrictions . . . ."

Such a community was located in Franklin County, Pennsylvania — a community of Old Baptist Brethren, or Dunkers. It included 298 persons . . . . This community, like the 54 others in the sect in the United States is composed of descendants of emigrants from the German Rhineland early in the eighteenth century. Formerly a larger sect, the German Baptist Brethren split into three denominations in 1881. Of these the community studied belongs to the smallest and most orthodox branch . . . .

[Among others, t]he genetic characters [of the Dunkers that were] analyzed included the ABO . . . blood groups; mid-digital hair; distal hyperextensibility of the thumb; [and] ear lobes . . . . In the case of the blood groups, comparisons could be made not only with the frequencies in the North American population but also with those in the Rhineland . . . . In the absence of genetic drift, it would of course be expected that the frequencies in the Dunker isolate would fall between those of the United States and the country of origin . . . . [But] ABO frequencies showed very striking departures from expectation. Thus the frequency of blood group A, which was expected to fall between 0.446 and 0.395, had risen to 0.593. The recessive group O, which one might expect to see increasing in frequency in an inbreeding community, had fallen below the value for West Germany instead of rising toward the characteristic American frequency . . . .

[Although f]requencies of the other traits studied could be compared only with frequencies in the American white population, the evidence . . . [indicates] that mid-digital hair types and ear lobes were . . . strikingly and significantly [different] from these in the control population. Distal hyperextensibility of the thumbs also differed strikingly [because of genetic drift].

Source: Bentley Glass, "Genetic Changes in Human Populations, Especially Those Due to Gene Flow and Genetic Drift" *Advances in Genetics,* Vol. 6 (1954), pp. 133–134.

## GENE FLOW

Another factor that can lead to genetic change in a population is **gene flow** — the infusion of the alleles from one population's gene pool into another's. This occurs when the members of two genetic populations in the same species that

Louis Riel
(1844–1885)

National Archives of Canada/C06688(D)

have developed their own characteristic traits in isolation from one another come into contact and mate. One of the most famous human examples of this process occurred in Canada, when European fur traders arrived in the Great Lakes region and began to mate with Indian women. By the mid-point of the nineteenth century there were an estimated 800 French *coureurs de bois* in the Upper Great Lakes region of the country, and, presumably, like the Europeans who had preceded them into the region, many had Cree and Ojibwa wives. The offspring of such unions and their descendants are neither Indian nor European, but **Métis**, a distinct Aboriginal Canadian population with their own history and culture. It was the Métis who became engaged in two military resistances that paved the way for Manitoba and Saskatchewan to enter Confederation in 1870 and 1905 respectively. The Canadian Parliament finally acknowledged this fact when it passed a resolution in March 1992 honouring Métis leader Louis Riel, who was hanged for treason in 1885 and long regarded by many Canadians as a traitor. The resolution recognized Riel's "unique and historic role as a founder of Manitoba and his contribution in the development of Confederation." See QQ 4.13.

## 4.13 GENE FLOW AND THE MÉTIS

Shortly before he was executed for leading the Métis resistances, Louis Riel wrote about the origin of the Métis and their culture.

. . .

The Métis have as their paternal ancestors the former employees of the Hudson's Bay and Northwest [fur trading] companies; and as their maternal ancestors Indian women belonging to different tribes . . . .

Appropriate as the . . . English expression "*Halfbreed*" was for the first generation of mixture of blood, now that European and Indian blood is mingled in all degrees, it is no longer adequate.

The French word Métis expresses the idea of this mixture . . . [more] satisfactorily . . . and thus becomes a proper name for the people.

One small observation in passing . . . . Sometimes a . . . person will say to a Métis: "You don't look like a Métis at all. Surely you haven't much Indian blood. Anyone would take you for a pure white." Disconcerted by the tone of the question, the Métis, who is proud of his origin on both sides, casts about for an answer . . . . [W]ords like these finally overcome his silence . . . . "True our Indian origin is humble, but it is right that we should honour our mothers as well as our fathers. Why should we be concerned about the proportion of our European and Indian blood? However little we have of each, gratitude and filial love command us to say: We are Métis."

Source: Louis Riel, "Les Métis du Nord-Ouest." in Thomas Flanagan, ed., *The Collected Writings of Louis Riel,* Vol. 3 (Edmonton: University of Alberta Press, 1985), pp. 278–279 (translated by the authors).

# MUTATION

## POINT MUTATIONS

Genetic drift and gene flow are not the only factors that are responsible for the genetic configurations upon which natural selection operates. Scientists have also determined that new genetic information is constantly being introduced into the gene pool of a population because of **mutation** — a change in the hereditary instructions that alleles provide. **Point mutations**, in which a single allele is affected, are an example. These can be caused by chemicals, radiation, and by mutator genes which initiate mutations, and since the hereditary information in an allele is determined by the structure of its DNA, point mutations arise when the sequence of the chemical bases in an allele's DNA is altered. Although no one is certain how many mutant alleles our gametes house, current estimates indicate that each of us likely possesses several such alleles, some of which may be transmitted to our offspring in the course of reproduction.

Unfortunately, the altered genetic instructions can sometimes be injurious. The mutant *ras* **gene** is an example; it may be responsible for upwards of 15 percent of the types of **cancer** (a group of diseases produced when a single somatic cell or a group of such cells begins to multiply and spread in an uncontrolled fashion) from which human beings suffer. What remains to be determined is how best to prevent the *ras* gene from instructing somatic cells to divide in a way that produces cancer rather than normal cellular growth and repair — an outcome that typically occurs when those cells have been assaulted by cancer-producing agents. See QQ 4.14.

## 4.14 THE MUTANT *RAS* GENE

Debra Black writes about the ongoing study of a mutant gene by a team of Canadian and American scientists at the Cold Stream Harbor Laboratory in New York; their research may soon yield dividends in the ongoing battle against cancer.

• • •

Until recently, despite billions of dollars spent internationally on basic cancer research . . . scientists did not understand much about the disease. They knew that cancer occurs when something goes wrong in the genetic instructions that govern cell processes. Somehow, normal cells begin to grow uncontrollably, eventually invading healthy tissue and destroying it. But why or how this happens were questions they could not answer . . . .

[However,] in 1981 [Dr. Michael Wigler] and his team successfully took a piece of DNA from a human bladder cancer cell, inserted it into a healthy mouse cell and produced cancer. The gene he located, called the *ras* gene, is believed by some to account for 15 percent of the 100 or so kinds of human cancer . . . .

Wigler's task to isolate one gene from the 100 000 genes found in one cell was analogous to walking into a library the size of the National Library of Canada to look for a book on Sir John A. Macdonald — except that the library has no card catalogue, no librarian and no titles on any of the books. Wigler took the DNA he found in the cancerous human bladder cell and catalogued it into a DNA library. Using recombinant DNA techniques [in which bits of DNA from one cell are cut out and spliced into the DNA of a second cell] he sorted his library into books, chapters and pages. Then, one by one, he took the pages, or short pieces of DNA, and introduced them into living mouse cells until he found the gene responsible for cancer.

In subsequent comparative studies, Wigler and his team found only one difference between the *ras* gene that had produced the cancer and a *ras* gene doing its normal job of controlling growth. The difference lay in a single error in one of the 5000-letter genetic codes of the two genes . . . .

[Once] the mutated *ras* gene [is better understood] . . . it may only be one step to finding a way to stop the erroneous message the cancer gene sends . . . to the cell.

Source: Debra Black, "Cracking the Cancer Code," *Equinox*, Vol. 4, No. 24 (1985), pp. 105–107.

## CHROMOSOMAL MUTATIONS AND NUMERIC CHANGES

Chromosomes are also subject to alteration, sometimes in terms of their structure and sometimes in terms of their number. Structural changes, which are known as **chromosomal mutations**, can be of a number of types, including duplications or deletions of chromosome sections, reversals of sections, or altered shapes such as ring chromosomes in which the ends of a chromosome are joined. **Numeric changes**, on the other hand, involve either the addition or the deletion of a complete chromosome, and this can lead to a major change in the genetic information the offspring inherits if it survives. From time to time this happens among humans when the sex chromosome of either parent is deleted. An extra chromosome results when the father's X-bearing and Y-bearing chromosomes do not separate, or when the mother's pair of X-bearing chromosomes do not separate. In such cases the offspring will possess an additional sex chromosome, with various possible consequences. See Table 4.5.

| TABLE 4.5 | Numerically normal combinations of the human sex chromosomes and selected abnormal combinations | | | |
|---|---|---|---|---|
| MOTHER'S CONTRIBUTION | FATHER'S CONTRIBUTION | SEX CHROMOSOMES | CONDITION OF OFFSPRING | DISTINGUISHING FEATURES |
| X | Y | XY | normal male | none |
| X | X | XX | normal female | none |
| X | XY | XXY | Kleinfelter male | long limbs; sparse body hair; underdeveloped genitals; low intelligence; sterile |
| X | XX | XXX | Triple-X female | either none or may be mentally retarded and sterile |
| X | YY | XYY | XYY-male | appear to be none |
| X | 0 | X0 | Turner female | short stature; webbed neck; underdeveloped genitals; impaired space perception |
| XX | Y | XXY | Kleinfelter male | see above |
| XX | X | XXX | Triple-X female | see above |
| 0 | X | 0X | Turner female | see above |
| 0 | Y | 0Y | | aborted because the X chromosome is critical for life |

Source: From pages 163–169 of *An Introduction to Medical Genetics*, 6/e by J.A.F. Roberts, 1973. By permission of Oxford University Press.

## The XYY Karyotype

Males with an **XYY** karyotype are especially interesting in this respect. Although it was once widely believed that they were extraordinarily aggressive and prone to violent anti-social and criminal behaviour, that hypothesis has never actually been confirmed. In fact, those with the karyotype seem to be as well adjusted socially as others (see QQ 4.15). Of course, the best way

## 4.15 THE XYY KARYOTYPE

Digamber Borgaonkar and Saleem Shah discusses the alleged "aggressiveness" of the XYY male

• • •

Our review of data on the XYY male does not provide firm or conclusive statements on a number of characteristics. However, it does allow us to conclude that many of the premature speculations about the behavioral predispositions of XYY males toward aggressive, impulsive, antisocial, and violent behavior are *not* supported by a careful analysis of the relevant studies. The premature and incautious speculations about the XYY phenotype might in fact serve to illustrate a number of other problems. Such dramatic and hasty conclusions have often been based on scientifically inadequate

and improperly analyzed data. The resulting reports in the scientific and professional literature as well as the publicity in the mass media appear to have facilitated — in fact, seem to have contributed to — public misunderstanding about the XYY phenotype [which occurs at a rate of about one in 1000 male births in the general population]. We have also noted some of the ill-conceived suggestions to formulate public policies on poorly understood issues. Fortunately, however, there have been no efforts to endorse officially nor to implement such formulations. Indeed, more rigorous and cautious scientific reports are being made with increased frequency and researchers are questioning some of the earlier speculations.

Source: Digamber S. Borgaonkar and Saleem A. Shah, "The XYY Chromosome Male — or Syndrome?" *Progress in Medical Genetics* Vol. X (1974), p. 206.

to determine the accuracy of the hypothesis would be to undertake a long-term comparative study of **XY** and **XYY** males. The hypothesis, however, remains in limbo; for fear that parents with **XYY** sons would rear them differently if they knew that their offspring possessed the karyotype, research on the problem has been suspended.

## Trisomy 21

Although not associated with the sex chromosomes, the defect known as "Down syndrome," or **Trisomy 21**, is another example of a malady that is caused by the improper separation of a chromosome pair. The pair in question is number 21, and it is usually the female gamete that contributes two copies of the three chromosomes that produce the disorder. About one in 800 persons is born with the syndrome, which is characterized by mental retardation, a short body with stubby fingers and toes, a round face, and a large, protruding tongue that makes coherent speech difficult. They also die at a much younger age than their counterparts in the general population, especially between one and nine years of age, when they are 17 times more likely to die than their age-mates in the general population. In addition, those who are born with the syndrome frequently suffer from congenital anomalies and from circulatory and respiratory ailments; in fact, up to the age of 30 these are the major underlying causes of death among such individuals. Research has also shown that the probability of producing a child with trisomy 21 increases with the age of the mother, and that once a woman has given birth to a child with trisomy 21, she is more likely to give birth to another such child than a woman of the same age who has not given birth to any offspring with the disorder. See QQ 4.16.

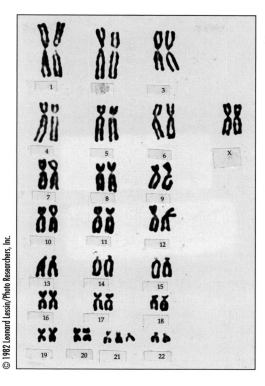

© 1982 Leonard Lessin/Photo Researchers, Inc.

▼

Trisomy 21 karyotype

▲

## QQ 4.16 TRISOMY 21

Patricia Baird, Chairperson of Canada's Royal Commission on New Reproductive Technologies, and Adele Sadovnick, her colleague from the Province of British Columbia's Ministry of Health, discuss the life expectancy and underlying causes of death of individuals who are born with Trisomy 21.

• • •

[W]e compared data on the U.C.O.D. [underlying cause of death] for a large, well-defined population of DS [Down syndrome] individuals and data on the U.C.O.D. for the age-matched general population. In general, an individual with DS is significantly more likely to die than is an individual in the age-matched general population, over all ages studied up to age 30. Below 1 year of age, a DS individual is six times more likely to die than is an individual in the age-matched general population. For ages, 1–9, the DS individual is more than 17 times more likely to die than others, and for ages 10–19 and 20–29, six times and four times more likely to die, respectively. These are very substantial differences in overall death rates. It is interesting that although the greatest absolute likelihood of dying is under 1 year, the age group with the greatest relative risk of dying is very definitely 1–9. This is new information, only available because it was possible to obtain comparative rates for death in DS and in the general population in the present study. In order, the three C.O.D. [cause of death] categories that have the greatest relative risk in DS are congenital anomalies, circulatory system, and respiratory system.

Source: P.A. Baird and A.D. Sadovnick, "Cause of Death to Age 30 in Down Syndrome" *American Journal of Human Genetics*, Vol. 43 (1988), p. 247. Reprinted by permission of University of Chicago.

## THE SICKLING ALLELE

Not all mutations are dangerous, at least not in the conventional sense of the term. Whether or not they are injurious often depends on the environment. The alleles that are responsible for the production of our **haemoglobin** — a complex iron-bearing protein in our red blood cells that plays an essential role in enabling those cells to transport oxygen from the lungs to the various cells of the body — illustrate the point. Three configurations are possible: we may have a pair of dominant (**AA**) alleles that produce haemoglobin A (**HbA**), a pair of mutated recessive (**aa**) alleles that produce haemoglobin S (**HbS**), or a pair of heterozygous alleles (**Aa or aA**) that produce both haemoglobin A and haemoglobin S. The latter configuration arises because the two alleles are **codominant**, which means that the effects of both alleles are expressed in the heterozygous state.

What is interesting about these configurations is their outcome. While those with haemoglobin A (**HbA**) produce red blood cells with a healthy, round shape, those with haemoglobin S (**HbS**) produce red blood cells with a crescent or sickle-like shape, and this prevents the cells from transporting sufficient oxygen to the other cells of the body. The result is a debilitating and often fatal form of anemia known as **sickle cell disease**. Meanwhile, those who are heterozygous stand in a different position. They have what is

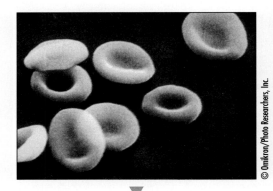

© Omikron/Photo Researchers, Inc.

Non-sickled red blood cells

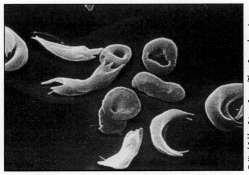

© Nigel Calder, Science Source/Photo Researchers, Inc.

Sickled red blood cells

known as the **sickle cell trait**, and, while they are somewhat anemic, the fact that they are heterozygous also has some benefits. This is particularly true when they are bitten by an *Anopheles* mosquito and their red blood cells are invaded by the parasite that causes malaria. When such persons are infected with malaria their red blood cells start to sickle rapidly because of the presence of haemoglobin S, and this prevents the parasite from reproducing during its 48 hour life-span. The result is that a more serious infection is prevented. Relatively healthy carriers consequently appear to be in a somewhat enviable position, and this is reflected by the fact that natural selection endowed human populations living near the equator, where malaria was widespread, with comparatively high frequencies (10 percent to 16 percent) of the HbS allele.

But while relatively healthy carriers have an advantage in areas where malaria is prevalent, where the disease has been eliminated that advantage is lost. As already mentioned, those with the sickle cell trait suffer from a mild form of anemia. In addition, although carriers themselves have healthy red blood cells, if two of them mate, then there is a 1 in 4 chance that their offspring will be stricken with sickle cell disease since: (**Aa**) x (**Aa**) = (**AA**) or (**Aa**) or (**aA**) or (**aa**). Thus, depending on the environment, the sickling allele may be either dangerous or beneficial. And what is true of the sickling allele is also true of other non-lethal mutations. See QQ 4.17.

## QQ 4.17 MUTATIONS

Isaac Asimov considers the causes and the evolutionary significance of mutations.

. . .

Any factor in the environment that increases the probability of a mutation is a *mutagenic agent*. Heat seems to be mutagenic . . . .

Another mutagenic agent is radiant energy, which includes both the ultraviolet of sunlight and X rays, as well as the various radiations produced by radioactive substances . . . .

There are also chemicals which . . . increase the rate of mutation . . . . [T]he best known are mustard gas, of World War I notoriety, and related compounds called "nitrogen mustards."

Even under the best and mildest of circumstances, mutations will take place, for mutagenic agents cannot be completely eliminated. There is sunlight, constantly showering life forms with ultraviolet light. There are the radiations emanating from the radioactive substances present in tiny quantities in soil, sea, and air. There are the cosmic-ray particles bombarding us from outer space. And there are always the workings of sheer chance during . . . [reproduction].

Accidents, in other words, will happen, and mutations will take place . . . [some of which produce disease and early death].

But mutations are not simply a matter of destructive error. Some changes may — through sheer chance — better fit an organism for its environment. It is upon this that the course of evolution through natural selection ultimately depends.

Source: From *The Genetic Code* by Isaac Asimov. Copyright © 1962 by Isaac Asimov. Used by permission of Dutton Signet, a division of Penguin Books USA Inc.

# SPECIATION

## GENETIC CHANGE, EVOLUTION, AND SPECIATION

As the above example shows, a genetic configuration that may be adaptive in one environment can be maladaptive in another. It is because of this fact that genetic drift, gene flow, and mutation play such an important role in the evolution of species. Were it not for the genetic change produced by these forces, the variability in the gene pool of a population would be reduced, and this would lessen the population's chance of acquiring the modified traits that would allow at least some of its members to adapt to new environmental conditions. Under the circumstances, the population would be less likely to survive when the environment changed. But precisely because there is ongoing genetic change, alterations in the environment do not necessarily disrupt the continuity of life. Instead, the altered environment in which a population lives is afforded an opportunity of preserving those forms whose modified traits make it possible for them to better reproduce than their ancestors. And if and when these new forms become reproductively distinct, a new species will have evolved from an old one. The process is called **speciation**.

## THE ALLOPATRIC AND SYMPATRIC MODELS

As far as experts are aware, in the majority of cases speciation conforms to the **allopatric** (*other place*) **model**. According to the model, new species arise when some members of a population are prevented from mating with others of their kind, either because they have migrated to the fringe of their territory, or else because geographic barriers such as waterways or mountains emerge that keep them apart. In such cases genetic information cannot be freely exchanged within the larger group. Instead, subpopulations develop, each suited to its own habitat and each with its own genetic make-up. As long as these allopatric populations are capable of interbreeding and producing fertile offspring, they belong to the same species. However, if their alleles become so different that they can no longer interbreed and produce fertile offspring, then speciation has occurred. Genetic changes favoured by natural selection will accelerate the process.

Genetic change also plays a role in the **sympatric** (*same place*) **model** of speciation, in which new species are said to arise from a continuous population despite the fact that the population is never completely split. In such cases genetic changes favoured by natural selection may produce breeds of a species whose members exploit their common habitat by behaving in alternate ways. Whether or not these behavioural differences lead to the production of new species depends upon the amount of gene flow between the alternate breeds. If their members continue to exchange genetic information on a regular basis, then the species will remain intact. Since research has shown this is almost always the case when different breeds of a species share at least a portion of the same territory, many scientists claim that the sympatric model rarely if ever conforms to events in the natural world. However, if the behavioural differences between the sympatric breeds become so pronounced that their members become reproductively isolated, then it is at least theoretically possible that the breeds will give rise to different species.

## ANAGENESIS AND CLADOGENESIS

The allopatric and sympatric models show how new species can arise directly from a parent population that is subject to genetic change and natural selection. In the long term, if one species evolves directly into another, without giving rise to any side branches, the process is known as **anagenesis** (see Figure 4.4a). This usually occurs when genetic changes favoured by natural selection produce a population that becomes increasingly better adapted to an ecological niche that remains relatively constant over time.

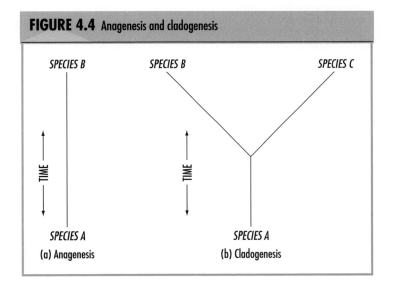

**FIGURE 4.4** Anagenesis and cladogenesis

SPECIES B    SPECIES B    SPECIES C

TIME    TIME

SPECIES A    SPECIES A

(a) Anagenesis    (b) Cladogenesis

In other situations, however, genetic changes favoured by natural selection may result in the production of a wide variety of new species that branch off from a common ancestor. This process is referred to as **cladogenesis** (see Figure 4.4b), and it usually occurs in situations where the environment provides a broad range of adaptive opportunities in the form of specific ecological niches. In turn, cladogenesis may result in the kind of adaptive radiation that characterized the evolution of Darwin's finches. As we shall soon see, both anagenesis and cladogenesis played an important role in the evolution of the primates — cladogenesis moreso when the first primates radiated into new species, and anagenesis more recently, after the first members of our genus had made their debut. See QQ 4.18.

**4.18 ANAGENESIS, CLADOGENESIS, AND PRIMATE EVOLUTION**

Alexander Alland, Jr., comments on anagenesis and cladogenesis and the roles these played in the evolution of the primates.

• • •

Sequential transformation and branching are both best explained as evolutionary adaptation to environment. In the first case (known as *anagenesis*) a single type develops greater and greater genetic specialization through time in areas where the environment has remained relatively constant . . . . This is the progressive development of goodness of fit. In the second case (known as *cladogenesis*) related isolated populations differentiate from one another as a response to differential selection pressures which arise in different microenvironments [through time] . . . .

There are countless examples of anagenesis and cladogenesis in both the plant and animal king-

doms . . . . [In fact, both played an important role in the evolution] of the primate order . . . . The common ancestors of these forms adapted more than seventy-five million years ago to life in the trees, in which stereoscopic vision and a pair of good grasping hands are advantageous. These are both major features of all but the oldest and most primitive members of the primate order. As the primates developed, they spread into different environmental zones. Some returned to a ground-dwelling existence and modified their earlier adaptations. Among these were the baboon and man, two species which have accommodated in different ways to terrestrial life. The human sequence shows increasing brain size, fully erect posture, and the loss of hands on the lower extremities.

Source: Alexander Alland, Jr., *Evolution and Human Behavior* (New York: Natural History Press, 1967), p. 20–23.

# KEY TERMS AND CONCEPTS

HEREDITY

SELF-FERTILIZING

CROSS-FERTILIZATION

ALLELES *(a-leelz)*

GENES

HOMOZYGOUS *(hoh-moh-zy'-gus)*

HETEROZYGOUS *(he-ter-oh-zy'-gus)*

CELLS

NUCLEAR DNA

MITOCHONDRIAL DNA
*(my-toh-kawn'-dree-uul + DNA)*

REPLICATION

MITOSIS *(my-toh'-sis)*

SOMATIC

MEIOSIS *(my-oh'-sis)*

GAMETES *(ga'-meets)*

ZYGOTE *(zy'-goht)*

CHROMOSOMES *(kroh'-moh-sohmz)*

PRINCIPLE OF SEGREGATION

PRINCIPLE OF RECOMBINATION

DOMINANT

RECESSIVE

GENOTYPE *(jee'-noh-typ)*

PHENOTYPE *(fee'-noh-typ)*

ALBINISM

POLYGENIC

KARYOTYPES *(ker'-ee-oh-typs)*

SEX CHROMOSOMES

GENE POOL

*148* CHAPTER FOUR

# SELECTED READINGS

## INTRODUCTION

Bowler, P. "Darwin's Concepts of Variation." *Journal of the History of Medicine*, Vol. 29 (1974), pp. 196–212.

Vorzimmer, P. "Charles Darwin and Blending Inheritance." *Isis*, Vol. 54 (1963), pp. 371–390.

## MENDEL'S VIEW

Dunn, L. *A Short History of Genetics.* New York: McGraw-Hill, 1965.

Orel, V. *Mendel.* Oxford: Oxford University Press, 1984.

Posner, E. and Skutil, J. "The Great Neglect: the Fate of Mendel's Classic Paper Between 1865 and 1900." *Medical History*, Vol. 12 (1968), pp. 122–136.

## ALLELES AND GENES

Carlson, E. *The Gene: A Critical History.* Philadelphia: Saunders, 1966.

Falk, R. "What is a Gene?" *Studies in the History of the Philosophy of Science*, Vol. 17 (1986), pp. 133–173.

## DNA

Alberts, B., *et. al. Molecular Biology of the Cell.* New York: Garland, 1989.

Crick, F. *Life Itself: Its Origin and Nature.* New York: Simon and Schuster, 1981.

McCarty, M. *The Transforming Principle: Discovery that genes are made of DNA.* New York: Norton, 1986.

Schulman, L. and Abelson, J. "Recent Excitement in Understanding Transfer RNA Identity." *Science*, Vol. 240 (1988), pp. 1591–1592.

Watson, J. *The Double Helix: A Personal Account of the Discovery of the Structure of DNA.* New York: Athaneum, 1968.

Watson, J. *The Molecular Biology of the Gene* (third edition). Menlo Park, California: Benjamin, 1976.

Winchester, A. and Mertens, T. *Human Genetics* (fourth edition). Columbus, Ohio: Charles E. Merril, 1983.

## MITOSIS AND MEIOSIS

de Grouchy, J. *Clinical Atlas of Human Chromosomes.* New York: Wiley, 1984.

Tijo, H. and Levan, A. "The Chromosome Number of Man." *Hereditas*, Vol. 42 (1956), pp. 1–6.

## SEGREGATION AND RECOMBINATION

Monaghan, F. and Corcos, A. "The Origins of the Mendelian Laws." *Journal of Heredity*, Vol. 75 (1984), pp. 67–69.

Olby, R. *The Origins of Mendelism* (second edition). Chicago: University of Chicago Press, 1985.

## DOMINANCE AND RECESSIVENESS

McKusick, V. *Mendelian Inheritance in Man* (seventh edition). Baltimore: Johns Hopkins University Press, 1986.

## HUMAN HEREDITY

Cummings, M. *Human Heredity: Principles and Issues* (second edition). St. Paul, Minnesota: West Publishing, 1991.

Hartl, D. *Our Uncertain Heritage: Genetics and Human Diversity* (second edition). New York: Harper & Row, 1985.

Mange, A. and Mange, E. *Genetics: Human Aspects.* Philadelphia: Saunders, 1980.

## THE COMPLEXITY OF HUMAN HEREDITY

Edwards, J. "The Mutation Rate in Man." *Progress in Medical Genetics*, Vol. 10 (1974), pp. 1–16.

Farley, J. *Gametes and Spores: Ideas about Sexual Reproduction, 1750–1914.* Baltimore: Johns Hopkins University Press, 1982.

Haas, A. and Haas, K. *Understanding Sexuality* (second edition). New York: Times Mirror/Mosby College Publishing, 1990.

Margulis, L. and Sagan, D. *The Origins of Sex.* New Haven: Yale University Press, 1986.

Weaver, R. "Changing Life's Genetic Blueprint." *National Geographic*, Vol. 166 (1984), pp. 818–847.

## POPULATION GENETICS

Bodmer, W. and Cavalli-Sforza, L. *Genetics, Evolution, and Man.* San Francisco: Freeman, 1976.

Cavalli-Sforza, L. and Bodmer, W. *The Genetics of Human Populations.* San Francisco: Freeman, 1971.

Glass, B. "On the Evidence of Random Genetic Drift in Human Populations." *American Journal of Physical Anthropology*, Vol. 14 (1956), pp. 541–555.

Hartl, D. *A Primer of Population Genetics* (second edition). Sunderland, Massachusetts: Sinauer Associates, 1987.

Stanley, G. *Louis Riel.* Toronto: McGraw-Hill Ryerson, 1963

Wallace, B. *Basic Population Genetics.* New York: Columbia University Press, 1981.

## MUTATION

Baird, P. and Sadovnick, A. "Life Expectancy in Down Syndrome." *Journal of Pediatrics*, Vol. 110 (1987), pp. 849-854.

Baird, P. and Sadovnick, A. "Underlying Causes of Death in Down Syndrome: Accuracy of British Columbia Death Certificate Data." *Canadian Journal of Public Health*, Vol. 81 (1990), pp. 456–461.

Croce, C. and Klein, G. "Chromosomal Translocations and Human Cancer." *Scientific American*, Vol. 252 (1985), pp. 54–60.

Crow, J. and Denniston, C. "Mutation in Human Populations." *Advances in Human Genetics*, Vol. 14 (1985), pp. 59–123.

Dickerson, R. and Geis, I. *Haemoglobin: Structure, Function, Evolution and Pathology.* Menlo Park, California: Benjamin/Cummings, 1983.

Ingram, V. "Gene Mutations in Human Haemoglobin: The Chemical Difference Between Normal and Sickle-Cell Haemoglobin." *Nature*, Vol. 180 (1957), pp. 326–328.

Livingstone, F. *Abnormal Haemoglobin in Human Populations.* Chicago: Aldine, 1967.

Muench, K. *Genetic Medicine.* New York: Elsevier, 1988.

Obe, G., ed. *Mutations in Man.* New York: Springer-Verlag, 1984.

Patterson, D. "The Causes of Down Syndrome." *Scientific American*, Vol. 257 (1987), pp. 52–61.

Smith, G., ed. *Molecular Structure of the Number 21 Chromosome and Down Syndrome.* New York: New York Academy of Science, 1884.

Sugimura, T., Kando, S., and Takebe, H., eds. *Environmental Mutagens and Carcinogens.* New York: Liss, 1982.

Wiesenfeld, S. "Sickle-Cell Trait in Human Biological and Cultural Evolution." *Science*, Vol. 157 (1967), pp. 1134–1140.

## SPECIATION

Dawkins, R. *The Extended Phenotype: The Gene as the Unit of Selection.* San Francisco: Freeman, 1982.

Eldredge, N. and Cracraft, J. *Phylogenetic Patterns and the Evolutionary Process.* New York: Columbia University Press, 1980.

Savage, J. *Evolution* (third edition). New York: Holt, Rinehart and Winston, 1977.

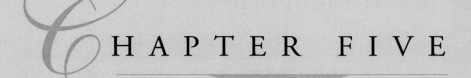

# CHAPTER FIVE

## EARLY PRIMATE EVOLUTION

*contents at a glance*

# INTRODUCTION

Inspired by the theories of Darwin and Mendel, which have been integrated into a single theoretical framework known as **neo-Darwinism**, anthropologists have made great progress in piecing together the story of human evolution. For instance, thanks to their efforts we now know that humans did not evolve from any of the living apes. Although we are closely related, the laws of heredity ensure that each new generation varies only slightly from its parents, humans producing human offspring and apes producing ape offspring. In other words, there is no "missing link" midway between apes and humans as some once believed.

The prevailing view is that the evolution of life began with the appearance of the first bacteria and algae about 3500 million years ago (mya). Thereafter, in concert with changes in the environment, natural selection favoured new genetic patterns that formed the basis for all subsequent plant and animal life. About 750 mya the first multicellular organisms appeared. These, in turn, gave rise to marine invertebrates, fish, amphibians, and reptiles, which appeared in successive waves between about 575 mya and about 230 mya. Later, between about 230 mya and about 65 mya, dinosaurs dominated the planet accompanied by the forerunners of birds and of mammals. More recently, about 65 mya, the **Age of Mammals** began; it was then that the mammals, including the primates, started to flourish. See Table 5.1

| **TABLE 5.1**  A simplified schema of the evolution of life | | |
|---|---|---|
| **MYA** | **DOMINANT LIFE FORMS** | **ALSO KNOWN AS** |
| 4500 – 3500 | none | |
| 3500 – 750 | bacteria and algae | |
| 750 – 575 | first multicellular organisms | |
| 575 – 500 | marine invertebrates | |
| 500 – 345 | fish | Age of Fish |
| 345 – 280 | amphibians | Age of Amphibians |
| 280 – 230 | reptiles | Age of Reptiles |
| 230 – 65 | dinosaurs | Age of Dinosaurs |
| 65 – today | mammals | Age of Mammals |

# THE RISE OF THE MAMMALS

According to palaeontologists, the first mammals were inconspicuous little animals that used their mammalian traits to great advantage. What has yet to be determined is which traits made the greatest contribution to the mammals' initial success. Assuming that dinosaurs were their main competitors, one possibility is that they owed their success to the fact they were warm-blooded. This may have been doubly important: first, it may have allowed the mammals to become nocturnal, which would have enabled them to avoid competing with cold-blooded diurnal dinosaurs, and second, the ability to regulate their internal body temperature may have allowed mammals to mature and reproduce more rapidly than dinosaurs. Another possibility is that it was lactation or suckling their young and parental care that set the mammals apart. This proposal is based on the idea that while young dinosaurs (like modern reptilian infants) were required but ill-equipped to fend for themselves from the moment they hatched, young mammals were suckled and cared for by their mothers until they were capable of surviving on their own. If this were the case, then the early mammals would have been far better equipped to survive and reproduce than the dinosaurs that they supplanted. Whatever the source of their evolutionary success, about 65 mya the heyday of the mammals began. See QQ 5.1.

 **5.1 THE RISE OF THE MAMMALS**

Alfred Romer discusses the rise of the mammals.

. . .

[Mammals are] the most intelligent and the most successful of land vertebrates. Arising, it seems, at about . . . [180 mya] from the mammal-like reptile stock, the group remained comparatively unimportant until the extinction of the great reptiles . . . [about 65 mya]. From that time on, however, mammals increased rapidly in numbers and diversification, so that now they include the greater part of the animals with which one is ordinarily familiar. They not only inhabit the surface of the earth but have invaded the air (bats) and returned to the seas (whales, seals, [and sea cows or] sirenians). In size they range from tiny shrews and mice to certain of the whales which are the largest of all known animals, exceeding even the great dinosaurs in size. Mammals are of particular interest to man, not only because they include many of his animal friends and enemies and much of his food supply, but also because he himself is a member of this group.

Source: Alfred Sherwood Romer, *Vertebrate Paleontology*, third edition, (Chicago: University of Chicago Press, 1966), p. 187.

# FOSSILS

The abovementioned theories about the rise of mammals are largely based on an analysis of **fossils**, which are the remains or traces of the remains of organisms that lived in the past. Sometimes organic materials such as bone, teeth, skin, hair, shell, seeds, wood, and charcoal are preserved, either in ice, in a tarpit, or in airless, wet acid media such as peat. The resulting fossils are composed of living cells. Far more often, however, fossils are formed through **petrification**, in which the hard parts of dead plants and animals are replaced either in whole or in part with minerals. This happens in one of two ways. In **permineralization** the pores or spaces in the hard parts are filled in with minerals such as silica, pyrite, and calcite that are absorbed from the local groundwater. Such fossils consequently consist of a mixture of organic and inorganic matter. In **mineralization** the hard parts dissolve completely but are replaced with minerals. These mineralized specimens are much like the originals except that they are composed entirely of stone. The more gradually mineralization occurs the more complete is the fossil. Fossils are also produced by imprints in sand, silt, and volcanic ash. If the imprint is not destroyed it can be filled with plaster to create a cast, the cast being a reproduction of the original. Although most organisms leave no trace of themselves after they die because mechanical, chemical, and biological processes destroy their remains, those fossils that do exist provide the special evidence that is required to reconstruct the natural history of species.

▼

Permineralized
bird eggs
from the
Cenozoic era

▲

The Raymond M. Alf Museum, Claremont, California

# THE GEOLOGICAL TIME SCALE

In order to make sense of the fossil record the age of the specimens is of the utmost importance, and, in order to call attention to their age, scientists frequently use terms such as "palaeozoic fossils," "mesozoic fossils," "cenozoic fossils," and so on. These terms refer to the position of fossils in the **geological time scale or calendar**, which divides the earth's history into a number of eras, periods, and epochs, each interval characterized by its own depositional features (see QQ 5.2). The divisions are based on the **Principle of**

**Superposition**, which states that the formations that cover the earth's core have been deposited in successive layers or strata — palaeozoic, for example, overlaid by mesozoic and mesozoic by cenozoic. It stands to reason that wherever the sequence has not been disrupted by folding or faulting, the ages of the strata and the fossils they contain increase with depth. Thus, as Table 5.2 shows, palaeozoic fossils are older than mesozoic, and mesozoic older than cenozoic.

## 5.2 THE GEOLOGICAL TIME SCALE

John Gribbin describes the first stage in the development of the geological time scale.

• • •

In order to understand how the history of the earth, and of life on earth, has unfolded, we need to know how old different geological samples are. The first stage in developing the geological "calendar" came as geologists determined which rocks are younger and which older among those available for inspection today. During the eighteenth and nineteenth centuries geologists, mainly based in Britain and Western Europe, built up their picture of the divisions of geological time and gave rocks of particular ages names (usually in Latinized or Greek form) corresponding to the regions of Europe in which they were found — so, for example, a particular time period [say, for instance, between about 570 mya and 500 mya] was given the name Cambrian from the Roman name for Wales, where many Cambrian rocks are found.

Source: John Gribbin, Genesis: *The Origins of Man and the Universe* (Toronto: Oxford University Press, 1982), p. 197.

**TABLE 5.2** Geological eras and periods

| ERAS | PERIODS | MYA |
|------|---------|-----|
| Azoic | | 4500 – 3500 |
| Proterozoic | | 3500 – 575 |
| Palaeozoic | Cambrian | 575 – 500 |
| | Ordovician | 500 – 430 |
| | Silurian | 430 – 395 |
| | Devonian | 395 – 345 |
| | Carboniferous | 345 – 280 |
| | Permian | 280 – 230 |
| Mesozoic | Triassic | 230 – 180 |
| | Jurassic | 180 – 135 |
| | Cretaceous | 135 – 65 |
| Cenozoic | Tertiary | 65 – 1.8 |
| | Quaternary | 1.8 – today |

# Determining the Relative Age of Fossils

## Stratigraphy

It was the principle of superposition that paved the way for anthropologists to develop the **Principle of Stratigraphy**, which, like its forerunner, is based on the order in which ancient materials lie buried. Generally speaking, according to the principle of stratigraphy, those fossils discovered at the bottom of an undisturbed excavation are older than those found near the surface, and those found in-between are intermediate in age. Thus, the respective age of any fossil in the sequence can be determined by virtue of its vertical proximity to the others in the deposit (see Figure 5.1). Were it not for this fact thousands of fossils that might otherwise have remained a mystery have been assigned a **relative age** — one that indicates whether one fossil is older than another, although not by how much. See QQ 5.3.

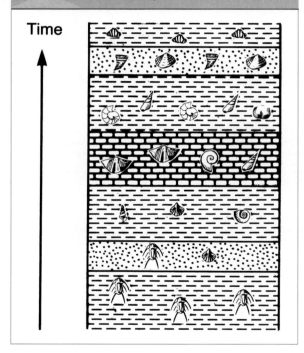

**FIGURE 5.1** Stratigraphic arrangement of fossils in an undisturbed sequence

Time

## QQ 5.3 STRATIGRAPHY

J. Forde-Johnston discusses the relationship between the earth's strata and the relative age of ancient remains.

• • •

[S]tratigraphy, borrowed . . . from the science of geology . . . [is based on the fact that t]he earth's crust is made up of a number of different layers or strata: granite, sandstone, limestone, chalk, etc., which quite clearly must have been laid in a definite sequence. Except where subsequently disturbed, by folding, for example, the lower layers of strata or rock must be earlier than the layers above, and this simple principle provides a time framework, a sequence of periods according to the number of layers involved. The lowest layer in any sequence is thus Period 1, the next layer above Period 2 and so on. It follows from this that anything found in the lowest layer (Period 1) is earlier than anything found in the next layer (Period 2) and so on, so that there is a very simple way of establishing the relative ages of all the things found in the different layers.

Source: J. Forde-Johnston, *History from the Earth: An Introduction to Archaeology* (New York: Phaidon Press Limited, 1974), p. 19.

## FLUORINE, URANIUM, AND NITROGEN DATING

Of course, where there has been folding or faulting the principle of stratigraphy is moot. However, when this is the case other relative dating techniques may be brought into play. Three of these are used to establish the relative age of ancient bone. Sometimes this is done by using the **fluorine test**, which is based on the fluorine content of bone that is buried in a moist sediment. When buried in such a sediment, bone absorbs fluorine from the groundwater. If two specimens are found in the same such location, the one with the more fluorine should be the older. The same principle governs **uranium dating**, for the longer a bone is buried in a moist sediment the more uranium it absorbs from the groundwater. **Nitrogen dating** is similar except that the amount of nitrogen in buried bone decreases over time. The older specimen consequently would be the one with less nitrogen.

# DETERMINING THE ABSOLUTE AGE OF FOSSILS

While the importance of stratigraphy and of the fluorine, uranium, and nitrogen methods cannot and should not be underestimated, none of these can generate the **absolute or chronometric age** of a fossil — the age of a fossil in years. To accomplish this goal anthropologists must rely on other dating techniques; their use makes it possible to determine, in fairly precise terms, when the life forms that are represented by fossils populated the planet.

## POTASSIUM-ARGON DATING

Foremost among these absolute dating techniques is the potassium-argon (**K-Ar) method**, which geochronologists use to calculate the age of compacted volcanic ash called **tuffs** that are between about one million years old and about five million years old, and rocks in the same age range that were formed through volcanic activity. These include basalts, obsidian, micas, potash, feldspar, and hornblende. The method is based on the fact that such volcanic materials contain an unstable radioactive element known as $^{40}$K (potassium 40), which breaks down into $^{40}$Ar (argon 40 or argon gas) at a constant rate after the material has cooled. The "half-life" of $^{40}$K is about 1300 million years, which means that over a 1300 million year span about one-half of the contents of $^{40}$K in a sample of volcanic ash or rock is transformed into $^{40}$Ar. Thus, by measuring the ratio of the two elements in a sample selected for its potassium content, the absolute age of the sample can be determined (see QQ 5.4). And when anthropologists discover a fossil sandwiched between two layers of volcanic materials that can be dated by the potassium-argon method, then they are generally safe in assuming that the absolute age of the fossil lies between the age range of the layers.

 **5.4 CALCULATING A POTASSIUM-ARGON DATE**

Frank Poirier lists the steps that are involved in calculating a potassium-argon date.

• • •

1  Measure the size of the sample and its richness in potassium.
    This is easily done by standard laboratory procedures, and shows that the sample contains 0.1 g of potassium.

2  Calculate the annual decay rate for a sample that size.
    It is known that the potassium-40 in 1 g of ordinary potassium decays to argon at a rate of about 3.5 atoms per second.
    Therefore:
    3.5 x 60 = 210 per minute x 60 = 12 600 per hour x 24 = 302 400 per day x 365 = 110 376 000 per year. So, 0.1 g of potassium yields 11 037 600 atoms of argon per year.

3  Boil off the sample, and send it to the mass spectrometer (along with any contaminating air that may be in the bottle).

4  Obtain mass spectrometer reading.
    This particular sample gives a reading of: 36 765 875 000 000 atoms of argon-40 (from the air and the sample)

27 070 000 000 atoms of argon-36 (from the air only)

5  Eliminate the atmospheric contaminant.
    Since there are 295.5 argon-40 atoms to every argon-36 atom in the atmosphere, multiply the argon-36 total by 295.5:
    27 070 000 000 x 295.5 = 7 991 850 000 000.
    This is the number of contaminating atoms of argon-40 still in the sample.
    So, from the total reading in the mass spectrometer, deduct the atmospheric contaminant:
    36 765 875 000 000 - 7 991 850 000 000 = 28 774 025 000 000
    This is the number of argon atoms released by the sample.

6  Calculate the age of the sample.
    Since the sample decays at a rate of 11 037 600 atoms a year, it is necessary to divide the number of argon atoms in the sample by that number:
    $$\frac{28\ 774\ 025\ 000\ 000}{11\ 037\ 600} = 2\ 606\ 909.5$$
    Answer: The sample is 2.6 million years old.

Source: Frank E. Poirier, An Introduction to *Physical Anthropology and the Archaeological Record* (Minneapolis, Minnesota: Burgess Publishing Company, 1982), p. 169.

# FISSION-TRACK DATING

Cross-checking the age of a layer of volcanic rock that has been determined via the potassium-argon method is also possible. This can be done by using the **fission-track method** (see Figure 5.2), which has frequently been employed to date volcanic rocks that are between about 100 000 years old and about three million years old. Such rocks house two relatively rare radioactive uranium elements: $^{238}U$ (uranium 238), which contains 99.27 percent of a volcanic rock's uranium burden, and $^{235}U$ (uranium 235), which accounts for .72 percent of the burden. Although uranium 238 decays into lead natu-

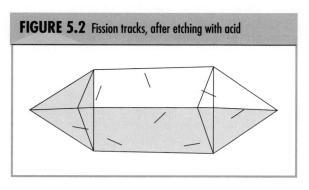

**FIGURE 5.2** Fission tracks, after etching with acid

rally, it also sometimes divides in half because of spontaneous fission, and when this happens the halves travel at extremely high speeds and collide with the structures in their path, creating a series of scars in the rock that are called "fission tracks." By counting the tracks and comparing the total with the number of tracks that are produced by intentionally generating fission in the uranium 235 in the sample, it is possible to estimate how much uranium 238 remains in the sample. This is because the ratio of uranium 235 to uranium 238 is known. And since the rate at which uranium 238 decays is also known, it is then possible to estimate the number of years that have elapsed since the sample was last superheated.

# THE MAGNETIC POLARITY TIME SCALE

Since the absolute age of a fossil that has been determined by the potassium-argon and fission-track methods may be anywhere within the range of ages of the layers of rock in which the specimen is sandwiched, whenever possible, anthropologists employ other techniques to more tightly define the absolute age of the fossil. One way of accomplishing this goal is to rely on the **Magnetic Polarity Time Scale** (MPTS). The scale is based on the fact that, from time to time, changes in the earth's core have caused the polarity of the earth's magnetic field (which currently runs from south to north) to run in the opposite direction. Such changes in polarity are called "reversals," and by carefully comparing the polarity of rocks with their K-Ar (potassium-argon) dates geochronologists have been able to construct the MPTS, an absolute time scale that plots the intervals between such reversals. Although the MPTS is by no means perfect — its accuracy diminishes as the age of the intervals increase and the intervals themselves are lengthy — it has proven to be quite useful in refining the absolute age of magnetized fossils that have previously been dated by the potassium-argon and fission-track methods. See QQ 5.5.

## 5.5 THE MAGNETIC POLARITY TIME SCALE

M. W. McElhinny comments on the development of the MPTS.

• • •

[In 1929 Matuyama] proposed that during the early part of the Quaternary Period [beginning about 1.8 mya] the earth's magnetic field was reversed and that this gradually changed over to the normal state. This result . . . has been amply confirmed by subsequent investigations.

If the field-reversal theory is correct, there must be a precise stratigraphic correlation of normally and reversely magnetized strata from all over the world. This approach is contingent on the condition that the duration of epochs during which the magnetic polarity

remains unchanged, must be sufficiently long to be resolved by the available geological techniques. For example, the classical techniques of palaeontology can hardly be used if polarity epochs lasted only 50 000 years. The development of the Potassium-Argon (K-Ar) dating technique has made it possible to use the method to date very precisely volcanic rocks whose magnetic polarities have been determined. From these measurements made at a number of laboratories around the world, a geomagnetic polarity time-scale [known as the MPTS] covering the past few million years has been built up . . . .

Source: M. W. McElhinny, *Palaeomagnetism and Plate Tectonics* (London: Cambridge University Press, 1973), p. 114.

## CARBON-14 DATING

Another important technique that anthropologists use to determine the absolute age of fossils is the **carbon-14 method**. It is used to determine the absolute age of carbon-bearing organic remains such as bone, wood, charcoal, hair, skin, shell, and seeds from the recent past, generally between about 500 years old and about 50 000 years old. The method is based on the discovery that all living organisms contain a relatively stable ratio of two **carbon isotopes**, that is, two carbon elements with the same atomic number but with different atomic weights: $^{14}C$, which is radioactive, and $^{12}C$, which is non-radioactive. These two elements mix in the upper atmosphere and become incorporated into carbon dioxide. Since green plants absorb carbon dioxide directly from the atmosphere, they are able to maintain the ratio of the two isotopes directly. Animals, on the other hand, maintain the ratio either by eating green plants or species that feed on such plants. In both cases, when an organism dies the ratio of the two elements changes. While the amount of $^{12}C$ remains constant, the amount of $^{14}C$ gradually decays by about one-half once every 5730 years, which is the "half-life" of the isotope. Consequently, by measuring the relative proportions of the two isotopes in a carbon-bearing organic sample, it is possible to estimate its absolute age. See QQ 5.6.

## 5.6 CARBON-14 DATING

Steward Fiedel describes the carbon-14 method.

• • •

[One of t]he most significant contribution[s] to . . . [dating ancient organic remains] was made by the physicist Willard F. Libby, who perfected the technique of carbon-14 dating in 1949. Carbon 14 is a radioactive isotope that is produced in the upper atmosphere when nitrogen atoms are bombarded by cosmic radiation. C14 behaves chemically just like normal non-radioactive C12. It is incorporated into carbon dioxide molecules which are absorbed from the air by green plants. When the plants are ingested by herbivorous animals, C14 passes into their tissues as well, and so on through the whole trophic [that is, nutritional] chain of life forms. When a plant or animal dies, it takes in no more C14, and the slow breakdown of the radioactive atoms within it begins. Libby discovered that this breakdown by the emission of beta particles proceeded at a constant rate. Half of the radioactive carbon atoms remained after 5568 years; half of these would decay in the next 5568 years, and so on. The "half-life" figure was later corrected to 5730 years.

Source: Steward J. Fiedel, *Prehistory of the Americas* (New York: Cambridge University Press, 1987), pp. 11–12.

# PANGEA, LAURASIA, AND GONDWANA

## PLATE TECTONICS

By employing the abovementioned dating techniques, and thanks to what they have learned from geologists about the movements of the earth's crust over its inner core, anthropologists have determined that the first primates lived in a world remarkably different from today.

About 230 mya, at the beginning of the Mesozoic era, the continents that we now know did not exist. Instead, there was only one supercontinent — **Pangea**, meaning *all lands* — which featured an arc of volcanoes in the south and three mountain ranges in the north. Slowly, over the next 60 million years, Pangea gave rise to two new continents that were separated by the Tethys Ocean: **Laurasia** in the north, which included what are now North America, Greenland, Europe, and Asia; and **Gondwana** in the south, which included South America, Africa, India, Antarctica, and Australia. Thereafter, Laurasia and Gondwana themselves began to break apart, so that by the beginning of the Cenozoic era (*c.* 65 mya), the configuration of the continents that we now know began to take shape (see Figure 5.3 on page 165). This rearrangement of the earth's surface was due to **plate tectonics**, the slow but continual movement of the semirigid plates on which continents and oceans rest. See QQ 5.7.

## PLATE TECTONICS AND THE EVOLUTION OF SPECIES

Plate tectonics had a marked effect on the evolution of untold species. The rearrangement of the earth's surface created new ecological niches that promoted the evolution of new mammalian forms. Among them were a num-

# 5.7 PLATE TECTONICS

**Jack Oliver looks at plate tectonics.**

• • •

According to the theory of plate tectonics, the surface of the earth is a mosaic, a dynamic global system of plates, some incorporating both continental and oceanic areas. The plates are like puzzle pieces set in excruciatingly slow motion, converging, diverging and slipping past one another over geologic time . . . .

The continents have been converging and colliding throughout much of the history of the planet. Some forty million years ago, for instance, the subcontinent of India, after drifting from its former position adjoining Antarctica across the intervening sea, entered into a long, drawn-out collision with Eurasia. The spectacu-

lar consequences of this episode include the highest mountain range on earth — the Himalayas — and the huge Tibetan plateau. That the process continues is evident from the seismic activity of China, which is periodically shaken by great earthquakes . . . . An imminent collision, on the geological time scale, is also underway in the southwestern Pacific, where Australia appears to be creeping inexorably toward Asia. Gathering and crumpling the islands and seas before it, Australia moves to plaster itself, along with remnants of smaller landmasses caught up in the collision, against the larger Eurasian continent.

Source: Jack E. Oliver, "The Big Squeeze: How Plate Tectonics Redistributes Mineral and Organic Resources," *The Sciences*, Vol. 31, No. 4 (1991), p. 24.

ber of small animals that fed on fruits and leaves and spent most of their time in the trees, in the huge subtropical and tropical forests that then circled the globe. At first glance it is hard to tell them apart from the forerunners of the **insectivora**, the mammalian order that includes modern insect-eating animals such as moles, shrews, and Old World hedgehogs. A closer examination reveals that they were the first primate-like mammals, quite unlike monkeys, apes, and humans, but equipped with the beginnings of the characteristics that all primates possess: flexible fingers and toes; nails instead of claws; mobile hands and feet; good hand-eye coordination; a small snout; a small number of teeth; and a relatively large brain. As the contents of Table 5.3 indicate, it was these forms that gave rise to the various primates that appeared during the Cenozoic era (*c.* 65 mya–today).

| **TABLE 5.3** | Major events in the evolution of the primates during the Cenozoic era | | | |
|---|---|---|---|---|
| **ERA** | **PERIOD** | **EPOCHS** | **YEARS AGO** | **MAJOR EVOLUTIONARY EVENTS** |
| Cenozoic | Tertiary | Palaeocene | 65mya–53mya | early primate-like mammals appear |
| | | Eocene | 53mya–37mya | early prosimian-like primates appear |
| | | Oligocene | 37mya–25mya | early monkey-like primates appear |
| | | Miocene | 25mya–5mya | early ape-like primates appear |
| | | Pliocene | 5mya–1.8mya | early human-like primates appear |
| | Quaternary | Pleistocene | 1.8mya–11 000 | humankind appears |
| | | Holocene | 11 000–today | |

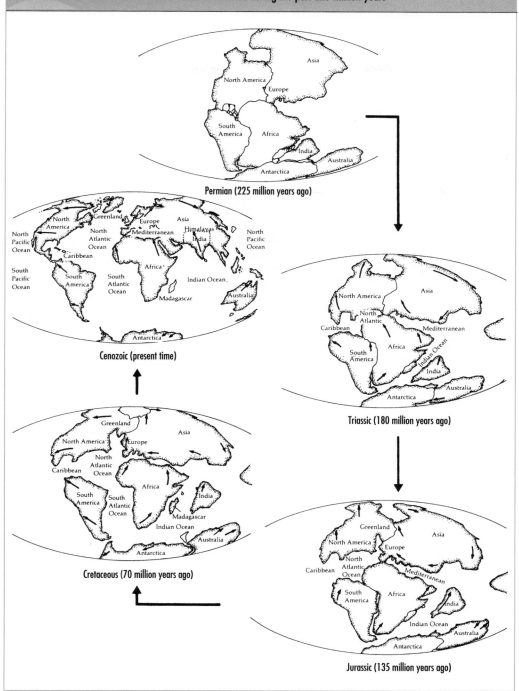

Permian (225 million years ago)

Triassic (180 million years ago)

Jurassic (135 million years ago)

Cretaceous (70 million years ago)

Cenozoic (present time)

# THE END OF THE AGE OF REPTILES

The survival of the first primate-like mammals was due to natural selection working in their favour and against the reptiles. By the start of the Cenozoic the heyday of the reptiles had passed. Many, including the dinosaurs, had already become extinct. Why this happened is still being debated. The prevailing theory is that their extinction was the result of an enormous asteroid strike at the end of the Mesozoic era. The strike apparently raised a huge dust cloud, so dense that it reduced the intensity of the sun's rays. This, in turn, upset the reptiles' food chain and natural selection took its toll. Recently, however, another hypothesis has been advanced which suggests that the demise of dinosaurs such as Tyrannosaurus, Triceratops, and others was due to disease (see QQ 5.8). The idea is based on the assumption that when the dinosaurs spread throughout the world the

Tyrannosaurus (left) and Triceratops (right)

Courtesy Department Library Services, American Museum of Natural History

## QQ 5.8 THE DOWNFALL OF THE DINOSAURS

Science reporter Virginia Morell explains why the dinosaurs may have become extinct.

• • •

Dinosaurs evolved early in the Triassic period, between 248 million and 213 million years ago. Their immediate ancestors were the theodonts . . . [but b]y 190 million years ago, dinosaurs had supplanted the . . . theodonts. They . . . remained the dominant land-dwelling animals until they died out at the end of the Cretaceous period, about 65 million years ago . . . . So what happened to them? . . .

The most popular theory . . . has a giant asteroid crashing into earth, kicking up a storm of dust, and casting a chilly pall over the planet. But . . . [some scientists] are dubious. "In the late Cretaceous there seems to have been a real dwindling of species," says [Phil] Currie

[head of palaeontology at Alberta's Dinosaur Provincial Park] . . . . So dinosaurs might already have been on the wane when the asteroid struck — or didn't strike.

Triceratops and Tyrannosaurus were two of the dinosaurs remaining at the end of the Cretaceous, and the triceratops herds . . . had spread to all continents . . . . Their dispersal . . . says [Robert] Bakker [adjunct curator of palaeontology at the University of Colorado Museum], coincided with the elimination of the geographical diversity and environmental pressures dinosaurs had previously experienced. "It wasn't a death star," Bakker insists. "It was far more messy than that . . . . [B]ecause the dinosaurs were already weakened by this geographical sameness, it wouldn't have taken much to finish them off. Disease, as a factor in extinction, is greatly underestimated."

Source: Virginia Morell, "The Birth of a Heresy," *Discover*, Vol. 8, No. 3 (1987), pp. 28–50.

environmental pressures that had promoted their diversity were eliminated, and that the sameness that resulted among them left them prone to disease.

# PALAEOCENE PRIMATES

## *PURGATORIUS*

The mammals were more fortunate than the dinosaurs. One interesting group of early mammals is represented by the genus ***Purgatorius***, whose fossils date back to the dawn of the Palaeocene, when the Mesozoic gave way to the Cenozoic. Their fossils, which consist of about four dozen teeth discovered in western North America, tell a puzzling story.

Although *Purgatorius* probably looked like an insectivore, with claws and a long snout, its molars were more like those of the living primates: smaller, less numerous, and less specialized than those of the insectivora, and more suited to a diet of fruits and leaves. The reason so much is made of the teeth is not only because they are all that remains of these ancient rat-size creatures, but also because they suggest that *Purgatorius* ate the foods that later primates preferred and lived in an

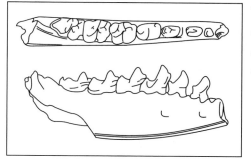

*Purgatorius teeth*

arboreal habitat. Whether or not *Purgatorius* was the first primate genus is nonetheless unclear. Since primates are characterized by a constellation of traits rather than by a single specialized feature, it is extremely difficult for palaeoanthropologists to determine when the first primates appeared. See QQ 5.9.

## 5.9 PRIMATES OR NOT?

**Elwyn Simons looks at *Purgatorius*.**

• • •

A recently described genus, *Purgatorius*, coined by Van Valen and Sloan [in 1965], is possibly the oldest known primate. *Purgatorius ceratops*, a species based on a single tooth, is of the same age as the dinosaur *Triceratops*. A later species of Paleocene age is based on some fifty isolated teeth from a single quarry. These teeth . . . show specific structural similarity to undoubted primate molars, but in the present state of knowledge (isolated teeth) this . . . [cannot be regarded as a clearcut] indication of phyletic [or evolutionary] relations . . . . To the extent that these animals are correctly associated with primate ancestry they suggest that . . . [they] arose from . . . Cretaceous mammals.

Source: Elwyn L. Simons, *Primate Evolution: An Introduction to Man's Place in Nature* (New York: The Macmillan Company, 1972), pp. 108–109.

## THE ARBOREAL THEORY

It is also unclear how the first primates — whoever they were — succeeded in adapting to an arboreal habitat. For many years the majority of anthropologists relied on the **arboreal theory** to explain this turn of events. The theory, which was developed around the turn of the century, holds that the first primates were the descendants of insectivores that were better equipped to live, move, and feed in the trees than their competitors because natural selection had equipped them with traits such as short faces, small snouts, close-set eyes, stereoscopic vision, clawless digits, grasping hands and feet, and teeth that were especially well-suited to feeding on fruits and leaves in the upper canopy of the tropical forest. However, while it is certainly true that these features enhance the ability to survive in an arboreal habitat, it is equally true that animals that do not possess this bundle of traits can also live, move, and feed in the trees. Opossums and tree squirrels are outstanding living examples; they move up the trunks and along the limbs of trees with great ease.

## THE VISUAL PREDATION THEORY

Given this fact, many anthropologists now favour the **visual predation theory** to account for primate origins. It holds that primate-like eyes and limbs are best regarded as adaptive devices that allowed those that originally possessed them to best survive by preying on insects that inhabited the slender vines and branches that formed the comparatively dark understorey of the tropical forest. Since the forerunners of opossums, squirrels, and similar living arboreal mammals presumably occupied a different part of the forest, this would have reduced the number of potential competitors in the ecological niche of the first primates and promoted their adaptive radiation. See QQ 5.10 and Figure 5.4.

## 5.10 PRIMATE ORIGINS

Matt Cartmill, who originally developed the visual predation theory, explains why he believes that it is superior to the arboreal theory.

• • •

Clawed digits, nonopposable thumbs and first toes, and wide-set eyes are primitive mammalian traits. For an arboreal mammal, the adaptive value of these traits is equal or superior to that of primate-like grasping extremities and closely apposed eyes. The loss of the primitive traits in the order Primates therefore cannot be explained by merely invoking the putative selection pressure per se. Visually directed predation on insects in the lower canopy and marginal growth of tropical forests is characteristic of many living prosimians, and also of small marsupials and chameleons. Primate-like specializations of the visual apparatus and extremities occur in all these groups. This suggests that grasping extremities were evolved because they facilitate cautious well-controlled movements in pursuit of prey on slender supports; and that optic convergence and stereoscopy in primates originally had the same adaptive significance they have in cats [that is, they allow

**FIGURE 5.4** Upper canopy, middle layer, and lower understorey of a tropical forest

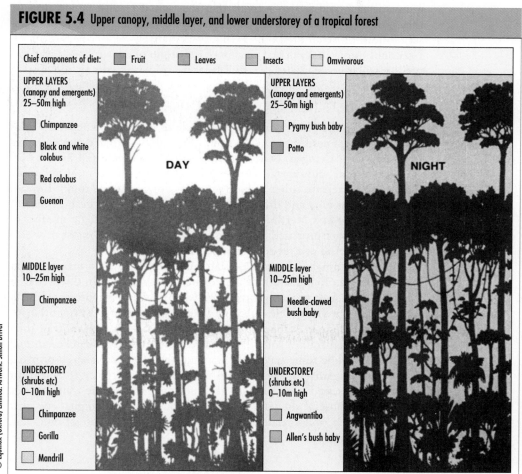

Chief components of diet: ■ Fruit ■ Leaves ■ Insects ■ Omvivorous

UPPER LAYERS (canopy and emergents) 25–50m high
■ Chimpanzee
■ Black and white colobus
■ Red colobus
■ Guenon

MIDDLE layer 10–25m high
■ Chimpanzee

UNDERSTOREY (shrubs etc) 0–10m high
■ Chimpanzee
■ Gorilla
■ Mandrill

DAY

UPPER LAYERS (canopy and emergents) 25–50m high
■ Pygmy bush baby
■ Potto

MIDDLE layer 10–25m high
■ Needle-clawed bush baby

UNDERSTOREY (shrubs etc) 0–10m high
■ Angwantibo
■ Allen's bush baby

NIGHT

## THE ANGIOSPERM RADIATION THEORY

There is, however, another way to account for primate origins. It is known as the **angiosperm radiation theory**, and, like the visual predation theory, it too holds that primate-like eyes and limbs made it possible for the first primates to live and breed successfully in the relatively dark understorey of the tropical forest. But instead of maintaining that the first primates preyed for the most part on insects, the angiosperm radiation theory holds that these animals were omnivores that subsisted primarily on angiosperms — flowering plants with protected seeds, ranging from herbs to trees, that began to flourish towards the end of the Mesozoic era, and are the dominant land flora of the modern world. The essence of the theory is that the adaptive radiation of the primates was promoted by the presence of angiosperms because these provided the first primates with foods such as flowers, fruits, gums, nectars, and seeds — foods that they could readily exploit because they possessed the visual acuity and hand-eye coordination that were necessary to feed on small food objects in a comparatively dark ecological niche. No doubt the accuracy of this version of the origin of our order and its competitors will become clearer as new Palaeocene primate fossils are discovered.

## THE PLESIADAPIFORMES AND *PLESIADAPIS*

Unfortunately, the picture during the middle of the Palaeocene (*c.* 60 mya) is also somewhat confusing. At the time, although the original configuration of Laurasia had changed, North America and Europe were still joined. Nor were the climate and vegetation of this huge northern land mass like those that are characteristic either of North America or of Europe today. Instead, the weather was mild and the plants were subtropical. It was in the forests of this ancient, massive North American-European continent that a number of long since extinct primate-like mammals made their debut, most of which were about the size of a modern domestic cat.

These mammals are called **plesiadapiformes**, and while none of their fossils have been found south of the equator, many have been discovered in the north. The best known genus is *Plesiadapis*, which is represented by at least fifteen species that ranged between what are now western North America and western Europe. It is in these two locations that most of their fossils have been found, some from Palaeocene sites in Saskatchewan and Alberta. But while some specialists maintain that *Plesiadapis* was a primate, others argue against this idea. And the debate is not easy to settle for *Plesiadapis* possessed both primate-like and non-primate-like traits: molars and ears like those of the lemurs, but claws on all of their digits, eyes that faced to the side, and a relatively large snout. It is this mixture of traits that makes the picture confusing.

## *PLESIADAPIS* AND DOLLO'S LAW

Even if the members of the genus *Plesiadapis* were primates, it is unlikely that they played a direct role in the subsequent evolution of our order. The reason is that, with the exception of lemur-like molars, their teeth were highly specialized and apparently best suited to a diet of stems. According to **Dollo's law**, which is otherwise known as the principle of the irreversibility of evolution,

once a species has acquired a highly conspicuous specialization it can never revert to the less specialized, ancestral condition. The much enlarged pincer-like incisors of *Plesiadapis* and the great reduction of their canines and premolars are a case in point (see Figure 5.5 and QQ 5.11). Such specializations are considered to be examples of **derived traits** as opposed to **ancestral traits**, that is to say, traits that reflect a relatively recent adaptation to a specific environment rather than ones that reflect the more longstanding general adaptive pattern of the ancestral group as a whole. Since modern primates lack the highly specialized teeth of the species that comprised *Plesiadapis*, it is almost certain that if the members of the genus were true primates, then they were terminal descendants of the common ancestor from which all primates evolved.

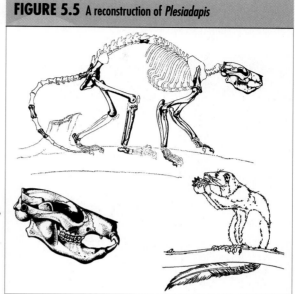

**FIGURE 5.5** A reconstruction of *Plesiadapis*

## QQ 5.11 A LESSON IN PALAEOANTHROPOLOGY

**Philip Gingerich discusses the dentition and diet of *Plesiadapis*.**

• • •

Any attempt to determine the diet of an extinct mammal must consider the morphology of all the teeth: incisors, canines, premolars, and molars. The molars of *Plesiadapis* show relatively great development of [the] buccal . . . [or cheek-side] cutting edges, as do the molars of living grazing herbivores. The incisor mechanism of *Plesiadapis* indicates a cropping [or shearing] mode of food acquisition. The reduction and loss of canines and premolars in *Plesiadapis*, leaving a large diastema [or space] between the incisors and the remaining premolars and molars, is also consistent with a cropping incisor mechanism. The canine and premolar teeth of many mammals are used to shear and puncture large bites of food when they are first ingested. Reduction and loss of the canines and premolars, with the resulting large diastema, is correlated with initial ingestion into the mouth of small pieces of food. Grazing mammals have a large diastema because they feed on food that is by its nature already in small pieces. Rodents have a large diastema because they gnaw a series of small pieces from a larger piece of food rather than ingesting the larger piece . . . . The great reduction of canines and premolars in *Plesiadapis* . . . thus suggests that *Plesiadapis* also ingested food in small pieces. The shape of the cutting edges on the incisors of *Plesiadapis* further suggests that the original food mass was predominantly in the form of stems.

Source: P. Gingerich, "Dental Function in the Palaeocene Primate *Plesiadapis*." In R.D. Martin, G.A. Doyle, and A.C. Walker, eds, *Prosimian Biology* (London: Gerald Duckworth and Company Ltd., 1974), p. 537.

The discovery of new fossils and the reinterpretation of existing finds may stimulate another exciting debate about our Palaeocene ancestors. In 1990 a research team from the University of Montpellier in France announced that they had discovered ten isolated cheek teeth at the foot of the High Atlas mountains in Morocco. The members of the expedition believe these teeth represent a previously unknown 60-million-year-old primate from North Africa. They named the find *Altiatlasius*, in honour of the High Atlas mountains, and its teeth, they claim, which are about the size of mustard seeds, are much more like those of living prosimians such as the Grey lesser mouse lemur (*Microcebus murinus*) of Madagascar and the Dwarf bush baby (*Galago demidovii*) of Africa than are the previously discovered teeth of *Plesiadapis*. Based on these albeit scant remains, the Montpellier team has suggested that *Altiatlasius* should replace *Plesiadapis* in our family tree. Some of those who agree with them have gone farther. They have argued that *Altiatlasius* is a direct descendant of the **microsyopids** — a group of primate-like Palaeocene mammals that were apparently well-suited to an arboreal habitat (see Figure 5.6). Rivalling *Purgatorius* in age, the microsyopids may thus be the stem from which *Altiatlasius* and all subsequent primates evolved. However, only time and additional fossil discoveries will reveal whether this new evolutionary scenario is correct. See QQ 5.12.

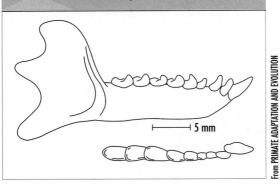

**FIGURE 5.6** Microsyopid jaw (and surface configuration of the teeth)

5 mm

From PRIMATE ADAPTATION AND EVOLUTION by John G. Fleagle, Academic Press, 1988

## QQ 5.12 A NEW SCENARIO

Philip Gingerich comments on the importance of *Altiatlasius* with respect to the geographical origin of the primates.

• • •

The fossil record of primates is one of the most intensively studied of all major mammalian groups. But until recently, fossils of the most ancient true primates were absent from Africa, where primates are a diverse faunal component today. This gap has now been filled with the discovery of the first true primate from the Palaeocene of Africa, by [Bernard] Sigé . . . [and his colleagues from] the University of Montpellier in France.

The new find, *Altiatlasius koulchii* . . . comes from late Palaeocene sediments . . . in . . . eastern Morocco, at the foot of the High Atlas mountains, a locality discovered in 1977 . . . .

The new primate is represented by ten isolated cheek teeth . . . but no teeth were found in association. Tooth size indicates that *Altiatlasius* . . . probably weighed no more than 50–100 grams.

# EOCENE PRIMATES

## THE ADAPIDAE AND THE OMOMYIDAE

By the beginning of the Eocene (*c.* 53 mya), which was much warmer and wetter than the Palaeocene, the evolution of the primates was in full-swing. Although the plesiadapiformes were dying out, (some survived into the middle Eocene) they were replaced by primates that were far better suited to a life in the tropical forest that now stood where *Plesiadapis* had lived. Among them were two remarkable families: the **Adapidae** and the **Omomyidae**, both of which lived in North America and Eurasia, which had continued to drift apart but were still connected in spots.

What makes the adapids and the omomyids stand out are the differences between them and their forerunners. Unlike the plesiadapiformes, the members of these two families tended to have a less prominent snout, eyes that faced more directly forward, nails rather than claws, and small, unspecialized incisors. Taken together, these features suggest that the adapids and the omomyids relied more on their eyes and their hands to survive than on their noses and their teeth. This not only foreshadowed the hand-eye coordination and the manual dexterity of the modern primates as a whole, but also allowed these Eocene primates to become better adapted to an arboreal habitat than their ancestors.

## THE LEMUR-TARSIER CONNECTION

The adapids and the omomyids are also important in another respect, and this is the resemblance between them and two varieties of living prosimians — lemurs and tarsiers. By looking closely at the fossils of North American adapids such as *Smilodectes* and *Notharctus*, and at the remains of an omomyid from western Europe known as *Necrolemur*, palaeoanthropologists have determined that whereas *Smilodectes* and *Notharctus* were essentially lemur-like, *Necrolemur* was much more tarsier-like. In fact, the resemblances are so striking that some palaeoanthropologists believe that

*Smilodectes* and *Notharctus* may have given rise to the Malagazy lemurs, and that *Necrolemur* may be the stock from which the tarsiers of southwestern Asia evolved. Although these hypotheses have been challenged — some believe that European adapids such as *Protoadapis* and *Adapis* are more likely the ancestors of lemurs, and that the proposed link between *Necrolemur* and the tarsiers must be based on firmer evidence — at least one thing is clear: by Eocene times, the evolutionary pathways that led to the modern prosimians were firmly established. See QQ 5.13.

## 5.13 ADAPIDS, OMOMYIDS, LEMURS, AND TARSIERS

Elwyn Simons discusses the evolutionary links between Eocene primates and lemurs and tarsiers.

• • •

The best-known examples [of North American Eocene primates] are species of two related lemur-like genera: *Notharctus* and *Smilodectes* . . . .

These New World primates resembled living lemurs both in their proportions and in their general structure. In contrast to the small-brained, snouty, side-eyed *Plesiadapis*, the skull of *Smilodectes* shows an enlargement of the front portion of the brain and a shifting of eye positions forward so that individual fields of vision can overlap in front. These features of the head, taken together with the animal's rather long hind limbs, suggest that in life *Smilodectes* looked rather like one of today's Malagasy lemurs . . . .

It is most unlikely, however, that either *Smilodectes* or *Notharctus* contributed to the ancestry of living lemurs. This honor can more probably be conferred on some members of a European genus, such as *Protoadapis* or *Adapis*, of equal Eocene age, if indeed the ancestors of modern lemurs were not already in Africa by this time . . . .

Unfortunately, none of these possible Old World precursors of living lemurs is sufficiently represented by fossils to provide the kind of detailed skeletal information we possess for their New World contemporaries.

This is also the case for a roughly contemporary European prosimian: *Necrolemur* . . . . In *Necrolemur* the evolutionary advances represented by *Notharctus* and *Smilodectes* have been extended. Enlargement of

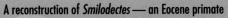

A reconstruction of *Smilodectes* — an Eocene primate

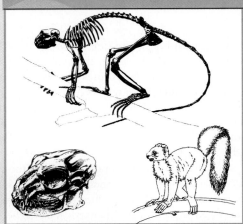

From "The Early Relatives of Man" by Elwyn L. Simons. Copyright © July, 1964 Scientific American, Inc. All rights reserved.

the forebrain and a further facial shortening are apparent. A forward shift of the eye position — with the consequent overlapping of visual fields and potential for depth perception — should have equipped *Necrolemur* for an active arboreal life in the Eocene forests. Actually this early primate, although it is probably not ancestral to any living prosimian, shows a much closer affinity for the comparatively advanced tarsier of southeast Asia than for the more primitive Malagasy lemurs.

Source: Elwyn L. Simons, "The Early Relatives of Man." *Scientific American*, Vol. 211, No. 1, p. 54. Copyright © 1964 by Scientific American, Inc. All rights reserved.

# OLIGOCENE PRIMATES

By the beginning of the Oligocene (*c.* 37 mya) most of the adapids and the omomyids had become extinct, almost certainly because of a major change in their environment. During the Eocene the climate in North America and in Europe was hot and damp, but by the beginning of the Oligocene temperatures had cooled and the air was much drier. This reduced the size of the tropical forest where most of the Eocene primates resided, and, although a few of them remained in the region, the heartland of primate evolution shifted to the south, to the tropical forests of South America and Africa, where a number of new, anthropoid-like primates made their initial appearance.

## SOUTH AMERICAN FINDS AND THE RAFTING HYPOTHESIS

Unfortunately, relatively few primate fossils have been discovered in South America. Despite years of painstaking research, only a handful of fragmentary remains have been found. The oldest of these are those of *Branisella*, a genus of extinct platyrrhine monkeys whose 26-million-year-old fossils were discovered in a late Oligocene deposit at La Salla, a site in Bolivia. While *Branisella* may represent the stock from which present day New World monkeys evolved, not enough fossils have been recovered to substantiate this view. See Table 5.4.

**TABLE 5.4**  The location and age of selected fossil platyrrhine genera discovered in South America

| COUNTRY | GENERA | AGE OF OLDEST REMAINS (MYA) |
|---|---|---|
| Bolivia | *Branisella* | *c.* 26 |
| Argentina | *Dolichocebus* | *c.* 20 |
| | *Tremacebus* | *c.* 20 |
| | *Homunculus* | *c.* 17 |
| | *Soriacebus* | *c.* 15 |
| | *Carlocebus* | *c.* 15 |
| Colombia | *Cebupithecia* | *c.* 15 |
| | *Nesosaimiri* | *c.* 15 |
| | *Stirtonia* | *c.* 15 |
| | *Micodon* | *c.* 15 |
| | *Mohanamico* | *c.* 15 |

Source: Based on information contained in: John G. Fleagle and Alfred L. Rosenberger, "Preface" *Journal of Human Evolution* (Volume 19, Numbers 1 and 2, 1990), pp. 2–3, and Bruce J. MacFadden, "Chronology of Cenozoic primate localities in South America," *Journal of Human Evolution* (Volume 19, Numbers 1 and 2, 1990), pp. 7–16.

How the first platyrrhine monkeys came to South America is likewise unclear. For many years it was thought that they were the descendants of North American omomyids that journeyed to South America during Eocene or Oligocene times. Nowadays, however, the majority of palaeoanthropologists favours the view that the ancestors of New World monkeys originated in Africa, and that they came to the New World across the Atlantic on floating rafts composed of driftwood and living plants, or else on natural floating islands (which can still be seen in the Atlantic today) — an idea that is known as the **rafting hypothesis** (see QQ 5.14). In either case, the first South American primates had to cross a water barrier to enter their new homeland. At the time the voyage was presumably made, South America was separated from both North America and Africa by open sea.

## 5.14  NEW WORLD MONKEY ORIGINS AND THE RAFTING HYPOTHESIS

R. L. Ciochon and A. B. Chiarelli describe how the ancestors of New World monkeys may have reached their homeland.

• • •

We . . . suggest that the last common ancestor of the Platyrrhini was derived from a precatarrhine African-based early anthropoid stock . . . [and that this early ancestor] made the . . . crossing to South America . . . across a much less expansive Eocene equatorial Atlantic Ocean . . . [than exists today].

The probable existence of an omomyid-derived preplatyrrhine/precatarrhine anthropoid stock in Africa during the late Eocene coupled with evidence derived from sea-floor spreading indicating a narrow equatorial Atlantic, the presence of tectonically active island chains and favorable east to west oceanic paleo-currents and winds in our opinion makes this scenario

for the origin and dispersal of the Platyrrhini the most parsimonious model. Of course, not . . . [everyone agrees with it]. Nevertheless, we feel it represents a reasonable working hypothesis that is subject to future testing and scrutiny. In its current form it is not clearly falsifiable . . . yet we are most certain that future pale-ontological discoveries, more complete paleogeo-graphic data, in-depth comparative anatomical studies, and new strides in the fields of genetic biology will turn this situation about. It is further hoped that the presenta-tion of this . . . model will not in any way lessen debate on the issue of platyrrhine origins . . . [but rather] *pro-mote* discussion on this subject.

Source: R.L. Ciochon and A.B. Chiarelli, "Concluding Remarks." In R.L. Ciochon and A.B. Chiarelli, eds. *Evolutionary Biology of the New World Monkeys and Continental Drift* (New York: Plenum Publishing Corporation, 1980), pp. 500–501.

## AFRICAN FINDS

Fortunately, much more is known about the Oligocene primates that lived in the Old World, especially those from a region in Egypt known as the **Fayum**, about 80 kilometres southwest of Cairo. Today the Fayum is located in Egypt's Great Western Desert, but during the Oligocene the region contained fresh water swamps bordered by grassland and woodlands. Many primate genera, possessing anthropoid-like rather than prosimian-like traits, lived there.

One was *Apidium*, a genus of squirrel-size creatures that possessed features that were similar to those of many living platyrrhines, and that may include the transatlantic voyagers that supporters of the rafting hypothesis are seeking. Judging by the structure of their eyes and their limbs, it is clear that the species that represented *Apidium* were agile arboreal quadrupeds. In addition, and once again like many modern platyrrhines, *Apidium's* teeth were arranged in a 2:1:3:3 pattern and were structurally suited to a diet of fruits.

Another genus, *Aegyptopithecus* (see Figure 5.7), which were fox-size animals, also possessed platyrrhine-like eyes and limbs and relied on a diet of fruits. *Aegyptopithecus*, however, had catarrhine-like teeth; they had a 2:1:2:3 dental formula, exactly the same as that of contemporary Old World monkeys, apes, and humans. Moreover, *Aegyptopithecus* had gorilla-like canines, and males were larger than females. Coupled with their other features, this has led some palaeoanthropologists to conclude that the genus gave rise to the great apes. Others, however, are skeptical, and they are probably right. Considering the overall appearance of the members of the genus, *Aegyptopithecus* was composed neither of monkeys nor of apes, at least not in the conventional sense of these terms. Instead, *Aegyptopithecus* is best regarded as a genus of primitive anthropoids, some of which were probably ancestral to both modern monkeys and apes, and, in this sense, foreshadowed what was about to happen in the Miocene.

**FIGURE 5.7** An artist's rendering of *Aegyptopithecus*

The Natural History Museum, London

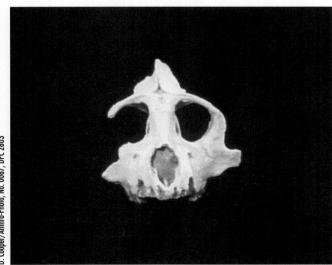

D. Cooper/Anthro-Photo, No. 0067, DPC 2803

▼

*Aegyptopithecus*
skull

▲

# MIOCENE PRIMATES

If the Eocene was the heyday of early prosimians, and the Oligocene the heyday of early anthropoids, then the Miocene (*c.* 25 mya to 5 mya) was the heyday of early **hominoids** — the common designation for the members of the superfamily to which apes and humans belong.

Like the prosimians and the early anthropoids that preceded them, the Miocene hominoids appeared in conjunction with changes in the environment, which, at the beginning of the Miocene, altered the landscape of East Africa. There, as today, the region was dominated by the **Great Rift Valley** — a long trough stretching from Syria down through eastern Africa for hundreds of kilometres to Mozambique (see Figure 5.8). It was in this location, in what was then a dense tropical forest, that the first hominoids lived. See QQ 5.15.

## 5.15 THE GREAT RIFT VALLEY

J. Desmond Clark describes the Great Rift Valley and explains why so many fossils have been found there.

• • •

The eastern part of Africa is split by a huge trough or fault, that runs nearly the whole length of the continent. This is the Great Rift Valley, which starts in Asia, in Syria, and continues southward down the Red Sea, though Ethiopia into Kenya, Uganda, and Tanzania, and finally loses itself beneath the alluvial [or sand and mud] sediments in the lower Zambezi valley in Mozambique. In Kenya and Uganda the trough splits into the Eastern Rift and the Western Rift; in a shallow basin on the plateau between the two branches lies Lake Victoria, Africa's greatest lake.

The bottom of the Great Rift lies at extremely variable elevations. In the Danakil section of the Ethiopian Rift the bottom drops in places to nearly 400 feet [or about 125 metres] below sea level, but southward the floor rises in a series of steps until in Kenya, in the Eastern Rift, the elevation is over 5000 feet [or about 1500 metres] . . . .

The Great Rift is a tectonically unstable zone where compression and tension of the earth's crust have pushed up the land bordering the trough into high ridges and mountains. The deeper portions of the trough are filled by great lakes, some of which, such as Lakes Tanganyika and Malawi, are among the deepest and longest in the world.

Although there exists now only one active volcano in the rift zone (Ol Doingo Lengai in Tanzania), there are numerous dormant and extinct ones, two of which — Mount Kilimanjaro (19 565 feet [or about 6000 metres]) and Mount Kenya (17 040 feet [or about 5200 metres]) — are perpetually snowcapped. Another huge snow-covered mass — this one a crystalline rock thrust of nonvolcanic origin — is the Ruwenzori range. Its highest peak, Mount Stanley, rises to 16 795 feet [or about 5100 metres]. The vegetation zones of the rift run in belts around the mountains, changing with altitude from rain forest at the foot to alpine tundra near the top, and the scenery is some of the most beautiful and varied in the world. The very rich fossil record preserved in the Great Rift Valley is due to the accumulation of deep sediments in the bottom of the trough and to the rapid burial of land surfaces by ash and dust from the volcanoes.

Source: J. Desmond Clark, "African Beginnings." In Alvin M. Josephy, Jr., ed. *The Horizon History of Africa* (New York: American Heritage Magazine, a division of Forbes Inc., 1971), pp. 19–20. © Forbes Inc., 1971.

**FIGURE 5.8** The Great Rift Valley

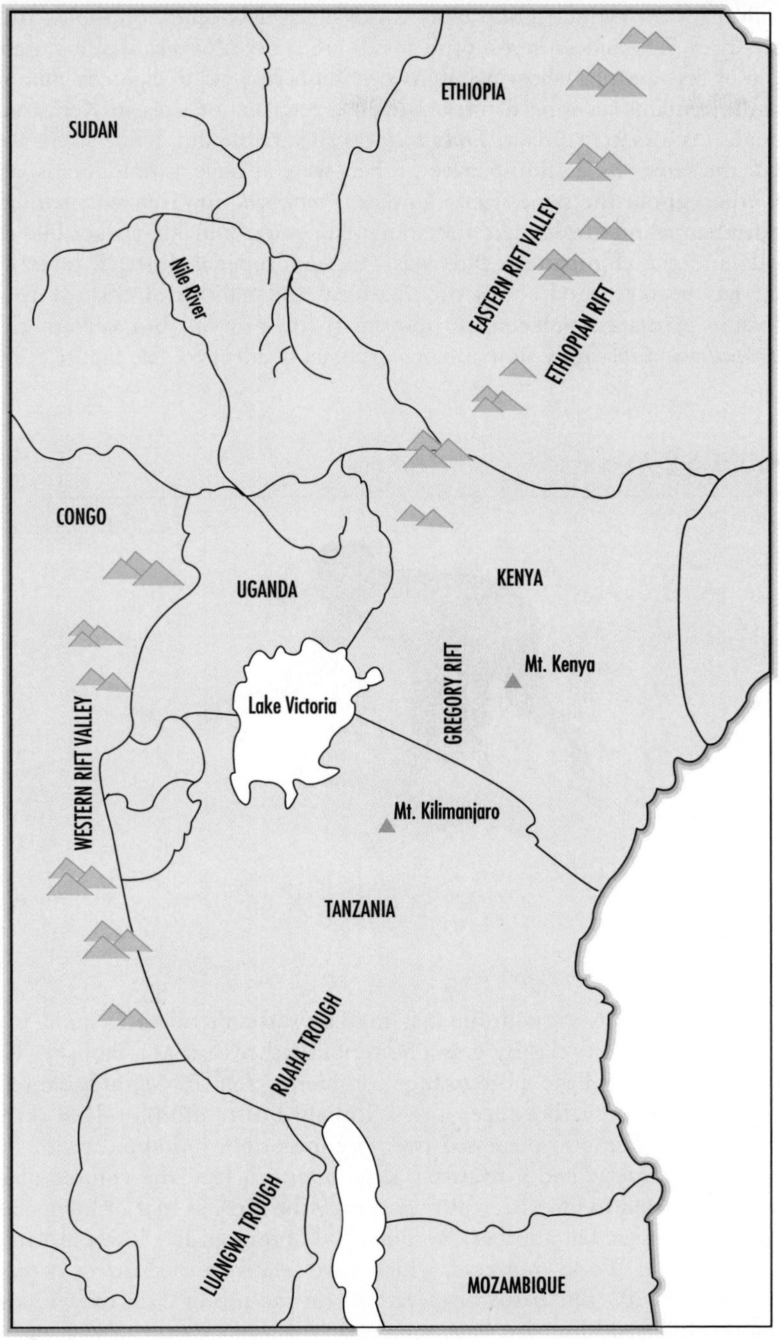

## PROCONSUL

The most ancient of these hominoids are **dryomorphs** — the common designation for an extinct group of forest-dwelling Miocene hominoids from East Africa. The oldest dryomorph fossils are those of *Proconsul*, a variable group of dryomorphs whose twenty-two-million-year-old to eighteen-million-year-old remains have been unearthed at several major sites in Kenya and Uganda. As a matter of fact, *Proconsul* was so variable that while some were about the same size as chimpanzees, others were as large as orangutans, and still others about the same size as gorillas. *Proconsul*, however, was neither a quadrumanual arborealist like the orangutan, nor a knuckle-walker like the gorilla and the chimpanzee. Nor was *Proconsul* bipedal. Instead, based on what has been learned about the anatomy and habitat of this group of Miocene primates, palaeoanthropologists are now all but certain that *Proconsul* was a relatively slow-moving arboreal quadruped. See Figure 5.9.

**FIGURE 5.9** A reconstructed skeleton of Proconsul

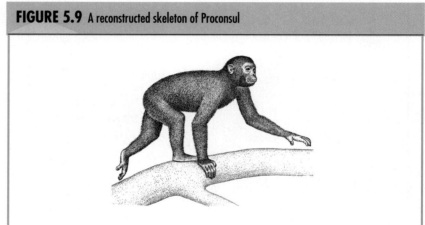

## PROCONSUL AND THE Y-5 PATTERN

Since *Proconsul* was a quadruped, it may be difficult to understand why palaeoanthropologists classify it as a hominoid rather than as a monkey. The designation seems all the more strange considering that *Proconsul* had a relatively flat, monkey-like face, and a lightly built, monkey-like skull. *Proconsul*, however, also possessed two very important hominoid traits. For one thing, *Proconsul* had a relatively large brain; in fact, the volume of its brain is estimated to have been almost exactly the same as that of living apes of comparable size. For another, *Proconsul* had hominoid-like lower molars.

Unlike Old World monkeys, which have relatively specialized, derived lower molars with four **cusps** or elevations on the top of each crown, contemporary apes and humans have more primitive, five-cusped lower molars whose origin dates back to the Oligocene. In fact, the lower molars of modern apes and humans are characterized by an ancestral trait that biological anthropologists refer to as the **Y-5 pattern**, that is, looked at from the top, the five cusps on the top of each crown are in the shape of a sideways Y with the arms of the letter facing the cheek. *Proconsul's* lower molars featured exactly the same pattern. Moreover, although some anthropologists once favoured the view that *Proconsul* was the direct ancestor of modern chimpanzees and gorillas, that view has recently been contradicted. *Proconsul*, it appears, was likely the last common ancestor of all subsequent hominoids, extinct and extant. See QQ 5.16.

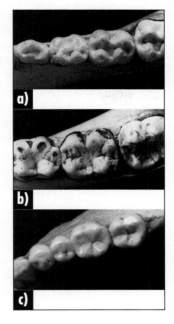

▼

The crowns of a lower molar of an

a) old world monkey,

b) an ape, and

c) a human

▲

---

 **5.16 PROCONSUL**

Alan Walker and Mark Teaford explain what has been recently learned about the extinct hominoid *Proconsul*.

• • •

The prehistoric ape *Proconsul* is now the best-known of our ancestors, yet its route from the obscurity of an excavation pit to fame in the scientific spotlight is as full of twists and surprises as any soap opera. It is a story of implausibilities, in which various pieces of important specimens, once unearthed, became separated and sent to distant lands until fortune brought them together again decades later. It is also a tale with a happy ending; recent expeditions to the excavation sites have revealed nearly 800 new specimens of hominoid primates — the superfamily of primates that includes the great apes, the lesser apes, and human beings. These have vastly increased the sample of *Proconsul* fossils, and the new finds show that *Proconsul* is a useful model of the last common ancestor of the great apes and man . . . [and not, as was once believed,] a specialized ancestor of the modern chimpanzee or the gorilla.

Source: Alan Walker and Mark Teaford, "The Hunt for *Proconsul*," *Scientific American*, Vol. 260, No. 1, pp. 76–82. Copyright © 1964 by Scientific American, Inc. All rights reserved.

---

## RAMAMORPHS

*Proconsul* was not the only relatively big-brained Miocene hominoid with the Y-5 pattern. While it was establishing a foothold in East Africa the continents continued to drift, and, about 16 mya, after being separated by the Tethys ocean for millions of years, Africa "docked" with Eurasia. In the

EARLY PRIMATE EVOLUTION 181

meantime, the climate in East Africa had become cooler and drier and patches of open woodland and grassland began to appear where tropical rain forests had stood. Taken together, this set the stage for the dryomorphs to migrate from their original East African homeland to Europe and Asia via the Arabian Peninsula, which they did between about 15 mya and about 12 mya, during the Middle Miocene. Those that settled in Europe and Asia — as well as some of *Proconsul's* African descendants — are known as **ramamorphs**, which is the common designation for an extinct group of woodland and savanna-dwelling Miocene hominoids from Africa, Asia, and Europe. While the ramamorphs were closely related to the dryomorphs, they also apparently spent some of their time on the ground.

## GIGANTOPITHECUS

Since the ramamorphs were widely distributed over a vast territory for millions of years, the fact that they came in a variety of shapes and sizes is hardly surprising. *Gigantopithecus* is an example of one of the largest ramamorphs. See Figure 5.10.

The species that represent *Gigantopithecus* were named after the enormous size of their jaws and teeth, which have been discovered in Late Miocene deposits in India and in Middle Pleistocene deposits in China and Vietnam. Based on these facial remains, which is all that is left of these unusual creatures, but which indicate that their jaws were huge and their molars five times the size of our own, some palaeoanthropologists believe that *Gigantopithecus* was the largest primate that every lived, weighing perhaps between 250 kilograms and 350 kilograms and standing between two metres and three metres tall. Other palaeontologists, however, claim that jaws and teeth alone are not a good basis on which to judge overall body size, and that *Gigantopithecus* may well have been about the same size as a modern gorilla. In either event, although there can be little doubt that this remarkable ramamorph spent much of its time on the ground, almost all palaeoanthropologists agree that it was too specialized to be a direct ancestor of living apes or humans (see QQ 5.17). It is possible, however, that our ancestors played a role in the demise of *Gigantopithecus*, since they co-existed during the Pleistocene. There are also some popu-

**FIGURE 5.10** An artist's rendering of *Gigantopithecus*

The Natural History Museum, London

## 5.17 GIGANTOPITHECUS

S. I. Rosen looks at how and where *Gigantopithecus* remains were discovered and comments on the evolutionary status of this perplexing primate.

• • •

For hundreds of years, oriental drugstores have sold ground-up fresh and fossil teeth and bones, so-called "dragon bones." These have been marketed as potions of all sorts. In the year 1935, the Dutch pale-ontologist G. H. Ralph von Koenigswald found a most interesting "dragon tooth," a molar belonging to some type of fossil hominoid. Von Koenigswald believed this to be [a] giant pongid, which he called *Gigantopithecus*. By 1939, he had found four such molar teeth but no jaws or other skeletal remains. The teeth are believed to be of middle Pleistocene age from fossil beds in Southern China . . . .

The 1950s saw the first discovery of *Gigan-topithecus* mandibles in China — the Kwangsi mandibles. In 1968 another lower jaw was found, this time in India and dating from the middle Pliocene. This indicates that primate populations of this type existed from perhaps five million years B.P. [that is, before the present,] up to possibly 250 000 years ago. The early part of the Pleistocene is known for megafauna — usu-ally large or giant animals . . . [whose modern descendants] are much smaller. The gigantopithecines of the Pleistocene likely represent one version of this gigantism . . . .

[Its gigantic jaws and molars, in fact,] led Dr. Franz Weidenreich . . . to deduce that this primate was in fact a giant hominid, which he called "Gigan-thropus." Weidenreich believed that man's early ancestors were giants. This idea has been refuted by the fossil evidence down to this day. Present popular dogma considers *Gigantopithecus* to be a giant gorillalike pongid who was ground dwelling and herbivorous [and an evolutionary dead end].

Source: S.I. Rosen, *Introduction to the Primates: Living and Fossil* (© 1974, Englewood Cliffs, New Jersey: Prentice-Hall Inc.), pp. 112–115, 167–170. Reprinted by permission of Prentice Hall.

lar writers who have argued that *Gigantopithecus* gave rise to Bigfoot, but this idea is considered to be rather far-fetched.

## RAMAPITHECUS AND SIVAPITHECUS

Other ramamorphs, however, are most definitely close relatives of present-day hominoids. One was **Ramapithecus** (see Figure 5.11), a group of twelve-million-year-old to seven-million-year-old ramamorphs whose remains were first discovered in India and named in honour of Rama, a Hindu prince. For many years all but a few palaeoanthropologists were convinced that these ramamorphs were the first **hominids** — the common designation for the bipedal members of the family to which humans belong — and that it was *Ramapithecus* that ultimately gave rise to humankind. They believed this to be true because of *Ramapithecus'* relatively short, flat, human-like faces, their relatively small, human-like canines, and their heavily enamelled molars — a trait that some insisted was proof positive that *Ramampithecus* spent consid-erable time on the ground since molars of this sort are associated with a diet of fibrous woodland plants that grow in the open at the fringe of the forest.

But with the recent discovery of additional and more complete ramamorph fossils in Asia, Europe, and Africa, the idea that *Ramapithecus* was a ground-dwelling human look-alike has been shown to be false. What the new fossils demonstrate is that *Ramapithecus* was much less human-like than was originally thought. Almost all palaeoanthropologists now believe that *Ramapithecus* was not only more or less arboreal, but also that it is actually part of *Sivapithecus* — a genus of relatively large hominoids (weighing 25 to 70 kilograms), that migrated from Africa to Eurasia during the Middle Miocene, and that

**FIGURE 5.11**  An artist's rendering of  *Ramapithecus*

The Natural History Museum, London

subsequently gave rise not to humans, but rather to the orangutan, which is, all things considered, much more like *Sivapithecus* than any other living primate. See QQ 5.18.

## QQ 5.18 FROM *SIVAPITHECUS* TO ORANGUTANS

David Pilbeam comments on the likely evolutionary relationship between *Sivapithecus* and orangutans.

• • •

The . . . newer specimens of Miocene hominoids allow us to paint the following phylogenetic picture. Somewhere between 17 and 12 M[y]a the species ancestral to large hominoids evolved, probably a generalized large bodied, very [sexually] dimorphic, arboreal frugivorous ape with a varied positional repertoire including arm-swinging, active climbing especially of vertical supports, and occasional bipedalism. It fed mainly in trees, but was a less restricted arborealist than *Pongo* [that is, the modern orangutan]. Sometime

between 18 and 12 M[y]a one or more large hominoid species migrated out of Africa. In the European and Asian forests and woodlands these hominoids underwent a modest adaptive radiation, including *Sivapithecus* from west and south Asia which can be linked to *Pongo* the sole living Asian great ape . . . . [T]he Asian and European hominoids largely disappeared following later Miocene climatic-habitat change, and *Pongo* is a highly derived descendant of *Sivapithecus*.

Source: D.R. Pilbeam, "Human Evolution," In G.A. Harrison, J.M. Tanner, D.R. Pilbeam, and P.T. Baker, *Human Biology: An Introduction to Human Evolution, Variation, Growth, and Adaptability*, third edition with corrections (New York: Oxford University Press, 1992), p. 100. By permission of Oxford University Press.

# SIVAPITHECUS, AND MODERN APES AND HUMANS

Finally, there is one other source of information that has a bearing on the evolutionary relationship between *Sivapithecus* and living apes and humans. That information was generated by constructing a **molecular clock**, an analytical device used to estimate when living species diverged from a common ancestor and began to evolve along separate lines. Given the fact that such species shared a common ancestor, it stands to reason that the differences in their blood proteins, amino acid sequences, fragments of DNA, and other molecular features have accumulated since they diverged.

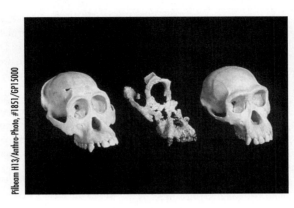

The skulls of an orangutan (left), *Sivapithecus* (centre), and a chimpanzee (right)

Assuming that these molecular differences are the products of neutral mutations, and that the mutations occur at the same rate in the species under consideration, it is then possible to estimate the evolutionary distance between them by comparing their molecular features — the rationale being that the more similar their molecular features the nearer to the present they diverged. However, no matter which feature is selected for comparison the molecular clock must be calibrated, and this is done by calculating the mutation rate in species whose divergence in time has previously been established via the fossil record.

To be sure, the molecular clock has been criticized. Some, for instance, argue that the mutations that are responsible for molecular change are not necessarily neutral, while others maintain that such mutations do not necessarily occur at the same rate even in closely related species. Notwithstanding these criticisms, the new research strategy has had a tremendous impact in palaeoanthropology. Those molecular clocks that have been constructed since the technique was introduced in the late 1960s have revolutionized the study of the evolution of the hominoids and the hominids alike. What such clocks indicate, and what many palaeoanthropologists now believe, is that gibbons began to evolve along independent lines about 12 mya, orangutans about 10 mya, and that gorillas, chimpanzees, and the first hominids diverged about 5 mya (see QQ 5.19). Equally important, given these estimated times of divergence, the idea that *Sivapithecus* was the direct ancestor of the orangutan is given further support — it was in the right location (Asia) at the right time (*c.* 10 mya) — as is the idea that *Sivapithecus* was not a hominid but hominoid. The hominids were yet to appear.

# 5.19 MOLECULAR CLOCKS

Glenn Conroy looks at molecular clocks and what they reveal about the divergence of gibbons, orangutans, and the ancestors of the African apes and ourselves.

• • •

A molecular clock relies on the premise that molecular differences between taxa [or taxonomic divisions] accumulate at a relatively constant rate when averaged out over geological time: a greater molecular difference implies an earlier divergence date. The clock must ultimately be calibrated against the fossil record, however. Suppose two taxa differ in some molecular parameter by $X$ units and that they diverged $Y$ million years ago as judged from the fossil record. Thus $X$ number of changes have occurred over $Y$ millions of years (assuming relatively constant rates of change). This relationship sets the clock. It is then a simple matter to calculate the divergence times of other taxa by measuring how much the taxa differ in the same molecular parameter. As an illustration, suppose two taxa that diverged 60 MYA differed by 20 units of some molecular parameter. Change in this parameter is thus assumed to occur at the rate of 1 unit per 3 million years. Consequently the divergence date of two taxa differing by 10 such units would be estimated at 30 MYA, 5 units at 15 MYA, and so on.

A number of such molecular clocks have been proposed for calculating divergence times within hominoid evolution. On average they suggest the following:

1. Separation of the gibbon from the great ape and human clade [whose members share the same common ancestor] occurred about 12 MYA.
2. Separation of the orangutan from the . . . [A]frican ape and human clade occurred about 10 MYA.
3. Separation of African apes from humans happened about 5 MYA.

Source: Glenn C. Conroy, *Primate Evolution* (New York: W. W. Norton & Company, Inc., 1990), pp. 259–261.

# KEY TERMS AND CONCEPTS

NEO-DARWINISM
AGE OF MAMMALS
FOSSILS
PETRIFICATION
PERMINERALIZATION
MINERALIZATION
GEOLOGICAL TIME SCALE OR CALENDAR
PRINCIPLE OF SUPERPOSITION
PRINCIPLE OF STRATIGRAPHY
RELATIVE AGE
FLUORINE TEST
URANIUM DATING

NITROGEN DATING
ABSOLUTE OR CHRONOMETRIC AGE
POTASSIUM-ARGON (K-AR) METHOD
TUFFS
FISSION-TRACK METHOD
MAGNETIC POLARITY TIME SCALE (MPTS)
CARBON-14 METHOD
CARBON ISOTOPES
PANGEA (*pan-jee'-uh*)
LAURASIA (*lor-ayzh'-uh*)
GONDWANA (*gahn-dwah'-nuh*)
PLATE TECTONICS

INSECTIVORA (*in-sek-ti-vor'-uh*)

PURGATORIUS (*pur-guh-tor'-ee-us*)

ARBOREAL THEORY

VISUAL PREDATION THEORY

ANGIOSPERM RADIATION THEORY

PLESIADAPIFORMES
(*plee-zee-uh-dap'-i-formz*)

PLESIADAPIS (*plee-zee-uh-dap'-is*)

DOLLO'S LAW

DERIVED TRAITS

ANCESTRAL TRAITS

ALTIATLASIUS (*al-tee-at-la'-see-us*)

MICROSYOPIDS (*my-kroh-sy'-oh-pidz*)

ADAPIDAE (*a-dap'-i-day*)

OMOMYIDAE (*oh-moh-my'-i-day*)

SMILODECTES (*smy-loh-dek'-teez*)

NOTHARCTUS (*noh-thahrk'-tus*)

NECROLEMUR (*nek-roh-lee'-mur*)

BRANISELLA (*bran-i-sel'-uh*)

RAFTING HYPOTHESIS

FAYUM (*fay-yoom'*)

APIDIUM (*a-pid'-ee-um*)

AEGYPTOPITHECUS (*ee-jip-tuh-pith'-i-kus*)

HOMINOIDS (*haw'-min-oydz*)

GREAT RIFT VALLEY

DRYOMORPHS (*dry'-oh-morfs*)

PROCONSUL (*pro-kahn'-suul*)

CUSPS

Y-5 PATTERN

RAMAMORPHS (*ra'-ma-morfs*)

GIGANTOPITHECUS (*jy-gan-toh-pith'-i-kus*)

RAMAPITHECUS (*ra-ma-pith'-i-kus*)

HOMINIDS (*haw'-min-idz*)

SIVAPITHECUS (*shee-va-pith'-i-kus*)

MOLECULAR CLOCK

# SELECTED READINGS

## INTRODUCTION

Little, C. *Terrestrial Invasion: An Ecophysiological Approach to the Origins of Land Animals.* London: Cambridge University Press, 1990.

## THE RISE OF THE MAMMALS

Benton, M. *The Rise of the Mammals.* London: Quatro Publishing plc, 1991.

Crawshaw, L., Moffitt, B., Lemons, D., and Downey, J. "Evolution of Endothermy: Histological Evidence." *American Scientist*, Vol. 69 (1981), pp. 543–550.

Kemp, T. *Mammal-like Reptiles and the Origin of Mammals.* New York: Academic Press, 1982.

## FOSSILS

Cvancara, A. *Sleuthing Fossils: The Art of Investigating Past Life.* New York: Wiley, 1990.

Raup, D. and Stanley, S. *Principles of Paleontology* (second edition). San Francisco: Freeman, 1978.

## THE GEOLOGICAL TIME SCALE

Eicher, D. *Geologic Time.* Englewood Cliffs, New Jersey: Prentice-Hall, 1976.

## Determining the Relative Age of Fossils

Oakley, K. "Analytical Methods of Dating Bones." In Brothwell, D. and Higgs, E., eds. *Science in Archaeology* (second edition). London: Thames and Hudson, 1969, pp. 35–45.

## Determining the Absolute Age of Fossils

Bishop, W. and Miller, J., eds. *Calibration of Hominid Evolution: Recent Advances in Isotopic and Other Dating Methods Applicable to the Origin of Man.* Edinburgh: Scottish Academic Press, 1972.

Cox, A. "Geomagnetic Reversals." *Science*, Vol. 163 (1969), pp. 237–245.

Dalrymple, G. and Lanphere, M. *Potassium-Argon Dating: Principles, Techniques, and Applications to Geochronology.* San Francisco: Freeman, 1969.

Libby, W. *Radiocarbon Dating.* Chicago: University of Chicago Press, 1955.

## Pangea, Laurasia, and Gondwana

Cachel, S. "Plate Tectonics and the Problem of Anthropoid Origins." *Yearbook of Physical Anthropology*, Vol. 24 (1981), pp. 139–172.

Hoffman, P. "Did the Breakout of Laurentia Turn Gondwanaland Inside-Out?" *Science*, Vol. 252 (1991), pp. 1409–1412.

Murphy, J. and Nance, R. "Mountain Belts and the Supercontinent Cycle." *Scientific American*, Vol. 266 (1992), pp. 84–91.

Seyfert, C. and Sirkin, L. *Earth History and Plate Tectonics: An Introduction to Historical Geology.* New York: Harper & Row, 1973.

## The End of the Age of Reptiles

Alvarez, W. and Asaro, F. "Extraterrestrial impact: accumulating evidence suggests an asteroid or comet caused the Cretaceous extinction." *Scientific American*, Vol. 263 (1990), pp. 76–84.

Bakker, R. *The Dinosaur Heresies: New Theories Unlocking the Mystery of the Dinosaurs and their Extinction.* New York: Morrow, 1986.

## Palaeocene Primates

Cartmill, M. "Arboreal Adaptations and the Origin of the Order Primates." In Tuttle, R., ed. *Functional and Evolutionary Biology of the Primates.* Chicago: Aldine, 1972, pp. 97–122.

Cartmill, M. *Rethinking Primate Origins.* Minneapolis, Minnesota: Burgess, 1975.

Clemens, W. "*Purgatorius*, an Early Paromomyid Primate." *Science*, Vol. 184 (1974), pp. 903–905.

Jones, F. *Arboreal Man.* London: E. Arnold, 1916.

Krause, D. "Paleocene Primates from Western Canada." *Canadian Journal of Earth Sciences*, 1978, volume 15, pp. 1250–1271.

Sigé, B, Jaeger, J.-J., Sudre, J., and Vianey-Liaud, M. "*Altiatlasius koulchii*, New Genus New Species: An Omomyid Primate from the Late Paleocene of Morocco, and the Origin

of the Euprimates [in French]."
*Palaeontographica Abt. A. Palaeozoologie-Stratigraphie*, Vol. 214 (1990), pp. 31–56.

Szalay, F., Rosenberger, A., and Dagosto, M. "Diagnosis and Differentiation of the Order Primates." *Yearbook of Physical Anthropology*, Vol. 30 (1987), pp. 75–106.

van Valen, L. and Sloan, R. "The Earliest Primates." *Science*, Vol. 207 (1965), pp. 435–436.

Watters, J., and Krause, D. "Plesiadapid Primates and the Biostratigraphy of the North American Late Paleocene." *American Journal of Physical Anthropology*, Vol. 69 (1986), p. 277.

Zimmer, C. "Distant Origins: Gliding Out of the Primate Order." *Discover*, Vol. 12, No. 1 (1991), p. 65.

## EOCENE PRIMATES

Gingerich, P. "New Species of Eocene Primates and the Phylogeny of European Adapidae." *Folia Primatologica*, Vol. 28 (1977), pp. 68–80.

Rasmussen, D. "Anthropoid Origins: a Possible Solution to the Adapidae-Omomyidae Paradox." *Journal of Human Evolution*, Vol. 15 (1986), pp. 1–12.

Swartz, J. and Tattersall, I. "The Phylogenetic Relationships of Adapidae (Primates, Lemuriformes)." *Anthropological Papers of the American Museum of Natural History*, Vol. 55 (1979), pp. 271–283.

## OLIGOCENE PRIMATES

Bown, T., *et. al.* "The Fayum Primate Forest Revisited." *Journal of Human Evolution*, Vol. 11 (1982), pp. 603–632.

Ciochon, R. and Chiarelli, A. "Paleobiogeographic Perspectives on the Origin of the Platyrrhini." In Ciochon, R. and Chiarelli, A., eds. *Evolutionary Biology of the New World Monkeys and Continental Drift.* New York: Plenum, 1980, pp. 459–493.

Conroy, G. "Primate Postcranial Remains from the Oligocene of Egypt." *Contributions to Primatology*, Vol. 8 (1976), pp. 1–134.

Delson, E. and Rosenberger, A. "Phyletic Perspectives on Platyrrhine Origins and Anthropoid Relationships." In Ciochon, R. and Chiarelli, A., eds. *Evolutionary Biology of the New World Monkeys and Continental Drift.* New York: Plenum, 1980, pp. 445–458.

Harrison, T. "The Phylogenetic Relationships of the Early Catarrhine Primates: a Review of the Current Evidence." *Journal of Human Evolution*, Vol. 16 (1987), pp. 412–80.

Kay, R. and Simons, E. "The Ecology of Oligocene Anthropoidea." *International Journal of Primatology*, Vol. 1 (1980), pp. 21–37.

Simons, E. "New Faces of *Aegyptopithecus* from the Oligocene of Egypt." *Journal of Human Evolution*, Vol. 16 (1987), pp. 273–290.

## MIOCENE PRIMATES

Andrews P. and Cronin J. "The Relationships of *Sivapithecus* and *Ramapithecus* and the Evolution of the Orang-utan." *Nature*, Vol. 297 (1982), pp. 541–546.

Ciochon, R. and Corruccini, R., eds. *New Interpretations of Ape and Human Ancestry.* New York: Plenum, 1983.

Goodman, M., Tashian, R., and Tashian J., eds. *Molecular Anthropology: Genes and Proteins in the Evolutionary Ascent of the Primates.* New York: Plenum, 1976.

Napier, J. *Bigfoot: the Yeti and Sasquatch in Myth and Reality.* London: Jonathan Cape, 1972.

Pilbeam, D. "Recent Finds and Interpretations of Miocene Hominoids." *Annual Review of Anthropology,* Vol. 8 (1979), pp. 333–352.

Pilbeam, D. "Hominoid Evolution and Hominoid Origins." *American Anthropologist,* Vol. 88 (1986), pp. 295–312.

Simons, E. and Ettel, P. "Gigantopithecus." *Scientific American,* Vol. 222 (1970), pp. 76–85.

Swartz, J. *The Red Ape: Orangutans and Human Evolution.* New York: Houghton-Mifflin, 1987.

von Koenigswald, G. "Gigantopithecus Blacki von Koenigswald, a Giant Fossil Hominoid from the Pleistocene of Southern China." *American Journal of Physical Anthropology,* Vol. 43 (1952), pp. 295–325.

Walker, A., Teaford, M., and Leakey, R. "New Proconsul Fossils from the Early Miocene of Kenya." *American Journal of Physical Anthropology,* Vol. 66 (1985), pp. 239–240.

Weidenreich, F. "Giant Early Man from Java and South China." *Anthropological Papers of the American Museum of Natural History,* Vol. 40 (1945), pp. 1–134.

Wolpoff, M. "*Ramapithecus* and Hominid Origins." *Current Anthropology,* Vol. 23 (1982), pp. 501–510.

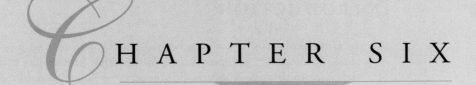

# CHAPTER SIX

# THE AUSTRALOPITHECINES

*contents at a glance*

# INTRODUCTION

If, as molecular clocks indicate, humankind has been evolving as a separate stock for roughly 5 million years, then the oldest human-like fossils should date back to the Late Miocene. But the story of human evolution between 8 mya and 4 mya is shrouded in mystery. Although the fossil record indicates that the ancestors of Old World monkeys radiated into new species and occupied new territory during this period, unfortunately, there is only a handful of human-like fossils that bridge the gap between the heyday of the dryomorphs and the ramamorphs and the appearance of the first undisputed hominids, in Africa, about 4 mya. All that remain are a few petrified teeth and some fossilized fragments of bone.

# THE GAP IN THE FOSSIL RECORD

## THE AQUATIC SCENARIO

Why the gap exists is a controversial subject. Those — and there are not very many — who support the **aquatic theory of human evolution** believe that the finds are sparse because the hominoids that gave rise to hominids began to spend the majority of their time in the water during the latter part of the Miocene (*c.* 8 mya to 5 mya), off the coastline of East Africa where their remains have since been destroyed. Proponents of this view also believe that it was the aquatic environment that fostered the development of upright posture, and that it was only after this posture was adopted that hominids took to the land. Originally formulated by marine biologist Sir Alister Hardy (1896–1985), the most outspoken champion of the aquatic scenario today is Elaine Morgan, a professional writer who has published several popular books on the subject. See QQ 6.1

## 6.1 THE AQUATIC APE

Elaine Morgan summarizes the aquatic theory of human evolution.

. . .

The aquatic theory of human evolution was first suggested by the marine biologist . . . Sir Alister Hardy . . . in an article in The New Scientist in 1960 . . . [titled "Was Man More Aquatic In The Past?"].

[Hardy noted] that . . . the arrangement of the hairs [on humankind] follows precisely the lines that would be followed by the flow of water over a swimming body. If the hair, for purposes of streamlining, had adapted itself to the direction of the current [of the water] before it was finally discarded, this is precisely what we should expect to find. [See part (a) of the accompanying figure.]

He showed how the cracking open of shellfish would foster the use of tools . . . [and] wrote that wading in water would explain not only our erect walk, but the increased sensitivity of our fingertips, through the habit of groping underwater for objects

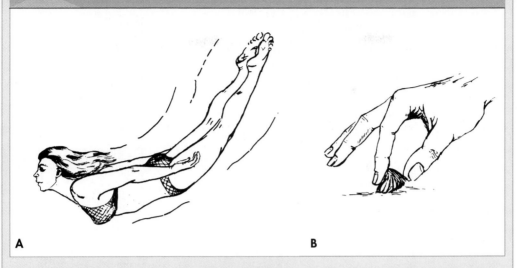

**Drawings used by Sir Alister Hardy to support his aquatic scenario. (A) The hairless, aquadynamic shape of the human body under water; (B) The use of the thumb and forefinger to locate and grasp food on the sea-bed.**

A

B

we could not clearly see . . . . [See part (b) of the accompanying figure.]

He pointed out that the best way of keeping warm in water is to develop a layer of subcutaneous fat [directly under the skin all over the body], analogous to the whale's blubber . . . and that Homo sapiens, alone among the primates, has in fact developed this . . . .

The Hardy theory also explains why, however far from the sea they may be found, the very earliest man-made tools . . . are always fashioned from "pebbles."

Above all, it gives a simple and adequate explanation of the . . . gap [in the fossil record].

Source: Elaine Morgan, *The Descent of Woman* (Scarborough House, 1985)

Yet as interesting as it may be, the vast majority of anthropologists are highly sceptical of the aquatic scenario. Although it does account for the gap in the fossil record, the fossils that represent hominoids on the far side of the gap, and hominids on the near side, indicate that both forms were adapted to the savanna. The fossils that represent the earliest known hominids are a case in point; they have been discovered at sites such as Taung, Sterkfontein, Kromdraai, Swartkrans, and Makapansgat in South Africa, and at Laetoli, Olduvai Gorge, Lake Natron, Chesowanja, Lake Turkana, Omo, and Hadar in East Africa (see Figure 6.1). Coupled with the fact that the albeit scant fossil remains from within the gap were recovered from what was once the grasslands of Africa, it is not surprising that almost all anthropologists abide by the **savanna theory of human evolution**, which holds that transition from hominoids to hominids took place on the savanna or grasslands of Africa.

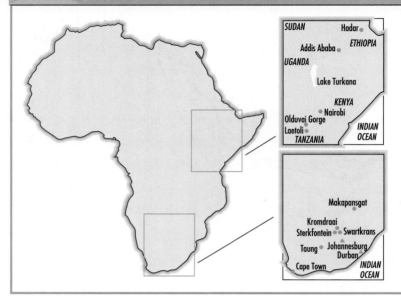

**FIGURE 6.1** The location of important sites that have a bearing on early hominid fossil discoveries in southern and eastern Africa

## THE CONVENTIONAL EXPLANATION

How then can the gap in the fossil record between 8 mya and 4 mya be explained? The conventional explanation is that the fossils were destroyed either by erosion, lava, or acid soils, or else remain buried in remote locations that have yet to be explored. Although this scenario is also hypothetical, the conditions that are necessary for fossils to be formed and those that prompt their discovery, make this explanation the one that anthropologists prefer. As already mentioned, fossils represent only a small proportion of previous populations and their discovery is often a matter of luck.

# AUSTRALOPITHECINES

Fortunately, Pliocene (*c.* 5 mya to 1.8 mya) fossils are more abundant. The climate became cooler and drier during this epoch. As a result, the tropical forests once again shrank giving way to more open woodland and grassland. Such was the case in South Africa and in the Great Rift Valley of East Africa, which supported animals such as elephants, rhinoceroses, hippopotamuses, giraffes, lions, leopards, horses, antelopes, hyenas, monkeys, and birds — some like their modern descendants and others remarkably different. It was there, too, near what were once lakes, rivers, and streams, that the first hominids lived. They are called **australopithecines**, a term that is used to refer to the extinct species that comprised the hominid genus *Australopithecus* (see Figure 6.2).

The members of the genus were not exactly like us. Nor should we expect them to be since they lived in different times and were adapted to different conditions. However, if we were able to observe them in their natural habitat, we could not help but be struck by their kinship with ourselves. See QQ 6.2.

**FIGURE 6.2** An artist's rendering of an australopithecine group

The Natural History Museum, London

## QQ 6.2 COMING UPON THE AUSTRALOPITHECINES

Glynn Isaac imagines what it would be like for us "see" the australopithecines in their natural habitat.

• • •

I imagine that if we had a time machine and could visit . . . [an australopithecine] site at the time of its original occupation, we would find hominids that were living in social groups much like those of other higher primates. The differences would be apparent only after prolonged observation. Perhaps at the start of each day we would observe a group splitting up as some of its members went off in one direction and some in another. All these subgroups would very probably feed intermittently as they moved about and encountered ubiquitous low-grade plant foods such as berries, but we might well observe that some of the higher-grade materials — large tubers or the haunch of a scavenged carcass — were being reserved for group consumption when the foraging parties reconvened at their starting point.

To the observer in the time machine behavior of this kind, taken in context with the early hominids' [possible] practice of making tools and equipment would seem familiarly "human." If, as I suppose, the hominids under observation communicated as chimpanzees do or perhaps by means of very rudimentary protolinguistic signals, then the observer might feel he was witnessing the activities of some kind of fascinating bipedal ape. When one is . . . reconstructing protohuman life, one must strongly resist the temptation to project too much of ourselves into the past. As Jane B. Lancaster of the University of Oklahoma has pointed out, the hominid life systems of two million years ago have no living counterparts.

Source: Glynn Isaac, "The Food-Sharing Behavior of Protohuman Hominids." *Scientific American*, Vol. 238, No. 4. Copyright © 1978 by Scientific American, Inc. All rights reserved.

# THE TAUNG CHILD

## DART'S FAMOUS DISCOVERY

In 1924, shortly after he had emigrated from England to South Africa, Raymond Dart, an anatomy professor at the University of the Witwatersrand in Johannesburg, was shown the fossil skull of a baboon by one of his students who had noticed it while dining at the home of an executive of the Northern Lime Company. The skull was from a limestone quarry at Taung, about 320 kilometres southwest of Johannesburg, and when Dart saw it he wondered whether similar fossils might be recovered from the site. He asked one of his colleagues to follow up on his suspicion, and when his co-worker paid a visit to Taung the manager of the quarry gave him several blocks of limestone with fossils embedded inside. These were shipped to Dart who immediately recognized that the collection contained not only the fossil skulls of two ancient baboons, but also the skull of a more human-like animal. Although it took Dart seventy-three days to chip away the cemented limestone that surrounded the fossil, when he had finished he had uncovered the skull of a three- to four-year-old infant that has since become known as the **Taung child**. See QQ 6.3.

## 6.3 DART'S FIRST LOOK AT THE TAUNG CHILD

**Raymond Dart describes his famous discovery.**

• • •

As soon as I removed the lid [of the box] a thrill of excitement shot through me. On the very top of the rock heap was what was undoubtedly an endocranial cast or mold of the interior of the skull. Had it been only the fossilized brain cast of any species of ape it would have ranked as a great discovery since such a thing had never before been reported. But I knew at a glance that what lay in my hands was no ordinary anthropoidal brain. Here in the lime-consolidated sand was the replica of a brain three times larger than that of a baboon and considerably bigger than that of any adult chimpanzee. The startling image of the convolutions and furrows of the brain and the blood vessels of the skull was plainly visible.

I stood in the shade holding the brain as greedily as any miser hugs his gold, my mind racing ahead. Here, I was certain, was one of the most significant finds ever made in the history of anthropology.

Source: Raymond Arthur Dart with Dennis Craig, *Adventures with the Missing Link* (Philadelphia: The Institutes for the Achievement of Human Potential, 1982), pp. 5–6.

Three features made the infant's skull stand out. One was the shape of its face and teeth, which were much more human-like than ape-like despite the fact that the face had the trace of a muzzle and the teeth were comparatively large. Another was the location of its **foramen magnum** — the large hole at the base of the skull where the spinal cord enters the brain. Its cen-

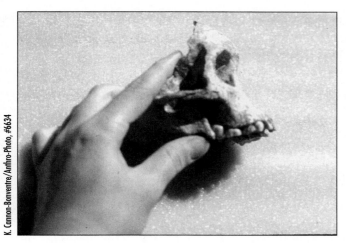

▼

Skull of the

Taung child

▲

tral position indicated that the Taung child was bipedal. However, what impressed Dart most about the skull was the volume of its brain — roughly 405 cm³ — about the same size as that of an adult chimpanzee, but about 25 percent larger than that of a chimpanzee of a similar size and age.

## REACTION

When Dart published a description of the Taung child in 1925 he dubbed it *Australopithecus africanus* (*A. africanus*), a term that is now used to refer to an extinct species of lightly built hominids that lived in South Africa between about 3 mya and about 2 mya, but which literally means southern ape of Africa. Dart, however, implied that the skull of the Taung child was that of a hominid not an ape, and those who read his account knew that this was pre-cisely what he meant. Given the con-troversy surrounding the "monkey trial" (discussed in Chapter 3), which was also held in 1925, the hostile reaction to the find by Christian fun-damentalists came as no surprise. But the scientific community of the day was equally upset. Although a few scholars accepted Dart's conclusion that the Taung child was more human-like than ape-like, the majori-ty of his colleagues greeted his opin-ion with scorn. Among them was Sir Arthur Keith, one of the leading palaeontologists of the day, who argued that the Taung child could

▼

Sir Arthur Keith

(1866–1955)

▲

not possibly be a hominid since the volume of its brain was less than 750 cm$^3$ — a minimum standard for "humanness" that he himself had proposed. Keith regarded the Taung child as nothing more than an unusual ape.

# GRACILE AUSTRALOPITHECINES

By 1947 Keith and most of Dart's other critics had changed their minds. What made them revise their opinion was the discovery of additional *A. africanus* remains, all of which are commonly referred to today as slender or **gracile australopithecines**. Although no other hominid fossils have ever been found at Taung, in 1936, Dr. Robert Broom, a Scottish physician living in South Africa and one of Dart's strongest supporters, recovered the remains of an adult gracile australopithecine from a cave site at the Sterkfontein lime-works near Johannesburg. Eleven years later, in 1947, Dart discovered the fragmentary remains of a gracile australopithecine skull in a cave at Makapansgat, which is also near Johannesburg, and, in the same year, Broom

Robert Broom (centre) looking over the Sterkfontein skull

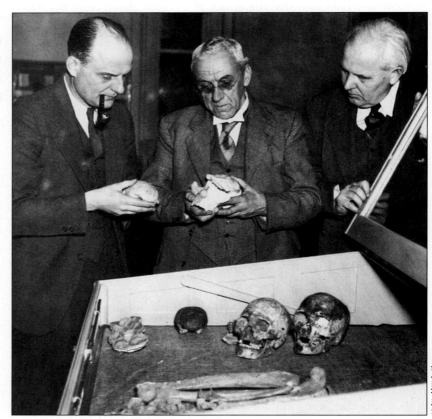

and his student, John T. Robinson, uncovered still more gracile australopithecine fossils at Sterkfontein. These included an almost complete 2.5-million-year-old to 3-million-year-old cranium nicknamed "Mrs. Ples," and several parts of the postcranial skeleton (vertebrae, ribs, pelvis, and thigh bone) of an equally old individual known as STS 14. It was fossils such as these that made Keith and his fellow skeptics change their minds. See QQ 6.4.

## 6.4 KEITH APOLOGIZES

In 1947 Sir Arthur Keith admitted that he was wrong and Dart was right.

. . .

When Prof. Dart of the University of the Witwatersrand, Johannesburg, announced in *Nature* the discovery of a juvenile *Australopithecus* and claimed for it a human kinship, I was one of those who took the point

of view that . . . [it was] near[er] akin to the living African anthropoids — the gorilla and chimpanzee . . . [than it was to humankind. But] I am now convinced . . . that Prof. Dart was right and that I was wrong; the Australopithecinae are in or near the line which culminated in the human form.

Source: Arthur Keith, "Australopithecinae or Dartians?" *Nature*, Vol. 159, No. 4037, p. 377. Reprinted with permission from *Nature*. Copyright 1947 Macmillan Magazines Limited.

Since 1947 hundreds of other gracile australopithecine fossils have been recovered, most from Sterkfontein and a few from Makapansgat. Like those discovered earlier, the more recent finds have also confirmed the accuracy of Dart's original conclusion about the hominid status of the Taung child. Moreover, based on the fact that gracile australopithecine fossils are found only in South Africa, and given the appearance and likely age of the fossils, it is now clear that the Taung child was part of a larger population of lightly built hominids that lived on the grasslands of South Africa between 3 mya and 2 mya. It is also clear that these hominids possessed human-like skulls and teeth; had a brain volume of about 442 cm³; weighed about 46 kilograms; stood about 1.4 metres tall; and habitually walked in an upright position.

John Reader/Science Photo Library/Photo Researchers, Inc.

▼

The skull of Mrs. Ples (upper left)

▲

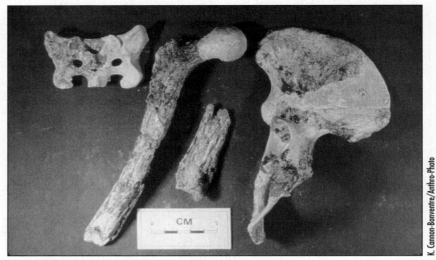

K. Cannon-Bonventre/Anthro-Photo

The partial post-cranial remains of STS-14

# ROBUST AUSTRALOPITHECINES

While scientists were debating the status of the Taung child, hominid fossils somewhat different from those of the gracile australopithecines were also discovered in South Africa. One of the first specimens came to light in 1938, when Robert Broom, the same man who discovered the first adult gracile australopithecine at Sterkfontein, purchased some teeth and skull fragments from a schoolboy who had found them in a cave at Kromdraai, less than four kilometres from Sterkfontein. See QQ 6.5.

## QQ 6.5 BROOM'S PURCHASE

Robert Broom describes how he acquired the Kromdraai specimen.

• • •

On the forenoon of Wednesday, June 8, 1938, when I met Barlow, he said, "I've something nice for you this morning"; and he held out part of a fine palate with the first molar-tooth in position. I said, "Yes, it's quite nice. I'll give you a couple of pounds for it." He was delighted; so I wrote out a cheque, and put the specimen in my pocket. He did not seem quite willing to say where or how he had obtained it; and I did not press the matter. The specimen clearly belonged to a large ape-man . . . .

I was again at Sterkfontein on Saturday, when I knew Barlow would be away. I showed the specimen to the native boys in the quarry; but none of them had ever seen it before. I felt sure it had not come from the quarry, as the matrix was different. On Tuesday forenoon I was again at Sterkfontein, when I insisted on Barlow telling me how he had got the specimen. I pointed out that two teeth had been freshly broken off, and that they might be lying where the specimen had been obtained. He apologized for having misled me;

and told me it was a school-boy, Gert Terblanche, who acted as a guide in the caves on Sundays, who had picked it up and given it to him. I found where Gert lived, about two miles away; but Barlow said he was sure to be away at school . . . .

The road to the school was a very bad one, and we had to leave the car, and walk about a mile over rough ground. When we got there, it was about half-past twelve, and it was play time. I found the head-master, and told him that I wanted to see Gert Terblanche in connection with some teeth he had picked up. Gert was soon found, and drew from the pocket of his trousers four of the most wonderful teeth ever seen in the world's history. These I promptly purchased from Gert, and transferred to my pocket . . . .

As the school did not break up till two o'clock, I suggested to the principal that I should give a lecture to the teachers and children about caves, how they were formed, and how bones got into them. He was delighted. So it was arranged; and I lectured to four teachers and about 120 children for over an hour, with blackboard illustrations, till it was nearly two o'clock. When I had finished, the principal broke up the school, and Gert came home with me. He took us up to the hill, and brought out from his hiding place a beautiful lower jaw with two teeth in position. All the fragments that I could find at the spot I picked up.

Source: Robert Broom, *Finding the Missing Link*, second edition (London: C. A. Watts & Co. Ltd., 1951), pp. 49–51.

In the course of excavations carried out between 1938 and 1950, fossils similar to the ones that Broom had purchased at Kromdraai were found there and also in a nearby cave at Swartkrans. Although Broom used the term *Paranthropus* (near human) to refer to those from Kromdraai, and called those from Swartkrans *Telanthropus* (distant human), the fossils are no longer known by these names. Instead, all of those from Kromdraai and the ones recovered from the lowest levels at Swartkrans are referred to as **robust australopithecines**, or, more technically, as ***Australopithecus robustus*** (*A. robustus*) — a taxonomic term that refers to an extinct species of ruggedly-built hominids that lived in South Africa between about 2 mya and about 1.5 mya. The designation also indicates that while the robust and the gracile australopithecines belonged to the same genus, they nonetheless were distinct species.

## DIFFERENCES AND SIMILARITIES

### DIFFERENCES

One of the things that makes palaeoanthropologists classify the robust and the gracile australopithecines as separate species is the age of their remains; whereas the fossils of the gracile australopithecines are between 3 million years old and 2 million years old, those of the robust australopithecines are between 2 million years old and 1.5 million years old. Given this timeframe the idea that the two forms represent different species makes good evolutionary sense. However, the most striking differences between them were the shapes of their faces, skulls, jaws, and teeth. Whereas the gracile australopithecines had comparatively protruding faces, smooth skulls, and moderately sized jaws and molars, their robust counterparts exhibited the opposite features. *A. robustus* was equipped with a less prominent muzzle, a sagittal crest, and comparatively large jaws and molars. See Figure 6.3.

**FIGURE 6.3** An artist's rendering of gracile (left) and robust (right) australopithecines

The Natural History Museum, London

## THE DIETARY HYPOTHESIS

At one time the differences between the skulls, jaws, and teeth of the gracile and the robust australopithecines were thought to be very important. Some palaeoanthropologists maintained that they indicated that the gracile and the robust forms ate different foods, or, more specifically, that the gracile australopithecines were omnivores and that the robust australopithecines were herbivores.

Over the years this **dietary hypothesis** has been modified: some claiming that *robustus* was a seed-eater who subsisted primarily on cereal grains, others arguing that *robustus* was a hunter and scavenger whose diet included a substantial amount of meat, and still others maintaining that both *robustus* and *africanus* may well have preferred to eat fruit and nuts. Today, however, while most palaeoanthropologists believe that the crested skulls and the comparatively large jaws and molars of the robust forms enabled them to grind up vegetable matter more effectively than their gracile counterparts, they also believe that both species ate a variety of foods, including fruits, nuts, seeds, roots, and small animals. See QQ 6.6.

## SIMILARITIES

Of course, since the gracile and the robust australopithecines were members of the same genus, it stands to reason that there were similarities between them, and this too is reflected in the fossil record. Based on the accumulation of gracile fossils from Sterkfontein and Makapansgat, and of robust ones from Kromdraai and the lowest levels at Swartkrans, it has become

## 6.6 THE DIET OF THE GRACILE AND THE ROBUST AUSTRALOPITHECINES

Richard Kay looks at the relationship between dental evidence and the diet of the gracile and the robust australopithecines.

• • •

Many theories have been advanced about the dietary adaptations of *Australopithecus* . . . . Some of these theories have utilized the evidence of australopithecine dental structure. J. T. Robinson . . . proposed that diets differed in the . . . species of *Australopithecus* [known as *A. africanus* and *A. robustus*]. The larger species, *A. robustus*, was primarily a vegetarian whereas the smaller species, *A. africanus*, ate meat as well as vegetable foods. Largely in agreement with Robinson, Jolly . . . envisioned primitive *Australopithecus*, as exemplified by *A. robustus*, shifting from a frugivorous diet to one centering on cereal grains. He saw "gracile" *Australopithecus* (*A. africanus*) as . . . [having] advanced over *A. robustus* in the direction of more meat eating, [and] thus [as having] evolved toward the condition in earliest *Homo*. . . . Szalay . . . countered that the dental specializations of *A. robustus*, argued

by Jolly as being adaptations for seed eating, make more sense if they are adaptations to meat tearing and bone crushing. He rejected Jolly's seed-eating hypothesis and argued that *"protohominids"* were hunters and scavengers. However the relatively small . . . vertically implanted incisors of *Australopithecus*, upon which Jolly's and Szalay's hypotheses were based, are found as well in a variety of leaf-eating primate species . . . . It remained for Kay . . . to point out that the same cheek-tooth morphology featured in Jolly's and Szalay's scenarios is found in living arboreal primates that eat hard-shelled fruits and nuts. Kay argued that these last items may also have figured prominently in the diets of australopithecines.

In view of . . . these findings, it seems clear that Jolly's and Szalay's dietary scenarios for early hominids may need modification. Simply stated, it is difficult to argue convincingly [on the weight of the dental evidence] that australopithecines were either seed-eaters or hunters and scavengers.

Source: Richard F. Kay, "Dental Evidence for the Diet of Australopithecus," *Annual Review of Anthropology*, Vol. 14 (1985), pp. 315–316.

---

increasingly clear that the robust australopithecines were only slightly larger than their gracile counterparts. Whereas, as already mentioned, adult graciles weighed about 46 kilograms, stood about 1.4 metres tall, and had a brain size of about 442 cm³, adult robust individuals weighed about 48 kilograms, stood about 1.5 metres tall, and had a brain size of about 530 cm³. Nor were the two forms alike only with respect to their overall size. Both lived in South Africa, and, like ourselves, both were bipedal.

# HYPERROBUST AUSTRALOPITHECINES

## *ZINJANTHROPUS BOISEI*

*Africanus* and *robustus* were not the only australopithecines who lived in Africa during the Pliocene. There were also australopithecines in East Africa. The discovery of their remains dates back to 1959, when Mary Leakey came across the teeth of what was clearly a hominid in **Olduvai Gorge** in

Tanzania, where she and her husband, Louis Leakey, had been searching for early humans for years.

The Main Gorge, which is 40 kilometres long and 90 metres deep, is the dried out basin of an ancient lake, and when one walks from the bottom to the top it is like travelling forward through the past two million years. At the bottom of the Gorge are four numbered beds that are between 1.9 million years old and 400 000 years old. Above these are the Masek, Ndutu, and Naisuisiu beds, which are between 400 000 years old and 100 000 years old. The teeth that Mary Leakey found were from Bed I — the deepest and most ancient bed in the Gorge — as was the nearly complete matching skull that she and her husband subsequently discovered (see QQ 6.7). The Leakey's named their find *Zinjanthropus boisei* (*Z. boisei*) — the human from East Africa, but it is most often referred to as "Zinj." See Figure 6.4.

## 6.7 THE LEAKEYS DISCOVER "ZINJ"

Louis Leakey explains how *Z. boisei* was found.

• • •

That morning I woke with a headache and a slight fever . . . . Reluctantly I agreed to spend the day in camp . . . .

Some time later . . . I heard the Land-Rover coming up fast to camp. I had a momentary vision of Mary stung by one of our hundreds of resident scorpions or bitten by a snake . . . .

The Land-Rover rattled to a stop, and I heard Mary's voice calling over and over: "I've got him! I've got him! I've got him!" . . . .

"Got what? Are you hurt?" I asked.

"Him, the man! *Our* man," Mary said. "The one we've been looking for. Come quick, I've found his teeth!" . . . .

I saw at once that she was right. The teeth were premolars, and they had belonged to a human. I was sure they were larger than anything similar ever found, nearly twice the width of modern man's.

I turned to look at Mary, and we almost cried with sheer joy, each seized by that terrific emotion that comes rarely in life. After all our hoping and hardship and sacrifice, at last we had reached our goal — we had discovered the world's earliest known human.

Source: L. S. B. Leakey, "Finding the World's Earliest Man." *National Geographic*, Vol. 118, No. 3 (1960), p. 431.

### *AUSTRALOPITHECUS BOISEI*

Although the Leakeys initially did not regard "Zinj" as a member of the same genus to which the gracile and the robust australopithecines belong, today most anthropologists consider the specimen to be a **hyperrobust australopithecine**, which is the common designation for an extinct species of especially ruggedly-built hominids that lived in East Africa between about 3 mya and about 2 mya. More technically, "Zinj" is regarded as a representative of either *Australopithecus boisei* (*A. boisei*) or *Australopithecus robustus boisei* (*A. robustus boisei*), depending on whether the hyperrobust forms are regarded as a distinct australopithecine species or as a subspecies of the robust australopithecines.

In either event, since "Zinj" was found the remains of other hyperrobust individuals have been discovered, not only in Olduvai Gorge but also at other East African sites, including Lake Natron in Tanzania, East Lake Turkana, West Lake Turkana, and Chesowanja in Kenya, and Omo in Ethiopia. Along with the fact that "Zinj" and his kind closely resembled *africanus* and *robustus* in terms of their cranial capacity and their overall size (the hyperrobust forms had a cranial capacity of about 515 cm³, weighed about 46 kilograms, and stood about 1.5 metres tall), these 2.5-million-year-old to 1.3-million-year-old fossils indicate that *boisei* was even more ruggedly built than *robustus*. The jaws and molars of the hyperrobust forms, for instance, were massive, and their skulls featured a prominent sagittal crest. In addition and most important, the same fossils have confirmed that *boisei* was bipedal. See Figure 6.5.

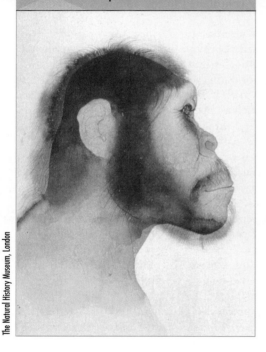

**FIGURE 6.4** Zinj's head reconstructed

The Natural History Museum, London

**FIGURE 6.5** An artist's rendering of *A. boisei*

© Ronald Bowen from *Origins* by Richard E. Leakey and Roger Lewin, C.P. Dutton, 1977

# THE ORIGIN OF BIPEDALISM

Why natural selection favoured bipedalism among forms such as *africanus*, *robustus*, and *boisei* has been the subject of considerable speculation. As already mentioned, one possibility is that upright posture was an adaptation to an aquatic environment. However, since the fossils that bracket the gap in the fossil record indicate the transition from hominoids to hominids took place on the land, this idea has little support.

C. Owen Lovejoy's **provisioning hypothesis**, which is part and parcel of the man-the-hunter scenario, accounts for the origin of bipedalism in a different way. Assuming that a bipedal gait was an adaptation to the African savanna, Lovejoy maintains that the primary adaptive advantage of bipedalism was that it allowed male hunters to carry the meat that they killed back to females and their infants who remained behind while the males were away hunting, and that this sort of provisioning contributed to the hominids' reproductive and hence evolutionary success.

Although enthusiasm for this idea has waned since support for the man-the-hunter scenario has collapsed, Lovejoy's underlying assumption that bipedalism contributed to hominid survival on the African savanna is still intact. In many anthropologists' minds, the single most important adaptive feature of upright posture was that it made it easier to spot and thereby avoid dangerous predators on the savanna such as saber-tooth cats and wild dogs, and to fend off these predators with a hand-held stick or a club when they attacked.

Two other possibilities are also worth noting, both of which focus on the diet of the australopithecines. One is Clifford Jolly's **seed-eating hypothesis**. It is based on the fact that several living monkeys and apes crouch on two feet and use both hands when they feed on small food objects on the savanna such as grass, tubers, and seeds, and that mobility increases access to such foods when these are widely dispersed. Jolly maintains that if the hominoids that gave rise to hominids were faced with a similar feeding situation, which he believes their dentition indicates, then the origin of bipedalism falls into place. According to the seed-eating hypothesis, natural selection favoured bipedalism because it facilitated two-handed feeding on small food objects on the grasslands of Africa.

Pat Shipman's **scavenging hypothesis** is equally intriguing. Based on her analysis of the cut-marks on the bones of ancient savanna-dwelling bovid or ox-like animals, Shipman maintains that while the hominids that lived in the vicinity of Olduvai Gorge between about 2 mya and about 1.7 mya were predominantly frugivores, they also derived a significant proportion of their food supply by scavenging meat and other food products from the carcasses of animals that had died as a result of natural causes or had been killed by non-human predators. The non-human predators' tooth marks on the bones in tandem with small tool marks on the same bones, Shipman says, indicate as much. Natural selection may well have favoured bipedalism, according to

Shipman, because it was the most efficient and the most effective way to locate dead animals, to fend off competitors that were intent on eating the same animals, and to secure access to other sources of food when scavenging failed. See QQ 6.8.

## 6.8 THE SCAVENGING HYPOTHESIS

Pat Shipman explains how the scavenging hypothesis may help us to better understand why natural selection favoured bipedalism.

• • •

The scavenging hypothesis proposes that the Oldowans [that is, the hominids that lived in the vicinity of Olduvai Gorge between about 2 mya and 1.7 mya] were poor hunters, infrequently capable of killing and defending their own prey. Instead, Oldowans relied mostly on scavenging to obtain meat, skin, or other substances from carcasses. Scavenging supplemented plant food foraging and did not provide the major proportion of dietary intake, since such a situation is unknown among living mammals . . . . The contribution to the diet is set at 33%, a figure chosen to indicate that scavenging is as significant a food-procurement strategy to Oldowans as it is to the most successful mammalian scavenger today, the spotted hyena. "Scavenging" refers both to obtaining meat or other substances from carcasses killed by other species and to carrion eating, or consuming partial or whole animals dead of nonpredatory causes . . . .

[W]hy bipedalism arose is a classic issue [that the scavenging hypothesis may help to resolve]. The striking congruity between the attributes of bipedalism . . . and the locomotor needs of scavengers might suggest to some that bipedalism is actually an adaptation . . . to scavenging.

Source: Pat Shipman, "Scavenging or Hunting in Early Hominids: Theoretical Framework and Tests." *American Anthropologist*, Vol. 88, No. 1 (1986), pp. 28–38. Reproduced by permission of the American Anthropological Association. Not for further reproduction.

The problem with Jolly's and Shipman's arguments, and in fact with all of the explanations that attempt to explain why bipedalism arose, is that the gap in the fossil record, which coincides with the advent of bipedalism, makes it impossible to determine which among them most accurately accounts for one of the most important evolutionary events in our biological history. Above all else, when it comes to reconstructing the natural history of the hominids, fossil evidence is required, and while there is circumstantial evidence that lends support to Jolly's and Shipman's arguments, there are simply no fossils that shed direct light on why natural selection worked in favour of a bipedal gait.

## HOW DID BIPEDALISM ARISE?

Anthropologists have encountered a similar problem in attempting to explain how bipedalism arose. Although there is no doubt that bipedalism became entrenched when the first bipedal hominids became reproductively distinct from their non-bipedal hominoid ancestors, it is still unclear whether this method of locomotion was a radical or a smooth departure from what had gone before. Perhaps the transformation was radical, or perhaps, as some have suggested, the advent of bipedalism was part of an ongoing evolutionary process in which hominoids that were preadapted for upright posture responded to a change in their environment by increasing their bipedal activity in an incremental way. Natural selection would then have produced the final biological changes that resulted in a fully bipedal gait. Once again, however, without direct fossil evidence, neither representation can be regarded as definitive. See QQ 6.9.

## QQ 6.9 THE EVOLUTION OF BIPEDALISM

Becky Sigmon, Professor of Anthropology at the University of Toronto, identifies what she surmises may have been the three principal stages in the evolution of bipedalism.

• • •

The emergence of erect bipedalism can be viewed in terms of three stages of development: a preadaptive, a behavioral, and a physical stage although the latter two are probably highly interrelated. During the preadaptive stage the pre-hominids would probably have been making use of bipedal locomotion in a number of circumstances (as . . . [is the case among] chimpanzee groups). During this period of development, however, the use of upright locomotion would not have been critical to the survival of the species.

Now let us assume that there is a change in the environment that is such as to make erect bipedalism of high adaptive value. This leads us to the second stage which is directly associated with the behavioral response of the organism . . . . With the change in

environment, the pre-hominids presumably responded behaviorally to increased selection pressure favoring bipedalism. The increased use of upright posture by pre-hominids was probably not merely a random event but improved their adaptation, hence their chances of survival. The capacity of the pre-hominids to make a behavioral response was probably a major factor contributing to their survival. Once the bipedal behavior became habitual, it influenced the natural selection pattern in a manner that favored the development and/or refinement of a physical type that further improved the behavioral character, bipedalism.

The third stage involved the physical adaptations that were necessary in changing a pronograde animal to an habitual biped. The most significant changes can especially be seen in the foot, pelvis, vertebral column, skull, and the corresponding muscular systems of these areas.

Source: Becky A. Sigmon, "Bipedal Behavior and the Emergence of Erect Posture in Man," *American Journal of Physical Anthropology*, Vol. 34, No.1, pp. 57–58. Copyright © 1971. Reprinted by permission of Wiley-Liss, a Division of John Wiley and Sons, Inc.

# BIPEDALISM AND TOOLS

Whatever their other features and no matter why and how it arose, bipedalism not only set *africanus*, *robustus*, and *boisei* apart from the apes, but also propelled them in a new direction — towards humankind. Among other things, bipedalism freed their hands, which may have allowed them to use perishable materials such as wooden sticks and clubs as hand-held tools. While there is no direct evidence to support this conclusion, the fact that chimpanzees occasionally use such materials in the wild to repel invaders suggests that the australopithecines also may have possessed this talent. But were perishable items the only implements they may have used, or was tool use among them widespread?

## OSTEODONTOKERATIC CULTURE

One person who thought that the South African australopithecines used a broad range of implements was Raymond Dart, and what led him to the conclusion was what he discovered at Makapansgat cave. There, in addition to the gracile australopithecine skull fragments he found, Dart discovered thousands of fossilized non-human animal bones. These bones, he argued, were tools. In fact, Dart believed that the South African australopithecines possessed an **osteodontokeratic culture**, by which he meant that they used "bones, teeth, and horns" as clubs, daggers, picks, scrapers, and so on. In other words, from Dart's point of view, Makapansgat cave was an arsenal that contained the weapons that the australopithecines had used to kill and butcher the antelopes, baboons, and other game that they had hunted on the grasslands of their South African homeland (see Figure 6.6). And Makapansgat cave is not the only location where such a collection has been found. Similar deposits have been discovered at in the caves at Sterkfontein, Kromdraai, and Swartkrans.

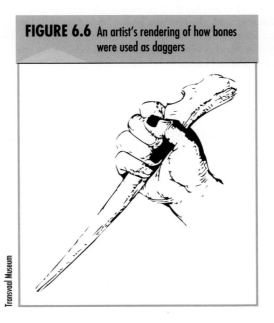

**FIGURE 6.6** An artist's rendering of how bones were used as daggers

Transvaal Museum

## ARDREY'S POPULARIZATION

While Dart made palaeoanthropologists aware that the South African australopithecines may have used "bones, teeth, and horns" as tools, it was the dramatist Robert Ardrey who brought the idea to the attention of the general public, and he did so in a sensational way. What Ardrey suggested was that osteodontokeratic culture signalled the beginning of "man's bloodthirsty history," when people began to kill for the sake of killing. See QQ 6.10.

In 1955 Robert Ardrey paid a visit to Raymond Dart and asked him about the "weapons" of the South African australopithecines.

. . .

For six years Dart and his students had been patiently developing the evidence that australopithecus had been a systematic, purposeful user of weapons. I examined the evidence and found it overwhelming. And now we sat in his office at the wrong end of the world while Dart looked out of his window at the thunderstorms chasing each other across the African sky . . . . There were skulls on his desk. In my hand was a jawbone of a twelve-year-old southern ape found a few years earlier at Makapan[sgat]. The jaw was broken on both sides. The front teeth were missing. There was a dark, smooth dent on the chin where the blow had landed; and the boy had died of it, for there had been no time for the bone to knit.

What if a weapon had done this deed? What if I held in my hands the evidence of antique murder committed with a deadly weapon a quarter of a million years before the time of man? What if the predatory transition should be susceptible to proof, and accepted as the way we came about? Could we afford to surrender, in such desperate hours as those we now lived in, our belief in the nobility of man's inner nature?

I asked Dart how he felt, from a viewpoint of responsibility, about putting forward such a thesis at such a time. I said that I understood his conviction that the predatory transition and the weapons fixation explained man's bloody history, his eternal aggression, his irrational, self-destroying inexorable pursuit of death for death's sake. But I asked, would it be wise for us to listen when man at last possessed weapons capable of sterilizing the earth?

Dart turned from his window and sat down at his desk . . . . And he said that since we had tried everything else, we might in last resort try the truth.

Source: Robert Ardrey, *African Genesis: A Personal Investigation into the Animal Origins and Nature of Man* (New York: Dell Publishing Co., Inc., 1961), pp. 32–33.

Yet despite Ardrey's convincing language, his suggestion that osteodontokeratic culture signalled the beginning of humankind's "inexorable pursuit of death for death's sake" is mistaken. Dart's claim that the South African australopithecines used bones, teeth, and horns as weapons has been effectively countered.

## THE LEOPARD HYPOTHESIS

C. K. Brain put the issue to rest. Brain's special interest is **taphonomy**, which literally means "the laws of burial" and is the study of what happens to the remains of animals between death and fossilization. In 1970 Brain published an article on the remains found in the Swartkrans cave in which he concluded that leopards, not australopithecines, were responsible for the bones in the cave. The idea is known as the **leopard hypothesis**, and what made Brain suspect that the fossils were the handiwork of leopards is what happens to the remains of their prey.

In order to protect their kills from hyenas, leopards often take their prey up into a tree, and when the meat has been eaten the bones fall to the ground. If a leopard feeds over and over in the same tree the bones will accumulate, and since what trees there are on a savanna often grow near moist fissures that lead into caves, the accumulation of bones in such caves is hardly surprising. This is how Brain accounted for what Dart believed were weapons in the Swartkrans cave. Equally important, Brain argued that the presence of australopithecine fossils in the Swartkrans cave was due to the same phenomenon. In other words, Brain proposed that the South African australopithecines were the hunted — not the hunters. See QQ 6.11.

## 6.11 THE LEOPARD HYPOTHESIS

**C. K. Brain looks back on the Swartkrans leopard hypothesis.**

• • •

Some years ago I speculated . . . on how the abundant australopithecine remains may have found their way into the Swartkrans cave. It was clear to me then that the very fragmentary hominid fossils probably represented carnivore food remains, and since leopards were well represented in the Swartkrans assemblage it seemed reasonable to assume that they had been involved. I had made the observation that when leopards are harried by spotted hyenas (whose remains were also known from Swartkrans) the leopards are obliged to take their prey into a tree or other inaccessible place. If they fail to this, they are likely to lose their meal to the hyenas, which are generally dominant in a competitive feeding situation. In woodland areas the leopards simply take their prey up into the nearest tree, but on the open highveld, which appears to have been largely a grassland in

Swartkrans, . . . trees were certainly less abundant. In the dolomitic [calcium magnesium carbonate] areas of the Transvaal highveld, large trees . . . are typically associated with the shaftlike openings of caves. In a generally treeless habitat, the cave entrances provide shelter from frost and fire to saplings that would not readily survive on the exposed hillsides. So the very trees available for leopard prey storage were those that overhung the shafts leading downward to the underground fossilization sites. I therefore speculated that, if in Swartkrans . . . leopards were preying on australopithecines and baboons, then they may well have fed on them in succeeding generations of trees that overhung the Swartkrans shaft. Food remains falling from the trees, fortuitously passing the waiting hyenas and gravitating into the subterranean cavern, ended up as the fossils by which the Swartkrans australopithecines are now known.

Source: C. K. Brain, *The Hunters or the Hunted?* (Chicago: University of Chicago Press, 1981), p. 266.

Since 1970 Brain has added some new twists to the leopard hypothesis. In 1981 he published a book in which his aim was, in his own words, "to analyze the fossil assemblages from the caves of Sterkfontein, Swartkrans, and Kromdraai in order to decide how these bones may have found their way into the caves and to draw conclusions about the behaviour of the hominids and other animals that interacted with them." Brain analyzed 19 487 bones representing at least 1331 animals, and, based on this painstaking work, Brain now believes that several species of wild cat along with hunting hyenas were responsible for the accumulation of bones in the three South African caves. He has also suggested that the caves may have been sleeping sites for the australopithecines, and that their fossil remains inside the caves may be the result of attacks by cats and hyenas inside as well as outside the caves. In any event, because of Brain's careful research, his view of the South African australopithecines as the hunted rather than the hunters now prevails.

## OLDOWAN TOOLS

Although the leopard hypothesis accounts for the bones in the South African caves, it does not account for the tools that may have been made by "Zinj" and australopithecines like him in East Africa. According to some anthropologists, these East African australopithecines may have been the first hominids to manufacture stone tools. Over the years, a number of these implements have been found in association with hyperrobust fossils. These are called **Oldowan tools**, in honour of Olduvai Gorge where many such tools have been found. They were made by striking one rock called a **hammerstone** against another called a **core**, the hammerstone being used to alter the shape of the core, and the core, after it had been modified, yielding one or more stone tools. The end product was either a baseball-size **core chopper** (see Figure 6.7), or a smaller **flake knife**; the former was likely used for skinning animals, cutting meat, and primitive woodworking, and the latter for finer butchering and sharpening sticks. See QQ 6.12.

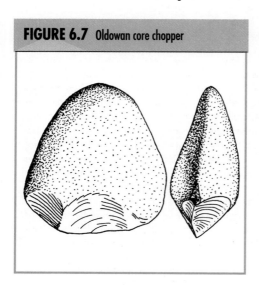

**FIGURE 6.7** Oldowan core chopper

## 6.12 OLDOWAN TOOLS

Mary Leakey discusses the Oldowan tools that she and others have discovered at Olduvai Gorge.

• • •

The principal tools of the Oldowan are choppers, usually made on waterworn, fist-size stones in which the rounded natural surface forms a butt that can be held comfortably in the hand. The working edges were roughly flaked from both [under]sides to form sharp, jagged cutting edges. Choppers are simple, primitive tools, but when newly made are surprisingly effective for a number of purposes such as skinning and cutting meat off a carcass as well as rough wood working. For a time choppers were thought to be the only tools of the Oldowan and they were certainly the most common and the most obvious, but excavations at Olduvai have shown that the Oldowan industry also contains a number of . . . sharp flakes that show evidence of use along the edges. These must surely have served for disjointing bones, sharpening wooden tools and other purposes for which the rather thick-edged choppers were unsuitable.

Source: Mary D. Leakey, "Olduvai Fossil Hominids: Their Stratigraphic Positions and Orientations," in Clifford J. Jolly, ed. *Early Hominids of Africa.* (London: Gerald Duckworth and Company Ltd., 1978), p. 7.

## WHO MADE OLDOWAN TOOLS?

But were the hyperrobust australopithecines responsible for such tools? Some palaeoanthropologists believe that the answer is yes. They claim that the East African australopithecines made Oldowan tools not only to hunt game, but also to protect themselves from enemies. All things considered, they say, there is little doubt but that *boisei* and perhaps other australopithecines were the first primates to use and make tools on a regular basis, and that it was these skills that foreshadowed the emergence of humankind.

Other palaeoanthropologists are skeptical. Some claim that it is highly unlikely that an animal with a brain no larger than an orangutan had the intelligence to make stone tools. In addition, there are those who think that Oldowan tools were manufactured by early members of the genus *Homo* — the genus to which we belong. Their argument hinges on the nature of the fossil record. So far, the tools that have been found in association with the East African australopithecines have also always been accompanied by more human-like fossils; the implication being that Oldowan tools were made by these more human-like forms. Thus, whether or not the East African australopithecines manufactured stone tools is unclear. And even if they did manufacture stone tools, whether or not they used these for hunting is still a matter of conjecture (see QQ 6.13). Many anthropologists currently believe

## 6.13 ROBINSON'S VIEW

John Robinson explains why he doubts that the East African australopithecines manufactured stone tools.

• • •

If it can be proved that australopithecines occur in direct association with a stone industry over a significant period of time *when no evidence whatever exists of the presence at that level of anything more advanced than the australopithecines*, then there will be a sound case for regarding the australopithecines as tool-makers. In fact there is evidence throughout the entire australopithecine period either proving or suggesting the presence of a more advanced form of hominid . . . . There is thus no period in the past in which australopithecines are known to occur in association with stone artefacts but about which all students are agreed that nothing more advanced than australopithecines occurs.

Courtesy of John T. Robinson

John T. Robinson

Source: John T. Robinson, "Australopithecines and Artefacts at Sterkfontein. Part I: Sterkfontein Stratigraphy and the Significance of the Extension Site," *South African Archaeological Bulletin*, Vol. 17, No. 66 (1962), p. 102.

that the bulk of the australopithecines' diet consisted of plant foods, and that what meat they ate was derived either from small animals that could be caught by hand, or else, as Shipman and others have argued, scavenged from animals that had died as a result of natural causes or were killed by non-human predators.

# THE OLDEST AUSTRALOPITHECINES

## LUCY AND THE FIRST FAMILY

Not too many years ago, a form of australopithecine even older than *africanus*, *robustus*, and *boisei* was discovered; one that lived in East Africa between about 4 mya and about 3 mya.

In 1974 Donald Johanson, an American palaeoanthropologist, and Maurice Taieb, a French geologist, unearthed a remarkable hominid fossil near **Hadar**, in the Afar Triangle of northeastern Ethiopia. What made the fossil stand out was that it was about 40 percent complete and at least 3 million years old. Based on the fact that the pelvis indicated that the specimen was female, Johanson and Taieb nicknamed the fossil **Lucy**, after the Beatles' song "Lucy in the Sky with Diamonds." See QQ 6.14.

## 6.14 HOW LUCY ACQUIRED HER NAME

Donald Johanson describes how Lucy was named.

• • •

We had recovered several hundred pieces of bone (many of them fragments) representing forty percent of the skeleton of a single individual. Tom's [Tom Gray, a coworker] and my original hunch had been right. There was no bone duplication.

But a single individual of what? On preliminary examination it was very hard to say, for nothing quite like it had ever been discovered. The camp was rocking with excitement. The first night we never went to

bed at all. We talked and talked. We drank beer after beer. There was a tape recorder in the camp, and a tape of the Beatles' song "Lucy in the Sky with Diamonds" was belting out into the night sky, and we played it at full volume over and over again out of sheer exuberance. At some point during that unforgettable evening [30 November 1974] — I no longer remember exactly when — the new fossil picked up the name of Lucy . . . .

Source: Donald C. Johanson and Maitland A. Edey, *Lucy, the Beginning of Mankind* (New York: Simon and Schuster, 1981), pp. 18–20.

One year later Johanson made another impressive find. Near the same location where Lucy was unearthed, he and his crew discovered the remains of an equally ancient group of thirteen individuals who have since become known as the **first family**.

## AUSTRALOPITHECUS AFARENSIS

Lucy and the first family are not the only hominid fossils that have been discovered at Hadar. Over the years, despite periodic interruptions on account of political and economic turmoil, excavations in this Ethiopian region have turned up the remains of between 36 and 65 equally ancient hominids. Today almost all palaeoanthropologists classify them as *Australopithecus afarensis* (*A. afarensis*), an extinct species of lightly-built hominids that lived in East Africa between about 4 mya and about 3 mya.

Lucy and her kind, however, are not regarded as a distinct species because of their weight, height, or brain size; in these respects they resembled *africanus*, *robustus*, and *boisei*. Nor is it strictly because their fossils are older. Instead, Lucy and her kind are allocated their own species because of the shape of their skulls and their teeth and the proportion of their limbs, which were more ape-like than those of the other australopithecines. In general, their skulls were less rounded, their canines were larger, and their arms were longer in relation to their legs. On the other hand, based on the shape of their postcranial skeletons, there is absolutely no doubt that although Lucy and her kind walked more slowly than we do today, they nonetheless had an erect posture and habitually walked in an upright position. It is for this reason that they are considered to be australopithecines. See QQ 6.15.

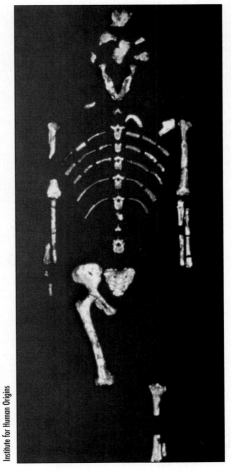

Institute for Human Origins

Lucy's
skeletal
remains

however, she often took to the trees, and climbed, as most primates do, using all four limbs.

Basic evolutionary principles provide one kind of verdict on the possibility. A species cannot develop detailed anatomical modifications for a particular behavior, such as bipedality, unless it consistently employs that behavior. For natural selection to have so thoroughly modified for bipedality the skeleton Lucy inherited, her ancestors must already have spent most of their time on the ground, walking upright . . . .

[Moreover, a] review of the rest of her skeleton and of other *Australopithecus* skeletons reveal[s] equally dramatic modifications that favor bipedality and rule out other modes of locomotion. The knee, for example, is adapted for withstanding greater stress during complete extension than the knee of other primates, and its design brings the femur [or thighbone] and the tibia [or inner bone of the lower leg] together at a slight angle, so that the foot can easily be planted directly under the body's center of mass when body weight is supported on one leg. The ankle is also modified for supporting the entire body weight, and a shock-absorbing arch helps the foot to cope with the added load. The great toe is no longer opposable, as it is in quadrupedal apes, but runs parallel to the other digits. The foot is now a propulsive lever for upright walking rather than a grasping device for arboreal travel. The arms have also become less suited to climbing: both the limb as a whole and the fingers have grown shorter than they are in apes.

Source: C. Owen Lovejoy, "Evolution of Human Walking," *Scientific American*, Vol. 259, No. 5 (1988), pp. 123–125.

# THE LAETOLI FOSSILS AND THE BLACK SKULL

## THE LAETOLI FOSSILS

Since Johanson and Taieb discovered Lucy and her kind at Hadar, two other collections of australopithecine fossils have been singled out for special attention. One includes several good specimens of hominid teeth and jaws and a few postcranial bones that were recovered from **Laetoli**, an early hominid site in Tanzania not far from Olduvai Gorge. Although the Laetoli fossils are similar to those from Hadar, not all palaeoanthropologists believe that they should be classified as *A. afarensis*. Many now maintain that the Laetoli fossils, and likely some from the Afar Triangle, are much more boisei-like than they are like Lucy and consequently should be classified in a different way.

Fossil footprints at Laetoli

John Reader/Photo Researchers, Inc.

No matter how they are classified, there is absolutely no doubt that the Laetoli hominids were bipedal. In 1977, nearby the fossils, Mary Leakey uncovered a 23-metre-long trail of fossil footprints that were made by two or perhaps three individuals who had strolled from north to south across the Serengeti Plain about 3.5 mya. See QQ 6.16.

## 6.16 FOSSIL FOOTPRINTS AT LAETOLI

Richard Hay and Mary Leakey discuss the significance of the Laetoli footprints.

. . .

The . . . footprints found in 1977 . . . are clear proof that 3.5 million years ago . . . [the] East African precursors of early man walked fully upright with a bipedal human gait.

An upright posture this early in the course of human evolution is of great importance. It freed the hands . . . for carrying and for toolmaking and tool use. In spite of diligent searching no stone tools have been found in the Laetoli beds. Hence it seems likely that the hominids who left their tracks . . . had not arrived at the stage of making stone tools. The fact remains that their upright posture gave them the full-time use of the first of all primate tools: unencumbered hands.

Source: Richard L. Hay and Mary D. Leakey, "Fossil Footprints from Laetoli," *Scientific American*, Vol. 246, No. 2 (1982), p. 56.

## THE BLACK SKULL AND ITS KIND

Another equally interesting collection of australopithecine fossils consists of a few teeth and lower jaws that were found at Omo, in Ethiopia, and an unusual fossil known as **KNM-WT 17000**, which was discovered in 1985 on the west shore of Lake Turkana, in Kenya. Otherwise known as the **black skull**, KNM-WT 17000 is a well-preserved 2.5-million-year-old cranium with an afarensis-like brain volume of about 410 cm$^3$ and a massively rugged, boisei-like shape. See QQ 6.17.

A. Walker, Johns Hopkins University School of Medicine, © National Museums of Kenya

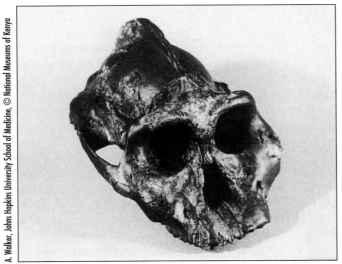

▼

The black skull
(KNM-WT 17000)

▲

Pat Shipman explains how her husband Alan Walker introduced her to the black skull and describes the specimen's unique constellation of traits.

• • •

. . . Alan led me to a . . . hill. . . . I was stunned as he showed me the large, blue-black pieces of a very heavily built skull he had just found . . . .

[W]hat . . . [he] had [discovered was] a very robust, heavily built australopithecine skull usually classified as either *Australopithecus robustus* or *A. boisei*. The age of the beds from which it came . . . was 2.5 million to 2.6 million years, so this skull was the oldest of its type ever found. It was accessioned under the number KNM-WT 17000 in the National Museums of Kenya . . . .

The extraordinary thing about WT-17000 is that it possesses primitive and highly specialized features in a combination that I, for one, would never have expected . . . . [Its] primitive features . . . found primarily in its braincase and jaw joint . . . are shared with *afarensis*, the most primitive hominid known. On the other hand . . . its [specialized traits] mostly in the face and teeth . . . are shared with later specimens of *boisei*, the most specialized australopithecine known.

Source: Pat Shipman, "Baffling Limb On The Family Tree." *Discover*, Vol. 7, No. 9 (1986), pp. 88–93.

In accord with this unusual combination of features, some palaeoanthropologists now maintain that the black skull and its potential counterparts from Omo represent either a primitive form of *A. boisei*, or a new species of australopithecine, which has been named ***Australopithecus aethiopicus*** (*A. aethiopicus*) — a term that refers to an extinct species of especially ruggedly built hominids that lived in East Africa about 2.5 mya.

# HOMINIZATION

As the new discoveries indicate, when it comes to the australopithecines there is much more that remains to be learned. This is especially true with respect to the role that they played in **hominization** — the process by which hominids became increasingly human. Although most anthropologists currently agree that *afarensis* — the oldest known hominid — was the stem from which all subsequent hominids branched off, the connecting links between them are still being debated. Five hypotheses are currently in vogue. See QQ 6.18.

Two hypotheses were formulated before the Laetoli fossils and the black skull were discovered. The first holds that *afarensis* gave rise to two distinct lineages: one that led to *africanus*, *robustus*, and *boisei*, and one that led to *Homo*. The second hypothesis arranges the remains in a different way. It proposes that *afarensis* evolved directly into *africanus*, and that it was *africanus* who gave rise to two lineages: one that led to *robustus* and *boisei*, and another that led to *Homo*.

Fred Grine diagrams the hypotheses that are currently being debated about the place of the australopithecines in human evolution.

• • •

Source: Fred E. Grine, "Australopithecus." In Ian Tattersal, Eric Delson, and John Van Couvering, eds. *Encyclopedia of Human Evolution and Prehistory* (New York: Garland Publishing, 1988), pp. 72–73.

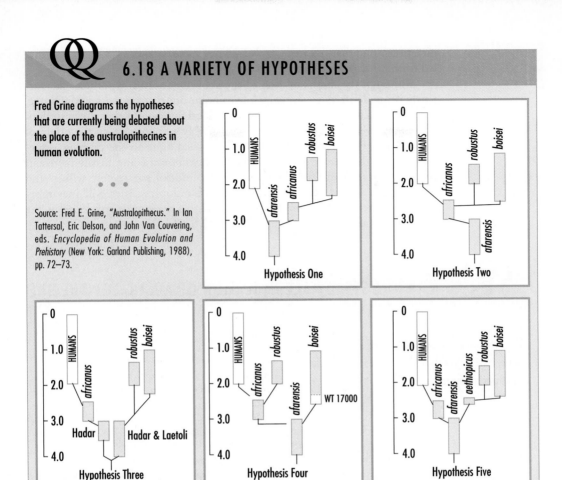

The third hypothesis, which was formulated more recently, focuses on the distinction that some palaeoanthropologists have drawn between the Lucy-like specimens from Hadar, and the non-Lucy-like fossils from Hadar and Laetoli. It states that the non-Lucy-like specimens evolved into *robustus* and *boisei*, and that the Lucy-like forms evolved into *africanus* who then evolved into *Homo*.

The last two hypotheses single out the black skull and its kind for special attention. The fourth assumes that the specimens do not represent a distinct species, but an early and primitive form of *boisei*. In accord with this view, proponents of the hypothesis claim that *afarensis* gave rise, on the one hand, to KNM-WT 17000 who ultimately evolved into *boisei*, and, on the other, to *africanus* who subsequently evolved into *robustus* and *Homo*. In contrast, the fifth hypothesis assumes that the black skull and its kind were a separate species. It contends that *afarensis* branched out in two different directions: one that led from *aethiopicus* to *robustus* and *boisei*, and one that led from *africanus* to *Homo*.

Although the merits of each of the abovementioned hypotheses are currently being debated, it is worthwhile to point out that no matter which one, if any, eventually proves correct, there are good indications that the australopithecines played at least some role, and possibly a major one, in the origin of humankind. Future research will no doubt shed light on the problem, especially given the steady accumulation of australopithecine fossils in recent years. As more fossils are discovered, the biosocial parameters within and between the species that comprised the australopithecines will inevitably become clearer. As the sample size increases, so does our confidence in the conclusions based upon the sample. See QQ 6.19.

## 6.19 THE STEADY ACCUMULATION OF AUSTRALOPITHECINE FOSSILS

At a symposium on human evolution held at the University of Alberta, Phillip Tobias talked about the steady accumulation of australopithecine fossils.

• • •

In the 25 years that have elapsed since the University of Alberta celebrated its golden jubilee [in 1957], the study of human paleontology has made enormous strides in Africa. In 1957, only seven sites in Africa had yielded fossils of what are today accepted as early hominids. [See Table A.] Five of these were in the Union of South Africa and they included Taung,

the site of the earliest discovery of an australopithecine. The other South African sites were Sterkfontein (1936 onwards), Kromdraai (1938 onwards), Makapansgat (1947 onwards) and Swartkrans (from 1948). The two remaining sites were in Tanganyika of those days, namely Olduvai and Garuysi (now known as Laetoli). The few fragments from these two East African localities represented at most five individuals, while the five South African sites represented a good sample of not fewer than 121 individuals and possibly as many as 157 individuals. To take the middle value, remains of some 144 early fossil hominid individuals were available from seven African sites a quarter of a century ago.

Devore 64-13-1A/Anthro-Photo

▼

Phillip V. Tobias
and students at
Sterkfontein
caves

▲

**TABLE A** African sites of early hominid discoveries

| 1924 | | 1957 | | 1982 | |
|---|---|---|---|---|---|
| COUNTRY | NUMBER OF SITES | COUNTRY | NUMBER OF SITES | COUNTRY | NUMBER OF SITES |
| South Africa | 1 | South Africa | 5 | South Africa | 5 |
| | | Tanzania | 2 | Tanzania | 3 |
| | | | | Kenya | 3–7 |
| | | | | Ethopia | 2 |
| | | | | Chad | 1 |

Source: Philip V. Tobias, "Hominid Evolution in Africa," *Canadian Journal of Anthropology*, Vol. 3, No. 2 (1983), p. 167.

**TABLE B** Twenty-five years of African early hominid discoveries

| 1957 | | | 1982 | | |
|---|---|---|---|---|---|
| NUMBER OF SITES | NUMBER OF INDIVIDUALS | MID-VALUE OF INDIVIDUALS | NUMBER OF SITES | NUMBER OF INDIVIDUALS | MID-VALUE OF INDIVIDUALS |
| 7 | 121–157 | 144 | 15 | 405–617 | 511 |

Source: Philip V. Tobias, "Hominid Evolution in Africa," *Canadian Journal of Anthropology*, Vol. 3, No. 2 (1983), p. 169.

By today, instead of seven we have 14 and possibly several more African sites that have yielded early hominids . . . .

[In addition, t]he number of individuals represented from the sites of South and East Africa has more then trebled. Instead of a minimum of 121 individuals, as in 1957, we have at our disposal an estimated minimum of 405 early fossil hominids and a possible maximum of 617 individuals, with a middle value of some 511 individuals . . . . [See Table B.]

Despite the shortcomings of the fossil hominid data, the almost explosive increase in the number of specimens in the past quarter of a century has given us a much clearer picture of the morphological nature of the hominids and of the pattern of hominid evolution.

Source: Phillip V. Tobias, "Hominid Evolution in Africa." *Canadian Journal of Anthropology*, Vol. 3, No. 2 (1983), pp. 167–168.

# KEY TERMS AND CONCEPTS

AQUATIC THEORY OF HUMAN EVOLUTION

SAVANNA THEORY OF HUMAN EVOLUTION

AUSTRALOPITHECINES
(*ahs-tray-loh-pith'-i-seenz*)

TAUNG CHILD (*tah-oong' + child*)

FORAMEN MAGNUM

*AUSTRALOPITHECUS AFRICANUS*
(*australopithecus + af-ra-kan'-us*)

GRACILE AUSTRALOPITHECINES

ROBUST AUSTRALOPITHECINES

*AUSTRALOPITHECUS ROBUSTUS*
(*australopithecus + roh-bus'-tus*)

DIETARY HYPOTHESIS

OLDUVAI GORGE (*ol'-doo-vy + gorge*)

*ZINJANTHROPUS BOISEI*
(*zin-jan'-throh-pus + boy'-zee-eye*)

HYPERROBUST AUSTRALOPITHECINE

*AUSTRALOPITHECUS BOISEI*

*AUSTRALOPITHECUS ROBUSTUS BOISEI*

PROVISIONING HYPOTHESIS

SEED-EATING HYPOTHESIS

SCAVENGING HYPOTHESIS

OSTEODONTOKERATIC CULTURE (*ah'-stee-oh-don-toh-keh'-ra-tic + culture*)

TAPHONOMY (*taf-on'-oh-mee*)

LEOPARD HYPOTHESIS

OLDOWAN TOOLS (*ol'-doh-wahn + tools*)

HAMMERSTONE

CORE

CORE CHOPPER

FLAKE KNIFE

HADAR (*hah-dahr*)

LUCY

FIRST FAMILY

*AUSTRALOPITHECUS AFARENSIS*
(*australopithecus + af-ahr-en'-sis*)

LAETOLI (*ly-toh'-lee*)

KNM-WT 17000

BLACK SKULL

*AUSTRALOPITHECUS AETHIOPICUS*
(*australopithecus + ee-thee-oh-pikus*)

HOMINIZATION (*haw-min-i-zay'-shun*)

# SELECTED READINGS

## INTRODUCTION

Hill, A. and Ward, S. "Origin of the Hominidae: the Record of African Large Hominid Evolution Between 14 My and 4 My." *Yearbook of Physical Anthropology,* Vol. 31 (1988), pp. 49–83.

Zihlman, A., Cronin, J., Cramer, D., and Sarich, V. "Pygmy Chimpanzee as a Possible Prototype for the Common Ancestor of Humans, Chimpanzees and Gorillas." *Nature,* Vol. 275 (1978), pp. 744–746.

## THE GAP IN THE FOSSIL RECORD

Morgan, E. *The Aquatic Ape.*
New York: Stein and Day, 1982.

Morgan, E. *The Scars of Evolution: What our Bodies Tell Us About Human Origins.* London: Souvenir, 1990.

Willis, D. The *Hominid Gang: Behind the Scenes in the Search for Human Origins.* New York: Viking, 1989.

## AUSTRALOPITHECINES

Jolly, C., ed. *Early Hominids of Africa.* New York: St. Martin's, 1978.

Zihlman, A. "A Behavioral Reconstruction of *Australopithecus.*" In Reichs, K., ed. *Hominid Origins: Inquiries Past and Present.* Washington, D.C.: University Press of America, 1983, pp. 207–238.

Zihlman, A. "Knuckling Under: Controversy over Hominid Origins." In Sperber, G., ed. *From Apes to Angels: Essays in Anthropology in Honor of Phillip V. Tobias.* New York: Wiley-Liss, 1990, pp. 185–196.

## The Taung Child

Dart, R. "*Australopithecus africanus:* the Man-ape of South Africa." *Nature*, Vol. 115 (1925), pp. 195–199.

Keith, A. "The Fossil Anthropoid Ape from Taungs." *Nature*, Vol. 115 (1925), p. 234.

Lewin R. "The Taung Baby Reaches Sixty." *Science*, Vol. 227 (1985), pp. 1188–1190.

Tobias, P. *Dart, Taung, and the "Missing Link".* Johannesburg: Witwatersrand University Press, 1984.

## Gracile Australopithecines

Butzer, K. "Paleoecology of South African Australopithecines: Taung Revisited." *Current Anthropology*, Vol. 15 (1974), pp. 367–426.

## Robust Australopithecines

Broom, R. "More Discoveries of *Australopithecus.*" *Nature*, Vol. 141 (1938), pp. 828–829.

Grine, F. "Evolutionary History of the "Robust" Australopithecines: a Summary and Historical Perspective." In Grine, F., ed. *Evolutionary History of the "Robust" Australopithecines.* New York: Aldine de Gruyter, 1988, pp. 509–520.

Shipman, P. "The Gripping Story of *Paranthropus.*" *Discover*, Vol. 10 (1989), pp. 66–71.

## Differences and Similarities

Jolly, C. "The Seed-Eaters: a New Model of Hominid Differentiation Based on a Baboon Analogy." *Man* (new series), Vol. 5 (1970), pp. 5–26.

Szalay, F. "Hunting-Scavenging Protohominids: a Model for Hominid Origins." *Man* (new series), Vol. 10 (1975), pp. 420–429.

## Hyperrobust Australopithecines

Hay, R. *Geology of the Olduvai Gorge: A Study of Sedimentation in a Semiarid Basin.* Berkeley, California: University of California Press, 1976.

Leakey, L. "A New Fossil Skull from Olduvai." *Nature*, Vol. 184 (1959), p. 491.

## The Origin of Bipedalism

Langdon, J. "Fossils and the Origin of Bipedalism." *Journal of Human Evolution*, Vol. 14 (1985), pp. 615–635.

Lovejoy, C. "The Origin of Man." *Science*, Vol. 211 (1981), pp. 341–350.

Lovejoy, C. "The Natural Detective." *Natural History*, Vol. 93 (1984), pp. 24–28.

Potts, R. and Shipman, P. "Cutmarks Made by Stone Tools on Bones from Olduvai Gorge, Tanzania." *Nature*, Vol. 291 (1981), pp. 577–580.

Shipman, P. "Scavenger Hunt." *Natural History*, Vol. 93 (1984), pp. 20–27.

Wheeler, P. "The Thermoregulatory Advantage of Hominid Bipedalism in Open Equatorial Environments: the Contribution of Increased Convective Heat Loss and Cutaneous Evaporation Cooling." *Journal of Human Evolution*, Vol. 21 (1991), pp. 107–115.

Zihlman, A. and Brunker, L. "Hominid Bipedalism: Then and Now." *Yearbook of Physical Anthropology*, Vol. 22 (1979), pp. 132–162.

## Bipedalism and Tools

Behrensmeyer, A. and Hill, A., eds. *Fossils in the Making: Vertebrate Taphonomy and Paleoecology.* Chicago: University of Chicago Press, 1980.

Brain, C. "New Finds at the Swartkrans Australopithecine Site." *Nature*, Vol. 225 (1970), pp. 1112–1119.

Dart, R. *The Osteodontokeratic Culture of Australopithecus Prometheus.* Pretoria, Republic of South Africa: Transvaal Museum, Memoir Number 10, 1957.

Potts, R. "Why the Oldowan? Plio-Pleistocene Toolmaking and the Transport of Resources." *Journal of Anthropological Research*, Vol. 47 (1991), pp. 153–176.

Susman, R. "Who Made the Oldowan Tools? Fossil Evidence for Tool Behavior in Plio-Pleistocene Hominids." *Journal of Anthropological Research*, Vol. 47 (1991), pp. 129–151.

## The Oldest Australopithecines

Boaz, N. "Status of *Australopithecus afarensis.*" *Yearbook of Physical Anthropology*, Vol. 31 (1988), pp. 85–113.

Johanson, D. "Ethiopia Yields First "Family" of Early Man." *National Geographic*, Vol. 150 (1976), pp. 791–811.

Johanson, D., Taieb, M., and Coppens, Y. "Pliocene Hominids from the Hadar Formation, Ethiopia (1973–1977): Stratigraphic, Chronologic, and Aleo-Environmental Contexts with Notes on Hominid Morphology and Systematics." *American Journal of Physical Anthropology*, Vol. 57 (1982), pp. 373–402.

McHenry, H. "Sexual Dimorphism in *Australopithecus afarensis.*" *Journal of Human Evolution*, Vol. 20 (1991), pp. 21–32.

Stern, J., Jr. and Susman, R. "The Locomotor Anatomy of *Australopithecus afarensis.*" *American Journal of Physical Anthropology*, Vol. 60 (1983), pp. 279–318.

## The Laetoli Fossils and the Black Skull

Clark, G. "Some Thoughts on the Black Skull: an Archaeologist's Assessment of WT-17000 (*A. boisei*) and Systematics in Human Paleontology." *American Anthropologist*, Vol. 90 (1988), pp. 357–371.

Coppens, Y., Howell, F., Isaac, G., and Leakey, R., eds. *Earliest Man and Environments in the Lake Rudolf Basin.* Chicago: University of Chicago Press, 1976.

Day, M. and Wickens, E. "Laetoli Pliocene Hominid Footprints and Bipedalism." *Nature*, Vol. 286 (1980), pp. 385–387.

Leakey, M. and Harris, J. *Laetoli: A Pliocene Site in Northern Tanzania.* Oxford: Oxford University Press, 1987.

Leakey, R. and Walker, A. "New *Australopithecus boisei* Specimens From East and West Lake Turkana, Kenya." *American Journal of Physical Anthropology*, Vol. 76 (1988), pp. 1–24.

Walker, A., Leakey, R., Harris, J., and Brown, F. "2.5-Myr *Australopithecus boisei* from West of Lake Turkana, Kenya." *Nature*, Vol. 322 (1986), pp. 517–522.

## HOMINIZATION

Delson, E. "Human Phylogeny Revisited again." *Nature*, Vol. 322 (1986), pp. 496–497.

Johanson, D. and White, T. "A Systematic Assessment of Early African Hominids." *Science*, Vol. 203 (1979), pp. 321–330.

Kimbel, W., White, T., and Johanson, D. "Implications of KNM-WT 17000 for the Evolution of the "Robust" Australopithecines." In Grine, F., ed. *Evolutionary History of the "Robust" Australopithecines.* New York: Aldine de Gruyter, 1988, p. 259–268.

McHenry, H. "New Estimates of Body Weight in Early Hominids and their Significance to Encephalization and Megadontia in Robust Australopithecines." In Grine, F., ed. *Evolutionary History of the "Robust" Australopithecines.* New York: Aldine de Gruyter, 1988, pp. 133–148.

Skeleton, R., McHenry, H., and Drawhorn, G. "Phylogenetic Analysis of Early Hominids." *Current Anthropology*, Vol. 27 (1986), pp. 21–43.

Tobias, P. *The Brain in Hominid Evolution.* New York: Columbia University Press, 1971.

White, T., Johanson, D., and Kimbel, W. "*Australopithecus africanus*: its Phyletic Position Reconsidered." *South African Journal of Science*, Vol. 77 (1981), pp. 445–470.

# CHAPTER SEVEN

# THE BEGINNING
# OF THE GENUS *HOMO*

*contents at a glance*

# INTRODUCTION

If australopithecines were our ancestors, then they gave rise to the genus *Homo*, the one that includes ourselves and the immediate forerunners of our kind (see Figure 7.1). Although the first members of our genus were more human-like than the australopithecines, they resembled the australopithecines quite closely; in fact, when the remains of early *Homo* and *Australopithecus* are found together, it can be extremely difficult to tell the two apart. In time, however, our genus included representatives who possessed much larger bodies and brains than the australopithecines, so much so that the late Isaac Asimov referred to the members of our genus as "super-australopithecines." See QQ 7.1.

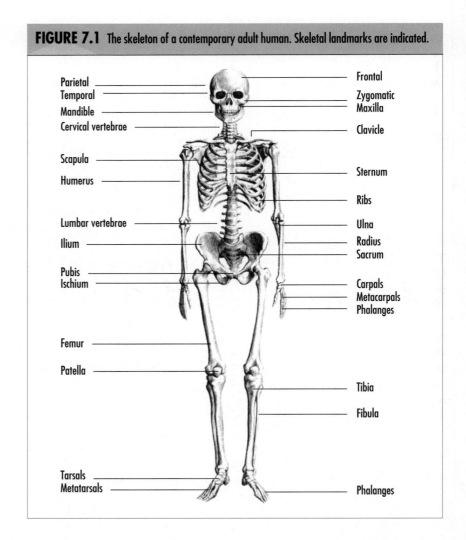

**FIGURE 7.1** The skeleton of a contemporary adult human. Skeletal landmarks are indicated.

Parietal
Temporal
Mandible
Cervical vertebrae
Scapula
Humerus
Lumbar vertebrae
Ilium
Pubis
Ischium
Femur
Patella
Tarsals
Metatarsals

Frontal
Zygomatic
Maxilla
Clavicle
Sternum
Ribs
Ulna
Radius
Sacrum
Carpals
Metacarpals
Phalanges
Tibia
Fibula
Phalanges

## 7.1 AUSTRALOPITHECUS AND *HOMO*

Isaac Asimov looks at the differences between *Australopithecus* and *Homo*.

• • •

If an australopithecine could reason as we do, we might ask it to describe its notion of a Super-australopithecine. It might answer thus:

"Well, first of all, I'd want it bigger and stronger than I am so it could defend itself better and be a better hunter than I am. I suppose that if it were, say, 1.6 meters tall, instead of 1.2 [to 1.5] as I am, and if it weighed about 70 kilograms, instead of 30 [to 40 kilograms] as I do, it would be tall enough and strong enough to be a Superaustralopithecine . . . ."

"[Also], it seems to me that if you really want a Superaustralopithecine . . . you . . . need . . . a larger brain; one that weighs, say 1.5 kilograms, or [three to] four times the size of mine. I realize that would mean Superaustralopithecine's skull would have to be huge and swollen and he would look ugly, but there's no way out, if you want someone who is super. A big-brained Superaustralopithecine — well, who can possibly tell what he would be capable of?"

The Australopithecines were extinct by about a million years ago, but by that time there were hominids that were larger and bigger-brained . . . a creature that was . . . [well on the way to becoming the] Superaustralopithecine described by our mythical australopithecine thinker . . . . We call him Homo.

Source: Isaac Asimov, "Introduction," In Isaac Asimov, Martin H. Greenberg, and Charles G. Waugh, eds, *Supermen* (New York: New American Library, Inc., 1984), pp. 8–9.

# *HOMO HABILIS*

## *HOMO HABILIS* AT OLDUVAI GORGE

The first undisputed representatives of the genus *Homo* were discovered in Olduvai Gorge by Louis Leakey and several of his colleagues. Encouraged by the discovery of "Zinj" in 1959, Leakey and his associates redoubled their efforts to find similar specimens in the Gorge. Instead, they found the remains of a more human-like animal. By 1964 the collection included some skull fragments and the lower jaw of a specimen referred to as OH 7, and the partial remains of two other individuals nicknamed "Cinderella" and "Olduvai George." Although the sample was small, Leakey and his colleagues concluded that the remains were those of *Homo*, and, acting on a suggestion made by Raymond Dart, they named the specimens ***Homo habilis*** (*H. habilis*) — "handy man" for short and "habilines" in colloquial terms. See QQ 7.2.

What convinced Leakey and his fellow researchers that OH 7, "Cinderella," and "Olduvai George" were more human-like than the australopithecines was that their faces, brains, and teeth were technically more like our own. For instance, while the specimens still had the trace of a muzzle, it was a smaller muzzle than the australopithecines possessed. With a

Louis Leakey, Phillip Tobias, and John Napier introduce *H. habilis.*

• • •

The recent discoveries of fossil hominid remains at Olduvai Gorge have strengthened the conclusions — which each of us had reached independently through our respective investigations — that the fossil hominid remains found . . . [between] 1960 [and 1964] at . . . Olduvai, did not represent a creature belonging to the sub-family Australopithecinae . . . .

An examination of these finds has . . . fully confirmed the presence of the genus *Homo* in the lower part of the Olduvai geological sequence, earlier than, contemporary with, as well as later than, the *Zinjanthropus* skull, which is certainly an australopithecine . . . .

We have come to the conclusion that, apart from *Australopithecus* (*Zinjanthropus*), the specimens we are dealing with from Bed I and the lower part of Bed II at Olduvai represent a single species of the genus *Homo* . . . [that we have named *H. habilis*, t]he specific name [*habilis*] . . . taken from the Latin, meaning able, handy, mentally skilful, vigorous.

Source: L. S. B. Leakey, P. V. Tobias, and J. R. Napier, "A New Species of the Genus *Homo* from Olduvai Gorge," *Nature,* Vol. 202, No. 4927, pp. 7–8. Reprinted with permission from *Nature.* Copyright 1964 Macmillan Magazines Limited.

cranial capacity of about 640 cm$^3$, the volume of their brains was also comparatively large; between 25 percent and 35 percent larger than that of the australopithecines. In addition, their teeth were smaller and narrower than those of the australopithecines and thus more like the teeth of our kind.

The mandibles of "Olduvai George" (left) and "Cinderella" (right)

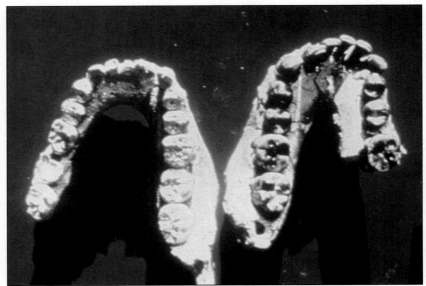

Courtesy of Robert Jurmain, San Jose State University

# ER 1470

The announcement of the new discoveries raised a number of important questions; one was whether the designation *H. habilis* was appropriate, another was whether the first members of our genus had appeared at the dawn of the Pleistocene or before. The answers to these questions remained cloudy until 1972, when a research team working under the direction of Richard Leakey — the son of Louis and Mary and a famous palaeoanthropologist in his own right — unearthed an almost complete *H. habilis* cranium at Koobi Fora on the east shore of Lake Turkana in Kenya. See Figure 7.2.

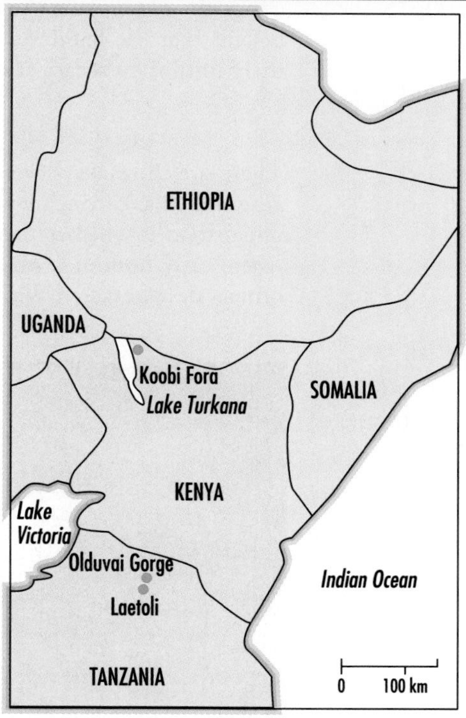

Known as ER 1470, two things made the fossil skull stand out. The first was the volume of its brain; estimated to have been about 775 cm³, ER 1470's cranial capacity was well outside the australopithecine range, and this helped to confirm the existence of *H. habilis* when its taxonomic standing was still in considerable doubt. The second striking feature of ER 1470 was its age. When it was first found it was suggested that the cranium was about 2.6 million years old, and this raised the ire of those who insisted that the first members of our genus had not appeared until the Pliocene had come to an end (*c.* 1.8 mya). Since then it has been determined that ER 1470 is upwards of 1.9 million years old, and while this is younger than was once thought, the age of the specimen has helped lay to rest the argument over the antiquity of our genus (see QQ 7.3). Today, based on what has been learned

**FIGURE 7.2** The location of Lake Turkana

---

## QQ 7.3 ER 1470

In 1973, one year after ER 1470 was discovered at East Lake Turkana (formerly Lake Rudolf), the find was heralded as an important event.

• • •

Leakey's report . . . [concerning the remains of ER or East Rudolf] 1470 . . . reiterates what more fragmentary and less dramatic material from the East Rudolf area has been suggesting for some time . . . . The comparatively large size of the skull

about *H. habilis* since the initial specimens were discovered, most experts believe that the habilines originated in Africa about 2.2 mya and persisted there until they became extinct about 1.5 mya.

## H. HABILIS AND HOME BASES

There are, however, several questions about *H. habilis* that have yet to be answered in a convincing way. One of the details of their lives that is still poorly understood is whether they congregated at **home bases** — selected campsites where early hominids may have manufactured stone tools, shared food, and otherwise interacted with one another on a regular basis. There are, for instance, several sites at Olduvai Gorge whose contents seem to indicate that home bases were part and parcel of habiline life. These sites, which are called **living floors**, contain substantial numbers of Oldowan stone tools and animal bones. One of these sites features a ring of stones about four metres in diameter inside which many Oldowan tools and animal bones were found, leading some to conclude that the outer ring of stones was once used as a foundation for a primitive structure.

But this site and others like it may not be home bases at all, despite the fact that the accu-

A 1.6 million-year-old living floor at Olduvai Gorge

Courtesy of Robert Jurmain and Harry Nelson

mulated debris creates the impression of communal activity. Some scholars have pointed out that the bone waste in such assemblages could have been deposited there because the sites were simply regular feeding areas where hominids rested briefly, rather than for prolonged periods, while they ate what they had scavenged from carnivore kills. The idea that the sites were home bases has also been challenged by those who have pointed out that it would have been dangerous for early hominids to remain stationary in open-air encampments for any length of time because of the presence of predators. Thus, whether or not the habilines established home bases is debatable.

## H. HABILIS AND OLDOWAN TOOLS

Nor is it known whether *H. habilis* (see Figure 7.3) invented the Oldowan tools that have been found at Olduvai Gorge. That the habilines made such tools is likely since their fossils have been found in association with Oldowan implements not only at Olduvai Gorge, but also at Omo in Ethiopia, at Koobi Fora in Kenya, and at Swartkrans in southern Africa. And given these associations, the idea that *H. habilis* invented Oldowan tools is appealing, particularly since it would allow anthropologists to draw a clear line of demarcation between *Australopithecus* and *Homo* on the basis of their tool-making skills. Unfortunately, however, the identity of the original Oldowan toolmakers at the abovementioned locations is in doubt. Since fossils of robust australopithecines have been found in the same deposits, it is possible that they, rather than the habilines, or that they *along with* the habilines, invented Oldowan tools. Still others have argued that, no matter who invented Oldowan tools, both the robust australopithecines and the habilines undoubtedly made use of such tools since both possessed the manual dexterity and the intellectual capacity to do so. In either event, it is important to remember that while Oldowan tools may have been used for butchering carcasses, splitting open bones to extract the marrow, and digging objects out of the ground, they may not have been used for hunting. Most experts hold that the early Oldowan toolmakers were far more adept at scavenging than they were at killing game. See QQ 7.4.

**FIGURE 7.3** An artist's rendering of *H. habilis*

The Natural History Museum, London

## QQ 7.4 OLDOWAN TOOLMAKERS

Randall Susman explains how anthropologists can determine whether early hominids manufactured stone tools, and identifies who he believes made the first Oldowan tools.

• • •

Contrary to the feeling of many archaeologists that it may be impossible to assess the capacity of tool behavior in early hominids (especially when more than one species is found at the same site), fossil evidence of the hand together with studies of comparative anatomy of extant humans and apes allows us to infer tool behavior in the same way we might deduce bipedalism, arboreality, diet, or other habits. The only prerequisites to the determination of hand function are (1) appropriate fossil material (for example, an entire set of wrist bones may be less diagnostic of precision grasp than a single distal phalanx [or terminal segment] of the thumb) and (2) determination of requisite and unambiguous behavioral indicators (viz., certain anatomy must be manifestly related to a given behavior as determined in living animals).

Different lines of evidence [i.e., archaeological, fossil, and primatological] suggest that both *Paranthropus* [i.e., *Australopithecus*] *robustus* (at least in South Africa) and *Homo habilis* were toolmakers. Furthermore, there is no definitive evidence, either artifactual or fossil, that earlier hominids such as *Australopithecus afarensis* or *A. africanus* were toolmakers.

Source: Randall L. Susman, "Who Made The Oldowan Tools: Fossil Evidence For Tool Behavior in Plio-Pleistocene Hominids," *Journal of Anthropological Research* Vol. 47, No. 2 (1991), p. 146.

## H. HABILIS AND H. ERGASTER

Finally, some palaeoanthropologists have expressed concern about the range of variation in the dimensions of habiline fossils. As the sample size has increased, it has become increasingly clear that the habilines were a highly variable group. Some, for instance, possessed straight jaws, large brains, and teeth that were smaller than those of the australopithecines but larger than our own; these are estimated to have stood between 1.3 metres and 1.6 metres tall and weighed about 48 kilograms. But there are also specimens whose faces were less robust, whose brain volume was only about 500 cm³, and whose teeth were somewhat smaller than the others although still larger than our own.

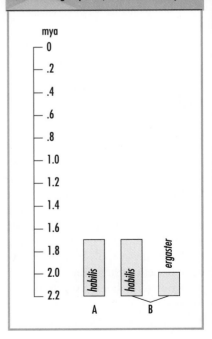

**FIGURE 7.4** Two views of early *Homo:* (a) as a single species, and (b) as two species

While there is little doubt that the smaller specimens are *Homo*, their tax-onomic position is unclear. As might be expected, some palaeoanthropologists say that the smaller individuals were females and that the larger-brained, more ruggedly constructed specimens were males; in other words, that the differ-ence in cranial capacity and in facial appearance was due to sexual dimor-phism. Others, however, maintain that habiline males were only 20 percent larger than habiline females at most, and that the only reasonable conclusion that can be derived from examining the habiline fossil record is that the small-er-brained individuals with the gracile faces represent a separate species. Some refer to this species as ***Homo ergaster*** (*H. ergaster*) — a name that was assigned to a lower jaw discovered at East Lake Turkana. See Figure 7.4 and QQ 7.5.

## 7.5 ONE SPECIES OF EARLY *HOMO* OR TWO?

Bernard Wood comments on the variation within *H. habilis*, which some palaeoanthropologists say indicates that there was more than one species of early *Homo*.

• • •

Opinions about *H. habilis* differ. Some see it as a vari-able, but well-defined, taxon, whereas others have come to acknowledge that the label *H. habilis* has been applied to such a heterogenous collection of material that the identity of the 'real' *H. habilis* is now all but obscured . . . .

Many attempts to sort within *H. habilis* have con-centrated on absolute size, but some studies, especially those which have looked at variation within early *Homo* from Koobi Fora, have also looked at shape dif-ferences . . . . Nonetheless, both approaches have suggested a broadly similar split of *H. habilis*, with one subgroup with large brains, orthognathic [or straight-jawed] faces and relatively large dentition . . . and the other being smaller-brained with more lightly-built faces . . . . It is natural to suspect that these two subgroups may be sexual dimorphs, but the pattern of variation within and between them does not suggest this. In summary, the cranial evidence . . . does little to unite the two morphs within *H. habilis*, nor does it par-ticularly help to resolve whether they represent intraspecific, and presumably sexual, differences, or taxonomic variation.

Source: Bernard Wood. "Who is the "real" *Homo habilis?*" *Nature*, Vol. 327, No. 6119, pp. 187–188. Reprinted with permission from *Nature*. Copyright 1987 Macmillan Magazines Limited.

# THE PLEISTOCENE EPOCH

## THE CLIMATE

The course of human evolution continued to move in our direction during the Pleistocene epoch — about 1.8 mya to 11 kya (thousand years ago). The epoch is also referred to as the **Ice Age**, because it was characterized by long-lasting periods of extremely cold weather, so cold that huge sheets of ice and debris called **glaciers** accumulated in the high-latitude and high-altitude regions of North America and northern Europe. How often such continental

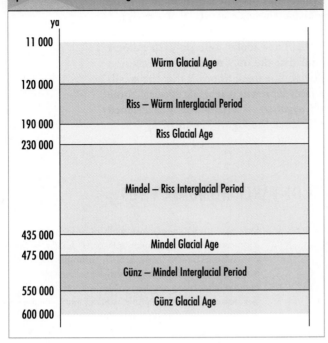

**FIGURE 7.5** A traditional view of the glacial ages and interglacial periods that occurred during the Pleistocene (c. 1.8 mya – 11 kya)

| ya | |
|---|---|
| 11 000 | Würm Glacial Age |
| 120 000 | Riss – Würm Interglacial Period |
| 190 000 | Riss Glacial Age |
| 230 000 | Mindel – Riss Interglacial Period |
| 435 000 | Mindel Glacial Age |
| 475 000 | Günz – Mindel Interglacial Period |
| 550 000 | Günz Glacial Age |
| 600 000 | |

ice sheets formed in the northern hemisphere during the Pleistocene is unknown. At one time scientists suspected that there were four major glacial advances during the epoch, each of which was halted by an intervening **interglacial period** during which warmer temperatures caused the ice sheets to melt (see Figure 7.5). Today it is estimated that there may have been as many as two dozen glacial episodes during the Pleistocene, and this has placed the original scheme in jeopardy.

## MAMMALIAN EVOLUTION

There is, however, absolutely no doubt that the Pleistocene was characterized by the rise and fall of many mammalian forms. While this was true everywhere, the impact in the Americas, in Europe, and in Australia was pronounced. On those continents the majority of large and medium-sized mammals became extinct just before the Pleistocene came to an end, partly as a result of human predation and partly because of environmental change (see QQ 7.6). At least this is how proponents of the **keystone herbivore hypothesis** explain this extraordinary turn of events. They maintain that Ice Age peoples in the Americas, in Europe, and in Australia hunted many large herbivores to the point of extinction, and that their demise, combined with climatic change, adversely affected the habitat of medium-sized herbivores condemning them to a similar fate. In the meantime, our genus had long since changed; the habilines had given way to hominids who were more like ourselves.

 **7.6 FAUNAL EXTINCTIONS**

Karl Butzer looks at the likely causes of faunal extinctions at the end of the Pleistocene.

• • •

The Pleistocene archaeological record suggests that over a span of some 2 million years, early hominids were transformed from their status of incidental or minor carnivores to modern humans, assuming the role of dominant world predators. It is therefore not only possible but also probable that the biomass [or the volume of organisms per unit volume of habitat] of preferred game was seriously altered in some habitats and biomes and that significant ecosystemic adjustments among other herbivores and their dependent carnivores took place. Several historical extinctions, near extinctions, and local extinctions can be directly attributed to hunters (e. g., the North American bison, the North African elephant, and the quagga zebra and blue antelope of South Africa). But the zooarchaeological record, however tantalizing, does not directly verify either decimation or extinction as the result of prehistorical hunting . . . .

A wave of animal extinctions marked the terminal Pleistocene on each continent, coincident both with the dispersal of efficient hunters and with environmental shifts from Pleistocene to Holocene equilibrium levels. The extinctions were not simultaneous within any one biome, and they affected several different kinds of animals, including a large range of birds . . . . This suggests complex ecosystemic responses to more than one causal factor. Just as prehistorical predation has not been directly linked to any one extinction, neither have environmental changes.

Source: Karl W. Butzer, *Archaeology as Human Ecology: Method and Theory for a Contextual Approach* (Cambridge, England: Cambridge University Press, 1982), p. 203. Reprinted with the permission of Cambridge University Press.

# HOMO ERECTUS

It was at the beginning of the Pleistocene, sometime between 1.8 mya and 1.7 mya, that the first hominids who overwhelmingly resembled ourselves made their debut. They are called *Homo erectus* (*H. erectus*) and they possessed an impressive constellation of traits, some reminiscent of the past but others clearly foreshadowing what was to come. For example, although their skulls were longer and lower than our own and their brow ridges were frequently well-developed, they were also relatively large-bodied and big-brained. *H. erectus* stood between 1.5 metres and 1.8 metres tall, weighed about 60 kilograms, and had a cranial capacity

**FIGURE 7.6** An artist's rendering of *H. erectus*

American Museum of Natural History

that ranged between 800 cm³ and 1000 cm³ in earlier forms, and between 1000 cm³ and 1300 cm³ in what some presume are later forms. Almost all anthropologists believe that *H. erectus* bridged the gap between the habilines and ourselves. See Figure 7.6.

# HOMO ERECTUS IN INDONESIA

## DUBOIS' DREAM

The first *H. erectus* fossils that came to light were unearthed in Asia, in the Dutch East Indies (now Indonesia) on the island of Java. They were found by Eugène Dubois (1858–1941), whose dream of discovering prehistoric human remains dated back to his high school years in Holland. After graduating in medicine at the University of Amsterdam, and then briefly teaching anatomy at the University of Leiden, Dubois set out to realize his dream. He enlisted as a medical officer in the Dutch East India Army, and in 1887 he was posted to Sumatra where he suspected that he would find the "Missing Link of Darwin" — a creature that possessed a combination of ape-like and human-like traits. See QQ 7.7.

## 7.7 DUBOIS' DREAM

**Edmund White explains the origin of Dubois' dream and how he set out to accomplish his childhood ambition.**

• • •

In the early 1870s a distinguished German scientist toured Holland, lecturing on the new theory of man's descent from the prehistoric apes. The talks stirred great interest, but they also provoked doubt and opposition. At Roermond, a young high school student named Eugène Dubois went to hear the lecture. It was a fascinating theory, the boy thought, even if there were no facts to prove it . . . .

In this atmosphere of often vituperous debate over man's origins, Eugène Dubois completed his schooling and began an academic career as instructor in anatomy at the Royal Normal School in Amsterdam. His professional specialty qualified him to take a more than casual interest in what was then the most prominent of scientific controversies, and he was fascinated by the many different family trees that were being published in learned — and popular — journals. But he also realized that armchair speculation would never really prove anything about man's distant past. To establish man's place in evolution, someone would have to find a fossil of a primitive creature that was the clear forerunner of man on earth.

At the age of 29 Eugène Dubois set out to find such a fossil. By then a lecturer at the University of Amsterdam, he dismayed his colleagues by taking off for the wilds of Sumatra, where, he said, he had reason to believe he would solve the mystery of man's origins. While his elders on the faculty sadly protested this touch of madness in an otherwise sober and promising young man, Dubois' departure in 1887 marked the start of the greatest manhunt in the history of science — the search for our ancestors.

Source: Edmund White, *The First Men* (New York: Time, Inc., 1973), pp. 33–34. From *Emergence of Man: The First Men.* By the Editors of Time-Life Books. © 1973 Time-Life Books Inc.

## PITHECANTHROPUS ERECTUS

During his first two years in Sumatra, Dubois was stationed at a small hospital in the interior of the island. Although his medical responsibilities were light, giving him plenty of time to search for fossils, he turned up nothing of interest. However, in 1890, after an attack of malaria, Dubois was placed on inactive duty in the Dutch colony of Batavia on the neighbouring island of Java, and it was there that he made his famous finds.

With the backing of the colonial government, which provided him with a team of local convict labourers and two Dutch guards, Dubois began a systematic search along the banks of the Solo River. His efforts were quickly rewarded. In 1891 Dubois instructed his labourers to dig into the bank of the Solo River near the village of Trinil (see Figure 7.7), where the river runs through a plain covered with volcanic debris. To Dubois' delight his workers unearthed a fragment of a lower jaw, a faceless skullcap, a left thighbone, and a right molar tooth — all of them somewhat ape-like but also unmistakably human. Based on these fragments Dubois concluded that he had discovered "the transition form which in accordance with the teachings of evolution must have existed between man and the anthropoids" — in other words, the "missing link" he was seeking. He named the Javanese fossils *Pithecanthropus erectus*, which means upright or erect ape-man: "pithecos" from the Greek for ape, "anthropos" meaning man, and "erectus" meaning upright or erect. Today the finds are classified as *H. erectus*, which means upright or erect human.

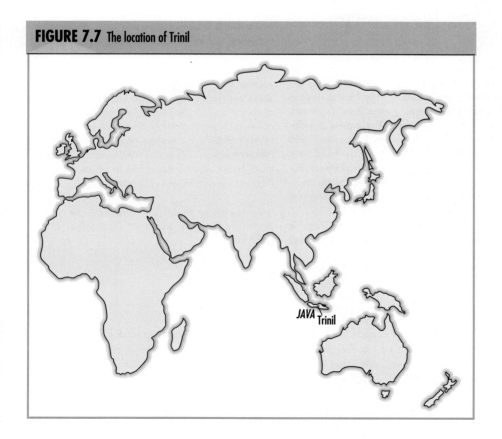

**FIGURE 7.7** The location of Trinil

## A COLD RECEPTION

If Dubois anticipated a hero's welcome when he returned to Europe from Java in 1895 to exhibit the fossils, he was sorely mistaken. Even before he set sail, Christian theologians denounced his find on the grounds that it supported Darwin's view of the origin of humankind. And while Dubois' fellow scientists were more sympathetic, few of them agreed that his discovery was a transitional form midway between apes and modern humans. Undaunted by the cold reception, Dubois attempted to gain support for his conclusion by displaying the fossils at scientific conferences and by describing them in scholarly journals. Yet despite his best efforts, the majority of scientists continued to reject his view, some arguing that Dubois' specimen was an ape, others that it was human, but scarcely anyone willing to agree with Dubois' interpretation. See QQ 7.8.

## 7.8 DUBOIS AND THE SCIENTIFIC COMMUNITY

Loren Eiseley comments on the harsh reaction to Dubois' discovery by virtually all of his scientific colleagues.

• • •

With the discovery in 1891 of *Pithecanthropus erectus* by Eugène Dubois, the first human type of genuinely low cranial capacity was revealed . . . . By this time much of the public outcry which had greeted Darwin's *Descent of Man* in 1871 had died down. The doctrine of evolution had been widely disseminated, discussed, and accepted in intellectual circles. The time would have seemed ripe for a clinching paleontological demonstration of the pathway of human descent. Unfortunately, however, the face of the Java hominid was missing and . . . distrust . . . at the . . . discovery emerged . . . .

At the Third International Zoological Congress which met in Leyden in 1895 Dubois exhibited and discussed his find. The zoologists present maintained that the skull was human and the human anatomists maintained it to be that of an ape . . . . We have the testimony of Marsh that in the beginning with the exception of Manouvrier in Paris and himself, no one took Dubois' claims at their full valuation. "Among a score or more of notices," he writes, "I do not recall a single one that . . . admitted the full importance of the discovery . . . ." "M. Dubois," Manouvrier ironically observes, "can congratulate himself on seeing placed in relief at Berlin the reasons according to which his Pithecanthropus could not be a man and, in England, much better reasons according to which the same Pithecanthropus could not be a monkey."

Source: Loren Eiseley, *Darwin's Century: Evolution and the Men Who Discovered It* (New York: Doubleday & Company, Inc., 1958), pp. 279–280.

## AN APOCRYPHAL TALE

Anthropological lore has it that Dubois was so disgusted by the negative reaction of his colleagues that he gathered his specimens, placed them in a strongbox, and then buried the strongbox under the floorboards of his dining room at his home in Haarlem, Holland. The fossils remained there, according to this version of events, until 1923 when Dubois finally agreed to let the remains be analyzed by palaeoanthropologists from the United

States. The story ends with the American palaeoanthropologists confirming that Dubois' original conclusion about the fossils was correct, but with Dubois announcing that the remains that he had travelled around the world to discover were actually those of a giant gibbon — an argument that he championed until he died in 1941.

The story is apocryphal. What actually happened was that the intransigence of his colleagues encouraged Dubois to turn his attention away from *Pithecanthropus* to other scientific matters, particularly the relationship between brain size and body size, a topic that the Dutch scholar pursued with zeal for the remainder of his scientific career. It was this pursuit that led Dubois to conclude that brain size doubles in relation to body size in conjunction with the appearance of more advanced evolutionary forms. Given this albeit mistaken conclusion, Dubois may have been speaking allegorically rather than literally when he suggested that his specimen was a giant gibbon. He may simply have meant to suggest that *Pithecanthropus* possessed an ape-like body and a human-like brain befitting its evolutionary status midway between apes and humans. In any event, *Pithecanthropus* was not an ape or an ape-man but a human, a fact that was confirmed by scholars such as Ralph von Koenigswald, who went to Java in 1937 and subsequently discovered the remains of about 40 *H. erectus* individuals from an area known as Sangiran, which is about 65 kilometres west of Trinil.

# HOMO ERECTUS IN CHINA

Another group of *H. erectus* fossils, this time from China, have an equally interesting history. The story begins in 1899 when a German physician, K. A. Haberer, who was an avid collector of fossil teeth, purchased a fossil tooth in a pharmacy in Beijing (Pekin or Peking), China, where such teeth, called "Dragon Bones" by the Chinese, were ground into powder and used for medicinal purposes. Haberer subsequently sent the tooth to the University of Munich for analysis where it was determined that the specimen was "a left upper third molar, either of a man or a hitherto unknown anthropoid ape," and this sparked an interest in the site where the specimen was found — **Zhoukoudian** (formerly spelled Chou Kou Tien and Choukoutien) — a cave site approximately 40 kilometres southwest of Beijing, the present capital of the People's Republic of China. See Figure 7.8.

## SINANTHROPUS PEKINENSIS

In 1924, at the same site, John Gunnar Andersson and Otto Zdansky made another important find. The two scholars had been sent to China by the Swedish Academy to search for fossils, and what intrigued them about Zhoukoudian was the presence of sharp pieces of quartz in the local limestone deposits. Andersson suspected that the quartz represented the remains of primitive tools that ancient humans had used in the cave in the distant past, so

**FIGURE 7.8** The location of Zhoukoudian

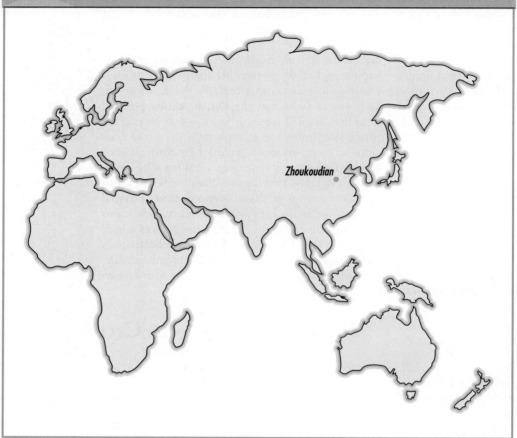

Zhoukoudian

he and Zdansky began to excavate. Within a few months they had filled several boxes with ancient materials, and in 1925 Zdansky returned with these to Sweden while Andersson remained behind.

When Zdansky inspected the collection in Sweden he came across two unusual teeth — a lower left first premolar and an upper right third molar, both of which seemed to be human but which were nonetheless quite primitive. Were these the teeth of an ape or had they belonged to a prehistoric human? Zdansky believed that the teeth were human, and in order to confirm his suspicion he sent the teeth back to China, to the person that he and Andersson believed was best equipped to make the ultimate determination. This was Davidson Black (1884–1934), a Canadian anatomist who had come to China in 1919, officially to join the staff of the Peking Union Medical College, but unofficially to search for the remains of ancient humans. In 1926,

# 7.9 ZHOUKOUDIAN

Chang Senshi and his colleagues from the Chinese Academy of Sciences describe Zhoukoudian.

• • •

At Zhoukoudian not far from Beijing . . . [there is a cave that] measures 140 meters in length . . . and up to 40 meters in breadth . . . . It is a veritable treasure-trove of man's early culture.

Vast changes have taken place about this site since liberation. On Longgushan Hill today stands . . . [a m]useum, surrounded by trees . . . . An exhibition center disseminating knowledge about the history of social development, this site attracts a stream of visitors from all over the world.

Source: Chang Senshi, *et. al. Atlas of Primitive Man in China* (Beijing: Science Press and Litton Educational Publishing, Inc., 1980), p. 32.

Neg. 33579, Courtesy Department Library Services, American Museum of Natural History

▼

The cave floor at Zhoukoudian

▲

shortly after he had received the specimens from Zdansky, Black also concluded that the fossil teeth were human. Black, however, was worried that no other fossils would be found at Zhoukoudian since Andersson was about to return to Sweden, so he appealed to and convinced the Rockefeller Foundation to finance more extensive research at the site. See QQ 7.9.

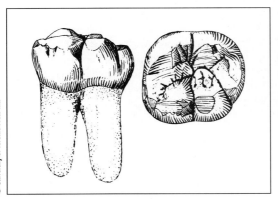

From page 32 of *Homo erectus: Papers in Honor of Davidson Black* edited by B.A. Sigmon and J.S. Cybulski, 1981. © University of Toronto Press Inc.

▼

An artist's rendering of the tooth discovered by Birger Bohlin (twice actual size)

▲

In 1927 another early hominid tooth was discovered at Zhoukoudian by Birger Bohlin, a young Swedish palaeontologist who was working under Black's direction. Bohlin gave the tooth to Black who immediately took the bold step of naming the new discovery *Sinanthropus pekinensis* — Chinese human of Pekin. Two years later, Pei Wen-Chung, another colleague of Black's, unearthed a nearly complete juvenile *Sinanthropus* skullcap at the site. Black released the news to the press. His bold step had paid off: like Dubois before him, Black too had correctly identified the remains of *H. erectus*. See QQ 7.10.

## 7.10 DAVIDSON BLACK AND *SINANTHROPUS PEKINENSIS*

Harry Shapiro writes about Davidson Black and *Sinanthropus pekinensis*.

• • •

In the history of the study of human evolution there are a series of associations that have become fixed in the minds of all . . . who know something about the subject. *Pithecanthropus erectus*, for example, and his discoverer, Eugene Dubois, are inevitably linked. Raymond Dart and *Australopithecus africanus* go together, as does Robert Broom with *Paranthropus* . . . and Louis Leakey with *Zinjanthropus* and *Homo habilis*. There is still another association that has a world-wide acceptance. This is Davidson Black and Peking Man.

In many, but not all of these associations, the linkage is partly the result of the acclaim that comes to the discoverer of an important fossil. In a few instances, however, it is the consequence of the recognition and identification of a fossil's true significance and the marshalling of the evidence to demonstrate this judgement. Davidson Black belongs in this category. He was not responsible for the discovery of Peking Man. That was the result of the efforts of J. Gunnar Andersson, who planned the explorations at Chou Kou Tien, and of Otto Zdansky, Birger Bohlin, and Pei Wen-Chung among others, who did the field work. But when Zdansky at Upsala, where the Chou Kou Tien fossils were sent for study and identification,

found among the miscellaneous fragments a couple of teeth that seemed to have primate or possibly hominid features, they were returned to Peking and referred to Davidson Black for his judgement. This was in 1925–26 . . . .

With only a few exceptions, the wisdom of turning to Black became immediately apparent to his colleagues. But in scientific centers around the world, where these matters were of prime importance, this recognition was not so prompt. Black published in 1926, in Nature and in Science, his conclusions that these teeth were hominid. Subsequently, he went even further to identify them as relics of a new type of early man, the oldest ever found in China or the Asiatic mainland, going back in age to the beginning of the Pleistocene, possibly, as dated then, to a million or more years ago. Considering the paucity of the evidence, it was with notable confidence that he ventured in 1927 to give it the distinctive name Sinanthropus pekinensis . . . .

But when in 1928 cranial and mandibular fragments and in 1929 a well preserved skull completely vindicated Black's identification, an . . . exciting wave of surprise circled around . . . [that] permeated with admiration for the perception and skill of Davidson Black.

Source: Harry L. Shapiro, "Davidson Black: An Appreciation." In Becky A. Sigmon and Jerome S. Cybulski, eds, *Homo erectus: Papers in Honor of Davidson Black* (Toronto: University of Toronto Press, 1981), pp. 21–23.

## An Unsolved Mystery

When Black died suddenly in 1934, the Rockefeller Foundation selected Franz Weidenreich to carry on the work at Zhoukoudian. Also an anatomist, and a refugee from Nazi Germany, Weidenreich arrived in Beijing in 1935. Anticipating the Japanese invasion of northern China in 1937, he plunged

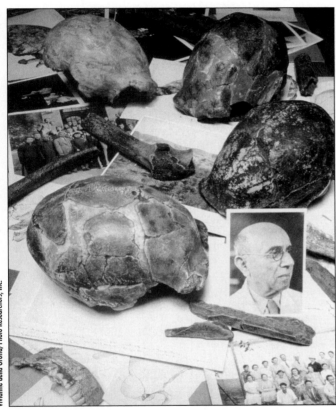

▼

Casts of *H erectus*
and a photograph of
Franz Weidenreich

▲

Vivianne della Grotta/Photo Researchers, Inc.

into the intensive work of making plaster casts of the fossils, drawing exact reproductions, and photographing and describing the specimens in detail. By the time Weidenreich left China for the United States, the Zhoukoudian finds included five skulls, nine skull fragments, six facial bone fragments, fifteen lower jaws, seven fragmented limb bones, and one hundred and fifty-two teeth — in all, the remains of forty-five individuals. In addition, thousands of fragments of stone and bone tools had been recovered.

It was fortunate that Weidenreich was so meticulous for, somehow, between 5 and 7 December 1941, on a train from Beijing to Chinwangtao, where the liner *President Harrison* was waiting to transport the specimens to the United States, they disappeared. When the small U. S. Marine detachment that was escorting the remains to Chinwangtao was arrested by Japanese soldiers on 7 December 1941, the fossils were nowhere to be found. What happened to them is still a mystery (see QQ 7.11). Although many leads have been followed up, the original Zhoukoudian specimens have never been recovered. Nevertheless, on the basis of Weidenreich's careful study, the Chinese *H. erectus* casts, which are housed in the American Museum of Natural History, constitute an important part of the fossil record.

## 7.11 AN UNSOLVED MYSTERY

There are many possible ways to explain the disappearance of the Chinese *H. erectus* fossils.

• • •

[W]hat happened to the bones is anyone's guess. And it has been almost everyone's guess. That they were actually put on a lighter [or loading boat], which capsized in the harbor. That local Chinese officials got them and ground them up for medicine. There is even an opium-scented story that they were smuggled safely away by sinister international merchants of the China coast and finally tracked down and purchased by an American medical man who lives in California and guards the hoard like Fafnir, the great worm . . . . [T]he press in Japan and in Hong Kong periodically erupted with developments of this tale for some years . . . .

So here we have Pekin Man, world traveler, associate of merchant pirates, or pawn of international

politics . . . . [However, i]t is unlikely that he ever left the environs of Chinwangtao. He is, if there is anything left of him, probably there today. The judgement of men who were in China in 1941 is likely to be the right one. Japanese officialdom was hot on the scent of the Pekin bones, but it was not Japanese officials who captured the Marines and looted the train with its secret freight at Chinwangtao. It was ordinary Japanese soldiers. And they doubtless did what ordinary Japanese soldiers, or ordinary American soldiers, would do in a captured train. Expecting something interesting and finding what looked like dog bones, they probably threw the whole lot onto the trash heap or over the dockside.

Source: William Howells, *Mankind in the Making: The Story of Human Evolution*, revised edition, (New York: Doubleday & Company, Inc., 1967), pp. 168–170.

# *HOMO ERECTUS* IN AFRICA AND POSSIBLY IN EUROPE

Because *H. erectus* fossils were first discovered in Asia, it was once thought that this is where the species originated. The idea is mistaken. *H. erectus* fossils have since been unearthed in Africa and also possibly in Europe, and the African specimens include the oldest found to date. The most ancient of these, which come from the Turkana basin in Kenya, are between 1.8 million years old and 1.7 million years old.

Nearly as old is the almost complete skeleton of a twelve-year-old boy from West Lake Turkana discovered in 1984 by Richard Leakey and his colleagues. Known technically as **KNM-WT 15000**, but often called the "Turkana boy," the specimen is about 1.6 million years old (see QQ 7.12). There is also a *H. erectus* braincase from Olduvai Gorge that is about 1.2 million years old. By way of contrast, the Asian fossils, which now include more specimens from sites in Indonesia and China, and others from sites in Vietnam and India, are estimated to be between one million years old and 200 000 years old. As for the European fossils and the tools that *H. erectus* may have left behind, although dates between 500 000 years old and 300 000 years old have been proposed, many of these are in dispute (see Table 7.1). Equally important, some palaeoanthropologists do not believe that the European discoveries should be classified as *H. erectus*; they contend that the finds represent ancient *H. sapiens*.

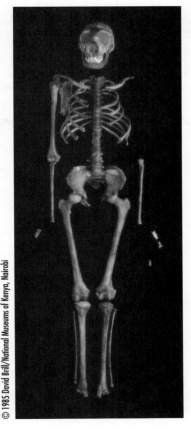

The "Turkana boy"

© 1985 David Brill/National Museums of Kenya, Nairobi

## QQ 7.12 KNM-WT 15000

Frank Brown, John Harris, Richard Leakey, and Alan Walker explain the evolutionary significance of the Turkana boy.

• • •

With the possibility of finding more parts when the excavation is continued and with further analysis, KNM-WT 15000 will provide a firmer foundation for the identification and understanding of unassociated postcranial bones and early hominid growth patterns. Except for those representing the latest stages of our evolution, it is also the first fossil hominoid, let alone hominid, in which brain size and body size can be measured accurately on the same individual. These two crucial variables, on which so much speculation about human origins and behaviour has been based, can now be determined for at least one individual early hominid.

Source: Frank Brown, John Harris, Richard Leakey, and Alan Walker, "Early *Homo erectus* skeleton from west Lake Turkana, Kenya," *Nature*, Vol. 316, No. 6031. Reprinted with permission from *Nature*. Copyright 1985 Macmillan Magazines Limited. pp. 792.

**TABLE 7.1**    The location and age of selected *H. erectus* remains discovered in Africa, Asia, and possibly Europe

| CONTINENT | COUNTRY | AGE OF OLDEST REMAINS (mya) |
|---|---|---|
| Africa | Kenya | 1.8–1.7 |
| | Tanzania | 1.2 |
| | Algeria | 0.4–0.7 |
| | South Africa | 0.5 |
| | Ethiopia | 0.2–0.5 |
| Asia | Indonesia | 1.0 |
| | China | 0.5 |
| | Vietnam | 0.4 (?)[1] |
| | India | 0.2 |
| Europe[2] | Hungary | 0.5 (?)[1] |
| | Spain | 0.5 (?)[1] |
| | Greece | 0.3 |
| | Germany | 0.3 |
| | France | 0.3 (?)[1] |

[1] (?) indicates a disputed age.
[2] Some palaeoanthropologists maintain that the European remains are not those of *H. erectus* but those of early *H. sapiens*. The reason is that the European fossils exhibit traits that are characteristic of both species.

# OUT OF AFRICA

Although *H. erectus* may not have resided in Europe, the locations where their fossils have been found make it clear that they lived not only in Africa but also in Asia. In fact, based on what is known about the distribution and age of the fossils, most anthropologists have concluded that after establishing a foothold in eastern and then in southern Africa during the first half of the Pleistocene epoch (*c.* 1.8 mya – 1.0 mya), *H. erectus* began to spread into the Middle East and from there moved eastward into Asia and perhaps northward into Europe. Although precise migration routes have yet to be established, the move was almost certainly accomplished by a process known as **fission**, in which small groups of individuals slowly drifted away from somewhat larger social units, some of the new groups surviving and others perishing. However they moved, *H. erectus* was the first hominid who ventured out of Africa. See QQ 7.13.

## 7.13 THE SPREAD OF HUMANITY

G. Philip Rightmire summarizes what is known about the movement of *H. erectus* out of Africa.

• • •

The earliest traces of *Homo erectus* have been uncovered in eastern Africa, where fossils from the Turkana basin are 1.7 to 1.8 million years in age. Later members of this species reached other parts of the continent, and probably *Homo erectus* had begun to disperse out of Africa before 1.0 million years ago. One route would have taken archaic people around

the eastern Mediterranean, to Europe and southwestern Asia. Assemblages of crude stone tools are found in Europe, although no human bones are known from these Early Pleistocene localities. The spread of *Homo erectus* eastward into Asia is poorly documented, but it is clear that populations were present in Java at an early date, and the famous caves at Zoukoudian were inhabited nearly 0.5 million years ago.

Source: G. Philip Rightmire, "The Dispersal Of *Homo Erectus* From Africa And The Emergence Of More Modern Humans," *Journal of Anthropological Research*, Vol. 47, No. 2 (1991), p. 177.

# LINGUISTIC SKILLS

Unfortunately, as is the case with the habilines, there are still many questions about *H. erectus* that remain unanswered. Consider their linguistic skills. One of the cultural breakthroughs that is sometimes associated with *H. erectus* is the use of spoken language to communicate. Already in *H. habilis* (although not among the australopithecines), there is evidence that the area of the brain that makes it possible for our kind to utter syllables and comprehend spoken words was present. Known as **Broca's area**, it is located in the left cerebral cortex towards the front. The same area was also present in *H. erectus*.

A specimen known as **Zhoukoudian cranium 5** has also shed light on the problem; the interior of its skull shows imprints of the same neural connections between the front and the back portions of the brain that make it possible for modern humans to receive, interpret, and send linguistic information in the form of spoken language. See QQ 7.14.

## 7.14 THE LINGUISTIC SKILLS OF *H. ERECTUS*

Milford Wolpoff looks at the linguistic skills of *H. erectus*, and, in particular, of the specimen known as Zhoukoudian cranium 5.

• • •

Human speech capacity is an extremely complicated behaviour, involving the development of tracts between the motor and associated speech areas and between the frontal and posterior parietal [or side and top] portions of the brain. The basis for speech ability would seem to lie more in neurological structures than in morphological ones. The importance of the frontal-posterior parietal tract is that it allows the frontal area to "make sense" out of the cross-modal associations of the posterior parietal area, while the motor-associated speech area tract provides a pathway for this information to reach the motor-speech area.

The evolution of language ability seems tied to the appearance of hemispheric dominance and asym-metry. The marked bilateral asymmetry in size and morphology in . . . Choukoutien cranium 5 provides one of the few direct morphological correlates with the ability to speak in humans. Thus, there is every reason to believe that *Homo erectus* was capable of human vocal language.

Source: Milford H. Wolpoff, *Paleoanthropology* (New York: Alfred A. Knopf, Inc., 1980), p. 206. Reproduced with permission from McGraw-Hill.

And if *H. erectus* did use spoken language to communicate, then the repercussions would have been enormous. For one thing, spoken language would have enabled them to share ideas ranging from how to make stone tools to how best to organize their social affairs, and this would have accelerated the development of culture. In addition, spoken language may have allowed *H. erectus* to "speak to themselves" as we are apt to do — a talent that some scholars say is intimately associated with the intellectual ability to distinguish between the self and the external environment — and this too would have accelerated the development of culture. But the evidence that *H. erectus* used spoken language is not incontrovertible. While it may be true that *H. erectus* possessed the biological attributes that are necessary to produce and understand spoken words, this does not mean that they employed spoken language to communicate. They may have used other means to communicate.

# TECHNOLOGICAL SKILLS

## ACHEULIAN TOOLS

The technological skills of *H. erectus* are likewise controversial. From about 2.5 mya until about 1.5 mya Plio-Pleistocene hominids and then early Pleistocene humans manufactured and used Oldowan tools. About 1.5 mya, however, these roughly fashioned core choppers and flake knives began to give way to more sophisticated stone implements. These are **Acheulian** (also spelled Acheulean) tools, large oval-shaped stone cores and flakes that are characteristically double-sided or bifacial. Made first in Africa and then in Asia and in Europe, Acheulian tools are typically found in abundance in sites that were or are closely associated with waterways or wetland environments. The tools were named after the village of St. Acheul in northwestern France, where examples were first unearthed in the nineteenth century.

Although Acheulian toolmakers fashioned picks and cleavers, by far the most remarkable tool they made was the **Acheulian hand axe** — an almond-shaped or teardrop-shaped stone implement between 12 centimetres and 15 centimetres long with sharp edges around the perimeter (see Figure 7.9).

Initially made by striking stone against stone, as time passed Acheulian hand axes were also manufactured by the **soft hammer technique**, in which flakes were carefully removed from a core by striking it with a "soft" hammer fashioned from either bone, antler, or wood. This technique allowed the toolmaker to manufacture the hand axe with an increasing eye to both beauty and precision.

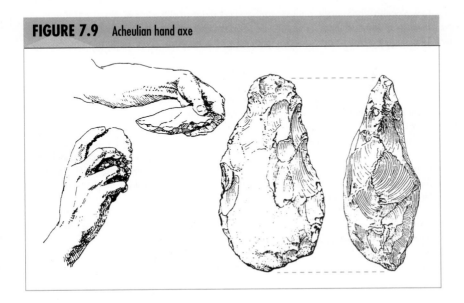

**FIGURE 7.9** Acheulian hand axe

How were these tools employed? Microscopic examination of the cutting edges on some hand axes indicates that they were used for butchering and woodworking, and experiments have shown that they could have been used as projectile weapons that were thrown either overhand like a baseball or sidearm like a discus (see QQ 7.15). Most anthropologists, however, believe that Acheulian hand axes were multi-purpose tools that were used for a wide variety of purposes including cutting, scraping, digging, and pounding.

 **7.15 THE ACHEULIAN HAND AXE**

Eileen O'Brien explains how she believes the Acheulian hand axe was used.

• • •

If we let the evidence speak for itself, the appropriate question is: What task would require force, call for a tool with a sharp edge around all (or most) of its perimeter but without a safe handhold, occur in or near water, and often result in the loss of a potentially reusable artifact? The possibility that occurred to me is that the hand ax was a projectile weapon. The idea, I have since discovered, has been thought of before . . . [but] has not taken hold, probably because it is not obvious how the larger hand axes could have been thrown . . . .

A few years ago, I decided that a practical experiment was what was needed. From my limited knowledge of track and field, I thought that for sidearm throwing, an analogy might be made between a hand ax and the Olympic discus.

The experiment took place in 1978, in the discus practice area at the University of Massachusetts, where I was then a student. Two student athletes participated . . . [one] a discus thrower, and . . . [the other] a javelin thrower . . . .

When grasped and thrown overhand, like a knife, the experimental hand ax performed like one, rotating symmetrically on edge in both ascent and descent. The average throw was just short of discus-style [that is, reaching a distance of about 30 metres], but more accurate, about . . . [500 centimetres] right or left of the line of trajectory [instead of about two metres]. It always landed on edge, but less often point first. Unfortunately, these results are the product of only six throws; owing to its weight and the ovate, broad point, the experimental hand ax was difficult to grasp and throw overhand . . . . This overhand style would probably be more suitable for lighter, more triangular hand axes [than the experimental one weighing about two kilograms]. In contrast, weight and shape were of no real concern when throwing the hand ax discus-style [which was done forty-five times]. Even a significant increase in weight might not have impeded the throwing motion, although it would have affected the distance of the throw.

Further testing is needed . . . but these first trials showed that a hand ax could perform appropriately as a projectile. The hand ax demonstrated a propensity to land on edge when thrown overhand or discus-style, a tendency to land point first, and a potential for distant and accurate impact. Its overall shape minimizes the effects of resistance while in flight, as well as at impact. This is not true of an unshaped stone or spheroid, for example. And despite its sharp edge, the hand ax could be launched without a safe handhold. The only apparent limitations to the hand ax's use as a projectile weapon are the strength, coordination, and skill of the thrower.

Source: Eileen M. O'Brien, "What Was the Acheulean Hand Ax? *Natural History* (Volume 93, Number 7, 1984), pp. 20–24.

Since Acheulian tools are often found in association with the remains of *H. erectus*, and since *H. erectus* undoubtedly made and used Acheulian tools, it might be assumed that *H. erectus* was the inventor. But the identity of the first Acheulian toolmakers, like the identity of the first Oldowan toolmakers, has yet to be confirmed: in the Turkana basin in Kenya, where the oldest *H. erectus* fossils have been discovered, there are no Acheulian tools; at Sterkfontein in southern Africa, Acheulian tools have been found in association with *H. habilis* but not with *H. erectus*; and at Olduvai Gorge in Tanzania, where the oldest Acheulian tools have been found, *H. habilis* and *H. erectus* are both present. Under the circumstances, it is impossible to determine who the original Acheulian toolmakers were.

## FIRE

Another technological breakthrough that is frequently attributed to *H. erectus* is the first use of fire, which also can be regarded as a tool. But despite the fact that the contents of some of the sites where *H. erectus* lived indicate that they used and controlled fire 300 kya or more, the evidence for first use is unclear. *H. habilis* and perhaps even the australopithecines may have used fire before *H. erectus*; there are Pliocene sites in Africa whose contents indicate that the use of fire may date as far back as 2 mya. Nor is it known how *H. erectus* captured fire. Some say that it was acquired from wildfires that were created by volcanic eruptions or by lightning strikes, while others maintain that it was taken from fires that toolmakers accidentally sparked in the debris that surrounded their workplaces. See QQ 7.16.

## 7.16 THE USE OF FIRE

J. Desmond Clark comments on the use of fire by *H. erectus* and perhaps also by earlier hominids.

. . .

It has . . . been suggested that it was the ability to make effective use of fire that enabled *H. erectus* and his suc-

cessors to live in the cold and temperate lands of Eurasia. Evidence of fire in the form of charcoals and burned bone occurs in Europe and the Far East by 300 000 BP [that is, before the present,] and not much later in Africa . . . . However, at more ancient open sites in the African tropics, where charcoal is not preserved . . . indirect evidence survives — burned earth or fire-

Some possible uses of natural fire by early hominids

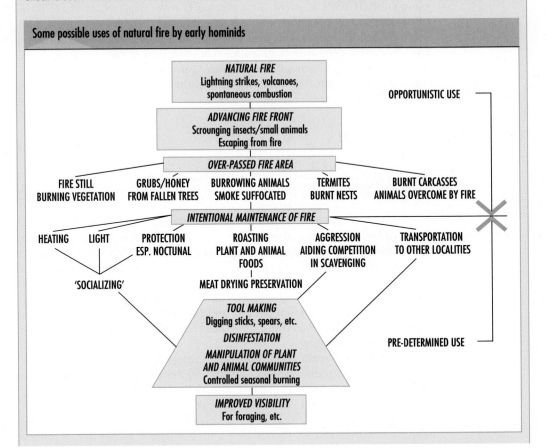

In either event, the controlled use of fire had a great impact on the way *H. erectus* lived. At the very least it would have allowed these ancient people to better defend themselves against predators. *H. erectus* also may have used fire to harden the tips of wooden spears and perhaps even to stampede big game animals such as prehistoric elephants into areas where these were killed, and such applications would have enhanced their ability to hunt. Fire also provided *H. erectus* with heat and light in a homeland whose northern limits bordered what were then the ice-bound regions of Eurasia, and where, in winter, the mean annual temperature was about three degrees Celsius lower than it is today. The ability to control fire may even have made it possible for *H. erectus* to venture out of Africa.

# SETTLEMENT PATTERNS AND SUBSISTENCE SKILLS

## TERRA AMATA

Finally, there is a quartet of sites whose contents have raised questions about the settlement patterns and subsistence skills of *H. erectus*. One of these is **Terra Amata** (beloved land), an alleged seasonal camp that was established sometime between 300 kya and 200 kya in what is now the city of Nice on the French Riviera (see Figure 7.10). Situated on the slopes of a hill that is now completely covered with modern buildings, Terra Amata was excavated by Henri de Lumley between January and July 1966, while the construction workers who had discovered the site waited for de Lumley to finish his work before resuming their own.

Although no human bones were found there, about 35 000 objects were recovered, including abandoned Oldowan and Acheulian tools and the bones of shellfish and fish, and of big game animals such as elephants, rhinoceroses, wild oxen, deer, and wild boars. The most striking find at Terra Amata, however, was what appeared to be the remains of 21 oval huts estimated to be between 8 metres and 15 metres long, and between 4 metres and 6 metres

**FIGURE 7.10** The location of Terra Arrata

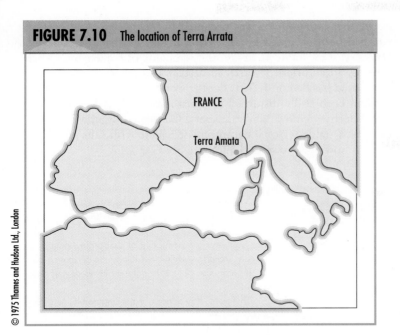

FRANCE

Terra Amata

wide. Almost all of the huts were equipped with a central hearth or fireplace and each was thought to be capable of housing between 10 and 20 people. It was also reported that the remains of the huts and the debris that was found in and around them indicated that *H. erectus* hunters had visited Terra Amata intermittently on an annual basis and had occupied the site off and on for more than a decade.

But what has been alleged and what has been proved about Terra Amata are quite different. Whether the site should be attrib-uted to *H. erectus* is ques-tionable; those who doubt that *H. erectus* resided in

Terra Amata: (top) excavated hearth: (middle) rendering of the hearth in use: (bottom) rendering of the exterior of the hut

Europe believe that Terra Amata was occupied by *H. sapiens*. And even if Terra Amata had been occupied by *H. erectus*, it may be that the bones of the big game that were found there are the remnants of scavenging rather than of hunting. In addition, research undertaken since Terra Amata was originally excavated has demonstrated that the stone tools and animal bones found there had been badly disturbed by horizontal and vertical displacement and by selective removal and transport. Consequently, the idea that Terra Amata was a stable seasonal encampment replete with huts equipped with fireplaces is now open to serious question. See QQ 7.17.

## 7.17 TERRA AMATA

Paola Villa explains what is and what is not known about Terra Amata.

• • •

[Terra Amata] is a site where material was found disturbed into a number of levels that probably correspond to several episodes of occupation. It appears, however, that in many areas of the site the levels were not stratigraphically isolated, and thus stratigraphic subdivisions are in large measure arbitrary. Vertical displacement of materials, as well as other site formation processes, have resulted in the partial mixing of residues of repeated occupations. The interpretive problem has been compounded by a complex stratigraphy, dissected with appropriate but time-consuming methods that, given the hurried pace of the excavation, contributed to the presence of understandable gaps in the records of a salvage dig. Some associations and structural features were preserved, but these are not traceable over the whole site. Therefore, spatial analysis of levels, as excavation has revealed them, would have to be treated with caution since

there is always the possibility of associating nonsynchronous materials while overlooking the attribution of contemporaneous objects to different levels during the excavation.

Therefore, it is difficult, if not impossible, to give precise answers to questions regarding (1) the overall layout of areas of different activities, (2) the association of tools either with structural features or between themselves, and (3) the length of occupation. Even the size of the living area at any point in time is uncertain. We know the maximum dimensions of the site, as these are indicated by the cumulative record of superimposed levels; however, the contemporaneity of different parts of the site at any given level is far from certain, since there are no long-distance links to demonstrate it. The postulated life-way of the Terra Amata hunters [that] has been described in detail by de Lumley . . . [and in some] textbooks . . . should be considered as largely speculative because they are based on ambiguous or inadequate data.

Source: P. Villa, "Conjoinable Pieces and Site Formation Processes." *American Antiquity,* Vol. 47, No. 2 (1982) p. 285.

## ZHOUKOUDIAN

The remains that have been found at Zhoukoudian, the cave home of *Sinanthropus pekinensis*, are likewise controversial. That Zhoukoudian was a seasonal camp is almost certain. There, intermittently between about 500 kya and about 250 kya, *H. erectus* lived in a cave heated and protected by fire. Since most of the non-human animal bones found in the cave were

▼

Workers unearthing
a *H. erectus* skull at
Zhoukoudian

▲

those of deer, it was once thought that the residents hunted mainly big game. In addition, it was alleged that they practised cannibalism since almost all of the human bones that were unearthed in the cave were broken, including the skulls, which may have been opened to extract the brain. However, a taphonomic study of the bones has cast doubt on these conclusions. While the study does not contradict the view that Zhoukoudian was a seasonal camp, it points to gnawing predators such as hyenas and wolves as the source of the broken human bones. Equally important, the study also suggests that the same predators were probably responsible for the non-human bones in the caves, calling into question what were once thought to be the formidable hunting skills of *H. erectus*. See QQ 7.18.

## 7.18 BIG GAME HUNTING AND CANNIBALISM AT ZHOUKOUDIAN

Lewis Binford and Chuan Kun Ho take exception to the claims that *H. erectus* at Zhoukoudian hunted big game and practised cannibalism.

• • •

In the half-century since Pei discovered the first skull of Beijing (Peking) man, the Beijing hominids have consistently been depicted as "primitive hunters who may have preferred to bring prey back to their cave and share the meat with others in the community rather than to consume it where it was killed."

. . . Not long after the reporting of these finds, the interpretation that this evolutionary ancestor had [also] been a cannibal began to appear. For instance, in a review of evidence for intrahuman killing in early time we read that "the lower cave of Choukoutien, Locality 1, may be termed the type site for evidence of murder and cannibalism in the Palaeolithic [or old stone age] as represented by popular literature and even by some prehistorians" . . . .

## TORRALBA AND AMBRONA

Two sites in north-central Spain, about 100 kilometres from Madrid, also deserve special mention in connection with the hunting skills of *H. erectus*. Separated by a narrow passageway through the Ambrona Valley — once a migration route for elephants, horses, and deer — the sites are named **Torralba** and **Ambrona** (see Figure 7.11). Unlike Zhoukoudian and perhaps Terra Amata, neither was a seasonal camp; instead, it was once widely believed that these were **kill sites** — places where big game was slaughtered, butchered, and made ready to carry back to temporary base camps where *H. erectus* slept. It was also reported that elephants were the primary target, and that these were dispatched by cooperating groups of *H. erectus* whose members used fire to stampede the elephants into what was then a boggy marsh. Once trapped in the mud, it was alleged, the elephants eventually tired and fell victim to the hunters' fire-hardened wooden spears and stone cutting tools. The remains of more than 50 elephants

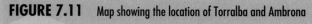

**FIGURE 7.11**   Map showing the location of Torralba and Ambrona

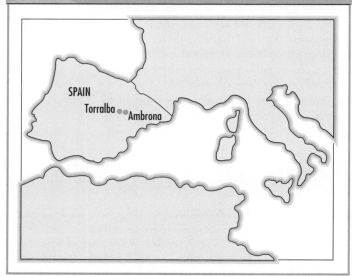

that were unearthed at Torralba and Ambrona seemed to indicate as much, as did the ashes and the Acheulian tools that were discovered at the sites.

However, based on a detailed microscopic examination of cutmarks on replicas of a substantial sample of bones that were recovered from Torralba and Ambrona, it has been determined not only that almost all of the bones have been so damaged by soil abrasion that cutmarks have been obliterated, but also that among those few bones that do exhibit cutmarks there is no significant difference in the number of those damaged by cutting tools and those damaged by carnivore teeth (see QQ 7.19). Thus, as is the case with respect to the animal bones that were found at Zhoukoudian, those discovered at Torralba and Ambrona likewise cannot be regarded as proof positive that *H. erectus* depended mainly on big game for their livelihood. Their hunting skills, like their linguistic skills, their technological skills, and their settlement patterns, remain to be revealed.

 **7.19 TORRALBA AND AMBRONA**

Pat Shipman and Jennie Rose summarize the results of their microscopic analysis of the cutmarks on the bones recovered from Torralba and Ambrona.

• • •

Torralba and Ambrona have been interpreted as butchery sites for many years, a contention recently challenged; natural death or carnivore activities are invoked as an alternative explanation. Scanning electron microscopy (SEM) of bone surface replicas distinguishes among hominid-produced cutmarks, carnivore tooth scratches, and other types of bone damage. A sample of 102 replicas, comprising the most likely cutmarks on a combined sample of roughly 3000 fossils from Torralba and Ambrona, were scanned to determine the major agent of damage. Microscopically verified cutmarks are present, but rare, occurring in less

than 1 percent of the bones in the combined sample. Carnivore tooth scratches are comparably rare. In contrast, evidence of sedimentary abrasion, which obliterates the diagnostic features of cutmarks, is present on nearly every bone from Torralba and Ambrona. It remains unresolved whether cutmarks were initially more common on these bones and were subsequently obliterated by abrasion, or whether the incidence of cutmarks was always low. These data demonstrate clearly that hominids and carnivores each damaged some bones at Torralba and Ambrona, but the frequency of each type of mark is too low to confirm strongly the interpretation of these sites as either butchery or carnivore remains.

Source: Pat Shipman and Jennie Rose, "Evidence of Butchery and Hominid Activities at Torralba and Ambrona; An Evaluation Using Microscopic Techniques." *Journal of Archaeological Science*, Vol. 10, No. 5 (1983), p. 465.

# KEY TERMS AND CONCEPTS

HOMO
HOMO HABILIS (*homo + ha-bil'-us*)
ER 1470
HOME BASES
LIVING FLOORS

HOMO ERGASTER (*homo + ur-gas'-tur*)
ICE AGE
GLACIERS
INTERGLACIAL PERIOD
KEYSTONE HERBIVORE HYPOTHESIS

HOMO ERECTUS (*homo + eh-rek'-tus*)

PITHECANTHROPUS ERECTUS
(*pith-i-kan'-throh-pus + erectus*)

ZHOUKOUDIAN (*zhoo'-kood-yen*)

SINANTHROPUS PEKINENSIS
(*sin-an'-throh-pus + pee'-kin-en'-sis*)

KNM-WT 15000

FISSION

BROCA'S AREA

ZHOUKOUDIAN CRANIUM 5

ACHEULIAN (*a-shoo'-lee-en*)

ACHEULIAN HAND AXE

SOFT HAMMER TECHNIQUE

TERRA AMATA (*terra + ah-mah'-tah*)

TORRALBA (*tor-ahl'-bah*)

AMBRONA (*um-broh'-nah*)

KILL SITES

# SELECTED READINGS

## INTRODUCTION

Blumenberg, B. and Todd, N. "On the Association Between *Homo* and *Australopithecus*." *Current Anthropology*, Vol. 15 (1974), pp. 386–388.

Butzer, K. and Isaac, G., eds. *After the Australopithecines: Stratigraphy, ecology and culture change in the Middle Pleistocene*. Paris: Mouton Publishers, 1975.

## HOMO HABILIS

Johanson, D., *et. al.* "New Partial Skeleton of *Homo habilis* from Olduvai Gorge, Tanzania." *Nature*, Vol. 327 (1987), pp. 205–209.

Lieberman, D., *et. al.* "A Comparison of KNM-ER 1470 and KNM-ER 1813." *Journal of Human Evolution*, Vol. 17 (1988), pp. 503–512.

Potts, R. "Home Bases and Early Hominids." *American Scientist*, Vol. 72 (1984), pp. 338–347.

Susman, R. and Stern, J. "Functional Morphology of *Homo habilis*." *Science*, Vol. 217 (1982), pp. 931–934.

Tobias, P. *Olduvai Gorge* (volume 4) *Homo habilis*. Cambridge: Cambridge University Press, 1989.

## THE PLEISTOCENE EPOCH

Benton, M. *The Rise of the Mammals*. London: Quarto Publishing, 1991,

Imbrie, J. and Imbrie, K. *Ice Ages: Solving the Mystery*. Cambridge: Harvard University Press, 1986.

## HOMO ERECTUS

Howells, W. "*Homo Erectus* – Who, When, and Where: A survey." *Yearbook of Physical Anthropology*, Vol. 23 (1980), pp. 1–23.

Howells, W. "*Homo erectus* in Human Descent: Ideas and Problems." In Sigmon, B. and Cybulski, J. eds. *Homo erectus: Papers in Honor of Davidson Black*. Toronto: University of Toronto Press, 1981, pp. 63–86.

Rightmire, G. *The Evolution of Homo erectus: Comparative Anatomical Studies of an Extinct Human Species.* Cambridge: Cambridge University Press, 1990.

Wood, B. "The Origin of *Homo erectus.*" *Courier Forschunginstitut Senckenberg,* Vol. 69 (1984), pp. 99–111.

## HOMO ERECTUS IN INDONESIA

Deacon, T. "Fallacies of Progression in Theories of Brain-Size Evolution." *International Journal of Primatology,* Vol. 11 (1990), pp. 193–236.

Jacob, T. "Solo Man and Peking Man." In Sigmon, B. and Cybulski, J. eds. *Homo erectus: Papers in Honor of Davidson Black.* Toronto: University of Toronto Press, 1981, pp. 87–105.

Pope, G. and Cronin, J. "The Asian Hominidae." *Journal of Human Evolution,* Vol. 13 (1984), pp. 377–396.

von Koenigswald, G. *Meeting Prehistoric Man.* New York: Harper & Brothers, 1956.

von Koenigswald, G. "Early Man in Java: Catalogue and Problems." In Tuttle, R., ed. *Paleoanthropology: Morphology and Paleoecology.* The Hague Mouton, 1975, pp. 303–309.

## HOMO ERECTUS IN CHINA

Black, D. "Tertiary Man in Asia: The Chou Kou Tien Discovery." *Nature,* Vol. 118 (1926), pp. 733–734.

Black, D., Teilhard de Chardin, P., Young, C., and Pei, W. *Fossil Man in China: The Choukoutien Cave Deposits with a Synopsis of our Present Knowledge of the Late Cenozoic in China.* Peiping, China: The Geological Survey of China, Geological Memoirs, Series A, Number 11, 1933

Hood, D. *Davidson Black: A Biography.* Toronto: University of Toronto Press, 1971.

Janus, C. with Brashler, W. *The Search for Peking Man.* New York: Macmillan, 1975.

Shapiro, H. *Peking Man: The Discovery, Disappearance and Mystery of a Priceless Scientific Treasure.* New York: Simon and Schuster, 1974.

Weidenreich, F. *Anthropological Papers of Franz Weidenreich,* 1939–1948 (compiled by S. Washburn and D. Wolffson). New York: Viking Fund, 1949.

Wu, R. "Hominid Fossils from China and Their Bearing on Human Evolution." *Canadian Journal of Anthropology,* Vol. 3 (1983), pp. 207–212.

Wu, R. and Lin, S. "Peking man." *Scientific American,* Vol. 248 (1983), pp. 86–94.

## HOMO ERECTUS IN AFRICA AND POSSIBLY IN EUROPE

Delson, E. "Paleobiology and Age of African *Homo erectus.*" *Nature,* Vol. 316 (1985), pp. 362–363.

Leakey, R. and Walker A. "*Homo erectus* Unearthed." *National Geographic,* Vol. 168 (1985), pp. 624–629.

Rightmire, G. "Stasis in *Homo erectus* Defended." *Paleobiology,* Vol. 12 (1986), pp. 324–325.

Wolpoff, M. "Stasis in the Interpretation of Evolution in *Homo erectus*." *Paleobiology*, Vol. 12 (1986), pp. 325–328.

## OUT OF AFRICA

Bar-Yosef, O. "Prehistory of the Levant." *Annual Review of Anthropology*, Vol. 9 (1980), pp. 101–133.

Bilsborough, A. and Wood, B. "The Nature, Origin and Fate of *Homo erectus*." In Wood, B., Martin, L., and Andrews, P., eds. *Major Topics in Primate and Human Evolution*. Cambridge: Cambridge University Press, 1986, pp. 295–316.

Isaac, G. "Casting the Net Wide: A Review of Archaeological Evidence for Hominid Land-Use and Ecological Relations." In Konigsson, L.-K., ed. *Current Arguments on Early Man*. Oxford: Pergamon Press, 1980, pp. 226–251.

## LINGUISTIC SKILLS

DuBrul, E. "Origins of the Speech Apparatus and its Reconstruction in Fossils." *Brain and Language*, Vol. 4 (1977), pp. 365–381.

Holloway, R. "Human Paleontological Evidence Relevant to Language Behavior." *Human Neurobiology*, Vol. 2 (1983), pp. 105–114.

Laitman, J. "The Anatomy of Human Speech." *Natural History*, Vol. 93 (1984), pp. 20–27.

Marshack, A. "Implications of the Paleolithic Symbolic Evidence for the Origin of Language." *American Scientist*, Vol. 64 (1976), pp. 136–145.

White, R. "Thoughts on Social Relationships and Language in Hominid Evolution." *Journal of Social and Personal Relationships*, Vol. 2 (1985), pp. 95–115.

## TECHNOLOGICAL SKILLS

Black, D. "Evidences of the Use of Fire by *Sinanthropus*." *Bulletin of the Geological Society of China*, Vol. 11 (1931), pp. 107–108.

Brain, C. and Sillen, A. "Evidence From the Swartkrans Cave for the Earliest Use of fire." *Nature*, Vol. 336 (1988), pp. 464–466.

Bordes, F. *The Old Stone Age*. New York: McGraw-Hill, 1968.

Gowlett, J. "A Case of Developed Oldowan in the Acheulian?" *World Archaeology*, Vol. 20 (1988), pp. 13–26.

James, S. "Hominid Use of Fire in the Lower and Middle Pleistocene." *Current Anthropology*, Vol. 30 (1989), pp. 1–11.

Klein, R. "The Archaeological Significance of Animal Bones from Acheulian Sites in South Africa." *The African Archaeological Review*, 1988, pp. 3–25.

Ohel, M. "Milking the Stones of an Acheulian Aggregation Locality in the Yiran Plateau in Upper Galilee. *Proceedings of the Prehistoric Society*, Vol. 52 (1986), pp. 247–280.

## SETTLEMENT PATTERNS AND SUBSISTENCE SKILLS

Binford, L. *Bones: Ancient Men and Modern Myths*. San Francisco: Academic Press, 1981.

Butzer, K. "Acheulian Occupation Sites at Torralba and Ambrona, Spain: Their Geology." *Science*, Vol. 150 (1965), pp. 1718–1722.

Cole, S. "A Spanish Camp of Stone Age Elephant Hunters." *New Scientist*, Vol. 390 (1962), pp. 160–161 and p. 482.

de Lumley, H. "A Palaeolithic Camp at Nice." *Scientific American*, Vol. 220 (1969), pp. 42–50.

Howell, F. and Freeman, L., Jr. "Ambrona: An Early Stone Age Site on Spanish Meseta." *The L. S. B. Leakey Foundation News*, Vol. 22 (1982), pp. 11–13.

# CHAPTER EIGHT

# ARCHAIC *HOMO SAPIENS,* NEANDERTALS AND OURSELVES

*contents at a glance*

# INTRODUCTION

In Latin, *Homo* means humankind, and *sapiens* means wise or intelligent. Taken together, these terms describe the descendants of *H. erectus* — ***Homo sapiens*** (*H. sapiens*) — our species. On the one hand, the terms bring to mind our kinship with other animals; they remind us that, however else we define ourselves, we are members of the Kingdom Animalia, the phylum chordata, the subphylum vertebrata, the class mammalia, the order primates, the suborder Anthropoidea, the superfamily Hominoidea, the family Hominidae, the genus *Homo*, and the species *sapiens*.

On the other hand, the designation that we use for ourselves also reminds us that we take special pride in our intelligence; it is the hallmark of our species. *H. sapiens* made their debut when natural selection began to favour individuals in the *H. erectus* population whose brains were larger and who were better able to reason than their fellow human beings. This juncture, as far as anthropologists are aware, was reached when our ancestors' unique mix of biological and cultural attributes had become more or less balanced. Thereafter humankind's capacity for culture continued to expand — we eventually came not only to dominate the planet, but also to understand its complexity. Of course, there are still gaps in our knowledge. This is certainly true with respect to our understanding of the evolution of our species. While there is no doubt that *H. erectus* was our ancestor, the evolutionary link between us continues to be hotly debated.

## Homo Erectus and Ourselves

Two models that depict the emergence of our kind in radically different ways are at the heart of the controversy. One is these is the **regional continuity or multiregional model**, which portrays the transition from *H. erectus* to *H. sapiens* as a gradual process beginning about 1.5 mya, when *H. erectus* spread from Africa to other parts of the world. Proponents of the model claim that the newcomers then gave rise to several regionally distinct *H. erectus* populations that independently evolved into *H. sapiens* in a step-like fashion in the regions that they occupied. The multiregional model consequently implies that the genetic roots of modern humans are relatively deep, and that these can be traced backward in time from existing regional populations through a graded series of ancestral regional forms to *H. erectus*. The other model is the **single origin** or **Noah's Ark model**, which represents the transition from *H. erectus* to *H. sapiens* as a relatively recent event that took place about 200 kya, when a regionally distinct African *H. erectus* population gave rise to a *H. sapiens* population that became differentiated in Africa and then spread from there throughout the world where modern regional features eventually became entrenched (see Figure 8.1). Thus, in contrast to the multiregional model, the single origin model promotes the view that the genetic roots of our kind are comparatively shallow, and that these can only be traced back with specificity to a relatively recent speciation event that

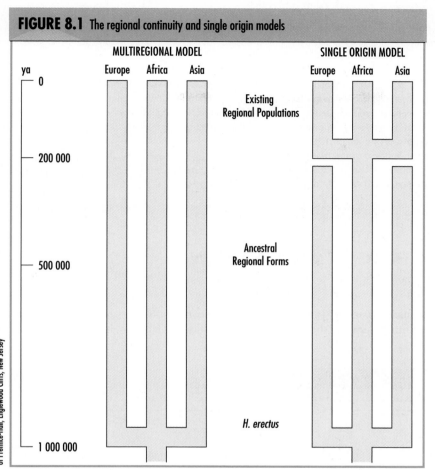

**FIGURE 8.1** The regional continuity and single origin models

occurred in Africa. If palaeoanthropologists agreed on how best to interpret the fossils that bridge the gap between *H. erectus* and ourselves, then the debate might be less acrimonious. However, there is no consensus and both models have their proponents. See QQ 8.1.

 **8.1 COMPETING MODELS**

C. B. Stringer and P. Andrews compare and contrast the multiregional and single origin models.

• • •

The two competing models for recent human evolution have been termed "regional continuity" (multiregional

origins) and "Noah's Ark" (single origin) . . . . In the multiregional model . . . recent human variation is seen as the product of . . . [a comparatively early] radiation of *Homo erectus* from Africa. Thereafter, local differentiation led to the establishment of regional populations which successively evolved through a series of evolutionary grades to produce modern humans in different areas of the world . . . .

According to the multiregional model . . . the appearance of *Homo sapiens* was thus primarily the result of a continuation of long-term trends in human evolution, and it has occurred mainly through the re-sorting of the same genetic material under the action of selection, rather than by the evolution and radia-tion of novel genetic material and morphologies . . . .

In contrast, the single origin model assumes that there was a relatively recent common ancestral popu-lation for *Homo sapiens* which already displayed most of the anatomical characters shared by living people. Proponents of this model have proposed Africa as the probable continent of origin of *Homo sapiens*, with an origin for the species . . . [in relative-ly recent times] followed by an initiation of African regional differentiation, subsequent radiation from Africa, and final establishment of modern regional characteristics outside Africa . . . .

Source: C. B. Stringer, and P. Andrews, "Genetic and Fossil Evidence for the Origin of Modern Humans." *Science*, Vol. 239, No. 4845 (1988), p. 1263. Copyright 1988 by the AAAS.

# ARCHAIC *HOMO SAPIENS*

## STEINHEIM

Whichever model prevails — many palaeoanthropologists believe that a com-promise between the two extremes best fits the facts — it is worthwhile to

---

**FIGURE 8.2** Map showing the location where Steinheim was discovered

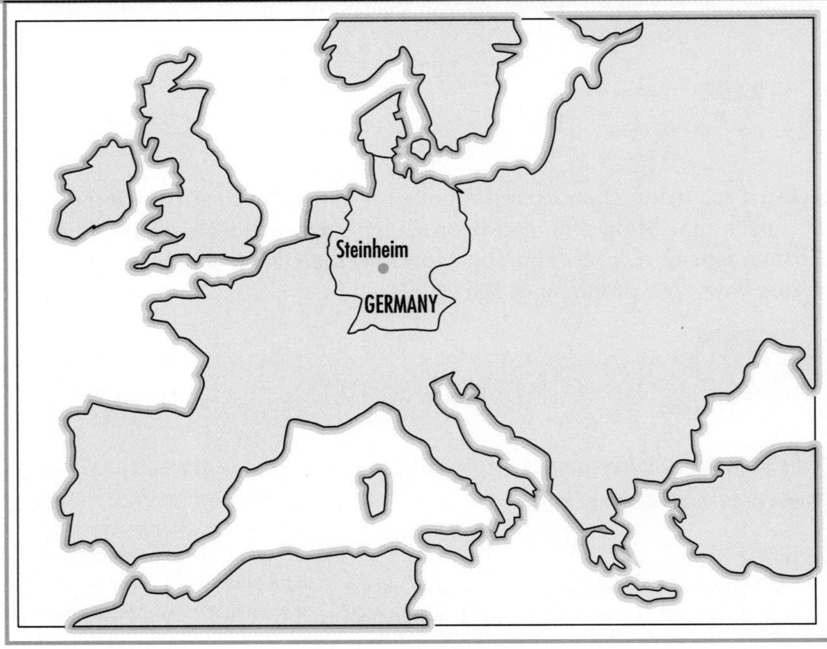

Steinheim
GERMANY

point out that the transition from *H. erectus* to *H. sapiens* is marked by the existence of a number of interesting fossils. One of these came to light in 1933 when Dr. F. Berckhemer, chief curator of the Natural History Museum of Württemberg in Germany, acquired a partly crushed petrified skull from the owner of a gravel pit in Steinheim, near the city of Stuttgart (see Figure 8.2). Called **Steinheim** because of where it was found, the skull was that of a young adult who had lived and died sometime between 250 kya and 200 kya, when the climate in Europe was warmer than today. The remains of an extinct species of elephant and the fossils of other warm-weather animals that were found in association with the

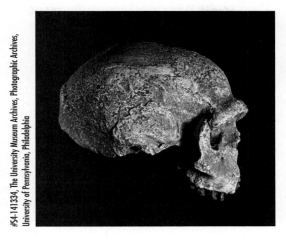

#54-141334, The University Museum Archives, Photographic Archives, University of Pennsylvania, Philadelphia

▼

The reconstructed skull of Steinheim

▲

Steinheim specimen attest to the warmer climate. In any event, it was not the associated remains that made the distorted and damaged skull stand out. It was, rather, the size of the brain that the skull had housed; estimated to have been between 1150 cm³ and 1175 cm³, Steinheim's cranial capacity was well above the current 900 cm³ minimum threshold for modern humankind. See QQ 8.2.

## QQ 8.2 STEINHEIM

Michael Day pinpoints the discovery of Steinheim and describes its morphological features.

• • •

**Site**   The Sigrist gravel pit, Steinheim on the river Murr, about 12 miles [or about 19 kilometres] north of Stuttgart, Wurttemberg, [formerly] West Germany.

**Found by**   Karl Sigrist, Jr., 24th July, 1933. Unearthed by F. Berckhemer . . . .

**Morphology**   . . . The specimen, which is distorted and damaged, consists of the cranial and facial skeleton of a young adult. The left orbit [eye socket], temporal and infratemporal regions [situated near the temple] and part of the left maxilla [or upper jaw] are missing, as is the premaxillary region [above the upper jaw] of both sides. The base of the skull is broken away around the foramen magnum [the hole at the base of the skull where the spinal cord enters the brain], but it is well preserved in its anterior portion. The upper right second premolar tooth and all of the molar teeth are in place.

The vault of the skull is long, narrow and moderately flattened; the supraorbital torus [or brow ridge] is pronounced, the nasal opening wide and the root of the nose depressed. In lateral view the degree of facial prognathism [or protrusion below the eyes] is small, the [nipple-like] mastoid processes [of the temporal bones behind the ears] small but well defined and the occipital region [at the back of the skull near the base] well rounded.

Source: Michael H. Day, *Guide to Fossil Man,* fourth edition (Chicago: University of Chicago Press, 1986), pp. 79–80.

## SWANSCOMBE

Not long after Steinheim was discovered, another equally ancient skull known as **Swanscombe** was also found in a gravel pit. This find was made at Swanscombe, on the Thames River in Kent, England, by Alvan Marston, a dentist and amateur archaeologist (see Figure 8.3). In 1935 Marston acquired a skull fragment that had been unearthed from the pit by local cement workers; one year later he recovered another fragment; and, in 1955,

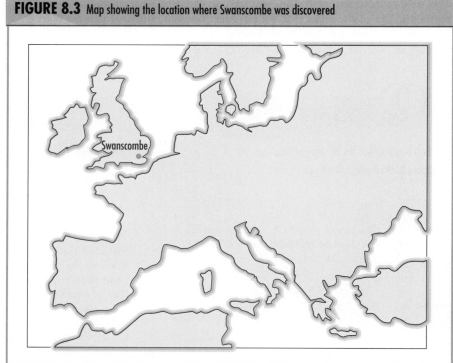

**FIGURE 8.3** Map showing the location where Swanscombe was discovered

Swanscombe

twenty years after the original find, a third fragment was discovered. Amazingly the three pieces were from the same individual, and, when reunited, they showed the top and the back of a faceless skull that otherwise was like the one that had been found at Steinheim. It too had housed a relatively large brain. Estimated to have been about 1275 cm$^3$, Swanscombe's cranial capacity was also well above the minimum threshold for modern humankind. Given these facts, many anthropologists initially contended that the two ancient individuals were *H. sapiens* like ourselves.

## REAPPRAISAL

But it is now almost certain that Steinheim and Swanscombe were not modern humans. True they had relatively large brains, and, like ours, their occipital bones (located at the back of the skull near the base) were rounded. Moreover, judging by the appearance of Steinheim, they both, again like us, had comparatively small faces and teeth. However, in 1964, with the aid of a computer, two palaeoanthropologists — J. S. Weiner and B. G. Campbell — compared the Swanscombe skull with those of other hominid fossils and with those of modern humans. They concluded that Swanscombe did not represent *H. sapiens*, at least not in the strict sense of the term. But neither was the skull exactly like that of *H. erectus*. Although the comparison revealed that Swanscombe had thick cranial bones, a low forehead, and a relatively long and low skull, in most other respects the specimen was quite different from *H. erectus*. It is on account of findings such as these that the majority of palaeoanthropologists now classify Swanscombe and the largely similar Steinheim as ***archaic H. sapiens***, an extinct variety of *H. sapiens,* reminiscent of both *H. erectus* and our species, who lived in Africa, Europe, and Asia between about 400 kya and 130 kya. These and other archaic *H. sapiens* are also commonly referred to as **transitional forms**. See QQ 8.3.

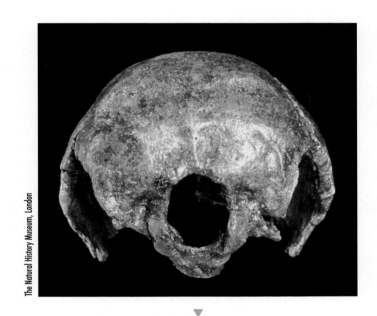

The Natural History Museum, London

The reconstructed skull of Swanscombe, as seen from the back

# 8.3 REVISITING SWANSCOMBE

J. S. Weiner and Bernard Campbell discuss what they learned by comparing the Swanscombe skull with the skulls of other hominid fossils and with those of modern humans.

• • •

The bones were examined with the intention of taking the maximum number of measurements . . . . It appeared possible to take 27 measurements in all between the various defined points, but these were reduced to 17, as the remaining 10 added no further information about the shape and size of the skull[s] . . . .

Our investigations of particular morphological characters and of the "total morphological pattern" of the skull bones have brought us to the conclusion that the features . . . do not "fall well within the ranges of variation exhibited by modern crania . . . ." On the basis of all the morphological and metrical features examined by us we would allocate Swanscombe not to the strict *sapiens* category . . . but to . . . [an] intermediate group.

Source: J. S. Weiner and Bernard G. Campbell, "The Taxonomic Status of the Swanscombe Skull." In Ovey, Cameron D., ed. *The Swanscombe Skull: A Survey of Research on a Pleistocene Site* (London: Royal Anthropological Institute of Great Britain & Ireland, 1964), pp. 182–202.

## OTHER TRANSITIONAL FORMS

Along with Steinheim and Swanscombe, palaeoanthropologists also classify a number of other fossils as transitional forms. Those found in Africa include the Bodo skull from Ethiopia, the Lake Ndutu skull from Tanzania, and the Kabwe or Broken Hill skull from Zambia. Asian finds include the Dali skull from China, the Solo skull from Java, and the Narmada cranium from India. Even more transitional fossils have been recovered from Europe, from sites such as Petralona in Greece, Arago in France, Mauer and Bilzingsleben in Germany, Atapuerca in Spain, and Vértesszöllös in Hungary (see Table 8.1). Although the physical features of these specimens are, like those of Steinheim and Swanscombe, reminiscent of *H. erectus*, they are also tantalizingly close to our own. Like us, the specimens were relatively big-brained, but, like *H. erectus*, their faces and skulls were more ruggedly constructed.

## THE LEVALLOIS METHOD

The intermediate status of the transitional forms is also reflected in their cultural skills. Although their toolkit was Acheulian, the transitional forms made a breakthrough (between about 200 kya and 100 kya) that foreshadowed the technological skills of our kind, one that resulted in the production of a new type of flake implement that was used for cutting, scraping, and butchering. Instead of shaping flakes after they had been struck from a core, archaic *H. sapiens* learned how to predetermine the shape of a flake by using the **Levallois method**. Named after the suburb of Levallois in Paris where such flakes were first unearthed, the method involved removing standardized flakes from a parent core that, after it had been worked, resembled a small tortoise shell. This was done by carefully trimming the core with a hammerstone to create a strik-

**TABLE 8.1** The location and age of selected archaic *H. sapiens* remains discovered in Africa, Asia, and Europe

| CONTINENT | COUNTRY | AGE OF OLDEST REMAINS (kya) |
|---|---|---|
| Africa | Ethiopia | 400 (?)[1] |
| | Tanzania | 400 (?) |
| | Zambia | 130 |
| Asia | China | 230 |
| | Indonesia | 250 (?) |
| | India | 100 |
| Europe[2] | Greece | 500 (?) |
| | Germany | 280 |
| | France | 215 (?) |
| | England | 250 |
| | Spain | 275 (?) |
| | Hungary | 185 |

[1] (?) indicates a disputed age.
[2] Some palaeoanthropologists maintain that the oldest European specimens are not those of archaic *H. sapiens* but those of *H. erectus*. The reason is that the specimens exhibit features that are characteristic of both species.

ing platform or working surface, and then directing a single crosswise blow to the edge of the striking platform with the hammerstone to remove the flake. Each core yielded from three to eight similar-sized flakes that were usually no less than six centimetres long and that were sometimes but not always refined by secondary flaking or retouching (see Figure 8.4). Although such tools were crude by modern standards, they were far more sophisticated than Oldowan flake knives or Acheulian axes. This was reflected in their design, which featured razor-sharp edges and a pointed tip. See QQ 8.4.

**FIGURE 8.4** The Levallois method

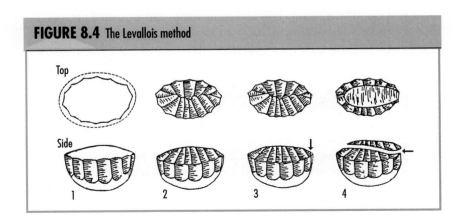

Top

Side

1  2  3  4

## 8.4 THE LEVALLOIS METHOD

Clive Gamble describes the Levallois method.

• • •

[One of t]he best known reduction strategies, which produced flakes . . . of desired dimensions, is named after the suburb of Levallois on the Seine river. It is here that this distinctive technique was first noted and where the shape of the flake . . . is predetermined by the careful preparation of the core *before* the flake is detached

. . . . The levallois technique produces broad flat flakes . . . [that] display the traces of previous flake removals on their dorsal surfaces and on occasion their striking platforms are faceted as a means of facilitating their removal from the core. The pattern of flake removals has resulted in the description of 'tortoise core' for the discarded core after the flakes have been removed.

Source: Clive Gamble, *The Palaeolithic Settlement of Europe* (Cambridge: Cambridge University Press, 1986), pp. 118–119.

## The Dilemma of the Transitional Forms

Equally important, the debris that has been found alongside the fossils of the transitional forms indicates not only that they were better able to fashion stone implements than *H. erectus*, but also that they included more plant and marine foods in their diet and that they built shelters to protect themselves from the elements. What remains to be determined is the precise role that these archaic *H. sapiens* played in the evolution of our kind. Were they our direct ancestors, or were they an evolutionary dead-end? Calling them *H. sapiens* does not answer the question, and grouping them together does not mean that the archaics were exactly alike. Detailed comparative studies have revealed that some transitional forms were more like *H. erectus*, others more akin to *H. sapiens*, and still others midway between *H. erectus* and our species, especially in terms of their cranial traits. Are these differences minor variations or do they indicate that there was more than one archaic species? And, if there were more than one species and the archaics are our direct ancestors, then who among them gave rise to our kind? And, if they were not our direct ancestors, then what caused their extinction and from whom did we evolve? Hopefully these questions will be answered as more specimens of transitional forms are found, especially in Asia and Africa where their remains are comparatively sparse and where the ages of the finds continue to be hotly contested.

# NEANDERTALS

Challenging questions have also been raised about **Neandertals** (also spelled Neanderthals), another descendant of *H. erectus*, who may have appeared as early as 130 kya but who flourished nearer the present, between about 80 kya and 40 kya. Like the transitional forms, their remains have also been found in Europe, Asia, and Africa. See Table 8.2.

**TABLE 8.2**  The location and age of selected Neandertal remains discovered in Africa, Asia, and Europe

| CONTINENT | COUNTRY | AGE OF OLDEST REMAINS (kya) |
|---|---|---|
| Africa | Sudan | 80  (?)[1] |
|  | Morocco | 50 |
| Asia | Israel | 70 |
|  | Iraq | 60 |
| Europe | France | 75 |
|  | Germany | 75 |
|  | Belgium | 75 |
|  | Gibraltar | 75 |
|  | Czech Republic | 45 |
|  | Bulgaria | 43 |
|  | (formerly) Yugoslavia | 40 |

[1](?) indicates a disputed age.

## EARLY FINDS IN GIBRALTAR AND GERMANY

In 1848, an unmistakably human but rugged-looking fossil skull was found in a cave at Forbes' Quarry in Gibraltar. Since Darwin had not yet published *On the Origin of Species*, it is not surprising that the find aroused little immediate interest. Eight years later a similar discovery was made. In 1856, in a small cave in the Neander Valley, near Düsseldorf in Germany, a fossilized skullcap similar to the one from Gibraltar was discovered by workmen who happened on the cave while quarrying for limestone (see Figure 8.5). The skullcap was accompanied by a collar bone, a shoulder blade, five arm bones, two leg bones, five ribs, and parts of a pelvis, and the owner of the quarry gave these to Johann Karl Fuhlrott, a science teacher in a nearby high school who was told that the bones were those of a bear. After examining the bones Fuhlrott came to a different conclusion. He believed that the remains were those of an ancient human. And so did Hermann Schaafhausen, a professor of anatomy at the University of Bonn, who had been shown the remains by Fuhlrott (see QQ 8.5). Although their conclusion sparked some debate about human antiquity, it was not until three years later, when Darwin published his book, that the Neander Valley fossils took on a special importance. Unfortunately, however, the status of these fossils, as well as the status of subsequent Neandertal finds, remained clouded by myths and misrepresentations for many years.

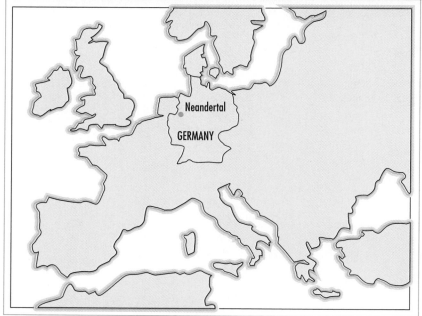

Neandertal

GERMANY

© 1975 Thames and Hudson Ltd., London

## 8.5 THE NEANDER VALLEY FINDS

George Constable recounts the story of the discovery of the Neander Valley finds.

. . .

[One of the first] Neanderthal . . . [finds] . . . turned up not far from the city of Dusseldorf, Germany, where a tributary stream of the Rhine flows through a steep-sided gorge known as the Neander Valley — *Neanderthal* in Old German. In 1856 the flanks of the gorge were being quarried for limestone. During the summer, workmen blasted open a small cave about 60 feet [or about 18.5 metres] above the stream. As they dug their pickaxes into the floor of the cave, they uncovered a number of ancient bones. But the quarrymen were intent on limestone; they did not pay much attention to the bones, and most of what was probably a complete skeleton of a Neanderthal was lost. Only the skullcap, ribs, part of the pelvis and some limb bones were saved.

The owner of the quarry thought that these fragments belonged to a bear, and he presented them to the local science teacher, J. K. Fuhlrott, who was known to be interested in such things. Fuhlrott had enough anatomical knowledge to realize that the skeletal remains came not from a bear but from a man — and a most extraordinary man at that [ — so] he called in an expert, Hermann Schaafhausen, professor of anatomy at the University of Bonn. Schaafhausen agreed that the bones represented one of the "most ancient races of man."

Source: George Constable, *The Neanderthals* (New York: Time, Inc., 1973), pp. 9–10. From *Emergence of Man: The Neanderthals*. By George Constable and the Editors of Time-Life Books. © 1973 Time-Life Books Inc.

# MYTHS ABOUT NEANDERTALS

## VIRCHOW'S PRONOUNCEMENT

One myth about the Neandertals involved the Neander Valley finds. When these were shown in Europe in the 1860s not all scientists were convinced that the fossils were ancient. While a few concluded that the bones were those of a recently deceased microcephalic idiot, others maintained that they had belonged to a Mongolian Cossack who had died in Germany in 1814, during Napoleon's retreat from Moscow. By far the most outspoken critic of the find, however, was Rudolf Virchow (1821–1902), the famous German scientist who had inaugurated the study of cellular pathology. Virchow's opinion was that the rugged-looking bones had belonged to a modern human who not only had suffered from bone disease, but who had also received several hard blows to the head. Although subsequent research showed that the bones were undoubtedly ancient and not misshapen because of disease or physical abuse, Virchow's pronouncement all but ended scientific speculation about the evolutionary significance of Neandertals for years.

The Bettmann Archive

Rudolph Virchow
(1821–1902)

## BOULE'S RECONSTRUCTION

Another myth which still lingers in the public imagination is that, compared to living humans, Neandertals were subhuman beasts. Among others, the French palaeontologist Pierre Marcellin Boule (1861–1942) was responsible for this idea.

In 1908, Boule, who was a professor at the Musée National d'Histoire Naturelle, acquired an almost complete Neandertal skeleton from three clergymen with an interest in archaeology who had discovered the specimen in a cave near the village of La Chapelle-aux-Saints, in southern France (see Figure 8.6). Although Boule correctly classified the fossil as an ancient human, his reconstruction of the skeleton, which is known as the **Old Man of La Chapelle**, was seriously flawed; partly because he was unaware that the individual whose skeleton he had reconstructed had suffered from arthritis of the jaws, spine, and likely the lower limbs, and partly because he believed that the Old Man represented a being who was radically different from ourselves. Instead of correctly portraying the Old Man as much more human-like than ape-like, Boule's faulty arrangement of the bones indicated the opposite. Unfortunately, it was not until the 1950s that the specimen was properly reconstructed. In the meantime, Boule's exaggerated account of the brutish appearance of Neandertals had been popularized in the press and in books. Such was the case, for example, in H. G. Well's *The Outline of History*, which sold over two million copies; in that book, Neandertals were described as hairy and repulsive creatures. See QQ 8.6.

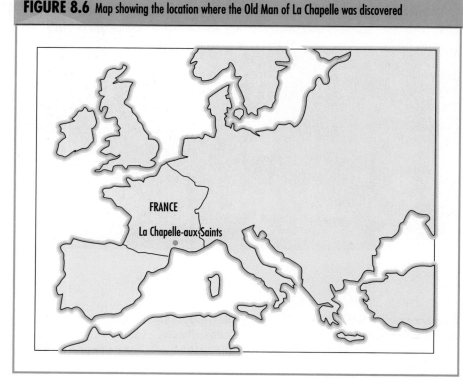

**FIGURE 8.6** Map showing the location where the Old Man of La Chapelle was discovered

FRANCE

La Chapelle-aux-Saints

# 8.6 UNDER THE INFLUENCE OF BOULE

In the following passage from *The Outline of History*, Boule's influence is evident.

• • •

The Neanderthal type of man prevailed in Europe at least for tens of thousands of years. For ages that make all history seem a thing of yesterday, these nearly human creatures prevailed . . . . We know nothing of the appearance of the Neanderthal man, but . . . an extreme hairiness . . . [can be assumed, as well as] an ugliness, or a repulsive strangeness in his appearance over and above his low forehead, his beetle [or overhanging] brows, his ape neck, and his inferior stature . . . . Says Sir Harry Johnston, in a survey of the rise of modern man in his *Views and Reviews*: "The dim racial remembrance of such gorilla-like monsters, with cunning brains, shambling gait, hairy bodies, strong teeth, and possibly cannibalistic tendencies, may be the germ of the ogre in folklore . . . ."

Source: H. G. Wells, *The Outline of History, Being a Plain History of Life and Mankind*. Garden City (New York: Garden City Publishing Co., Inc., 1929), pp. 65–70.

Field Museum of Natural History, Neg. #66830, Chicago

▼

Neandertal mistakenly reconstructed as a brute

▲

# THE APPEARANCE AND CULTURE OF THE NEANDERTALS

## BIOLOGICAL FEATURES

Since these myths were exposed, anthropologists have been able to reconstruct a much more accurate portrait of the Neandertals. With the steady accumulation of their fossils, it has become clear that, like many of our kind, they stood about 1.6 metres tall, weighed about 65 kilograms, and had a cranial capacity that ranged between 1300 cm$^3$ and 1750 cm$^3$. Neandertals, however, were more robust than modern humans. The shape of their postcranial skeletons indicate as much, as do the markings that are found on the bones of the upper body, torso, and limbs. These impressions made by their muscles, which are found in Neandertal men, women, and children, indicate that they were heavily muscled and extraordinarily strong. In addition, their skulls, faces, and teeth were ruggedly built. Their skulls, for instance, featured a prominent brow ridge; their midfacial region protruded; and their jaws, incisors, and canines were exceptionally large. It is also interesting to note that while Neandertal incisors and canines show extreme wear, the pattern is unlike that produced by chewing food; instead, the form of the wear suggests that Neandertals may have used their teeth to clamp down on skin, wood, and stone while these were being worked. See QQ 8.7.

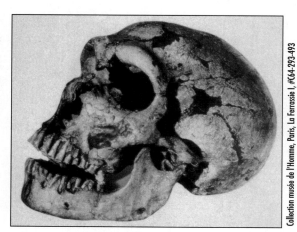

The skull of a Neandertal discovered at La Ferrassie, France

Collection musée de l'Homme, Paris, La Ferrassie I, #C64-293-493

# QQ 8.7 NEANDERTAL TEETH

Erik Trinkaus explains why he believes that Neandertals used their large front teeth to clamp down on materials.

• • •

Among the Neandertals . . . [the front teeth, the incisors and the canines,] are larger than those of virtually all living humans . . . . Why the Neandertals had large front teeth is still a matter of controversy, but it appears that they were using their incisors and canines for more than just chewing food. A number of the Neandertals . . . show extensive wear on their front teeth. The crowns of their teeth were being worn off completely, so that the stubs of the roots were functioning as the chewing surface, by the age of 35 to 40 years. And rather than having the chewing surface worn in the normal way, parallel to the plane of chewing, they were rounding off the front, or outer, margins of their teeth. Microscopic inspection of the chewing surfaces of these teeth shows crushing along the edges and small scratches running from within out, as though they were biting down on hard objects and then pulling them outward. They were probably using their teeth as a vise for holding a variety of objects, including skins, wood, and flint tools, while working on them.

Source: Erik Trinkaus, "New Light on the Very Ancient Near East." *Symbols*, Summer 1980, p.11.

## MOUSTERIAN CULTURE

Also, based on the non-human bones and tools that have been found in the caves and in the rock shelters where Neandertals lived, it is now known that they hunted big game such as mammoth, woolly rhinoceros, bison, cave bear, reindeer, and wolf, and that they designed new implements to better cope with their surroundings. Their toolkit is referred to as **Mousterian**, named after the village of Le Moustier, in France, where Neandertal implements were first discovered in the 1860s. What is impressive about Mousterian tools is not only the large number that have been recovered, but also the wide variety. Made mostly of stone, but sometimes fashioned from wood, bone, and antler, Mousterian implements were functional and designed for specific tasks. The Mousterian toolkit included **scrapers** — small tools with a steeply bevelled working edge that were used to clean hides and carve wood, bone, and antler; **notches** — sharp-edged stone flakes with angular indentations that were used as knives for cutting meat and hides; **burins** — chisel-like implements that were used for planing, trimming, and engraving wood, bone, and ivory; **borers** — icepick-like implements that were used to punch holes in leather and other soft materials; **drills** — bit-like drilling implements that were used to create holes in hard materials; and **denticulates** — saw-like stone implements that were used to cut and shred wood. In addition, the Mousterian toolkit included improved hand axes that were used for chopping, and stone-tipped poles that were used for spearing. The Neandertals who fashioned Mousterian implements — and whose culture is likewise known as Mousterian — may also have been the first humans to cook their food on a regular basis. See QQ 8.8.

▼

Mousterian
side scraper (A)
and point (B)

▲

A

B

C. Loring Brace describes the toolkit of the Neandertals and the way they cooked their food.

• • •

The term "Mousterian" comes from the village of Le Moustier in southwestern France where the type site is located. Tools of Mousterian form are distributed throughout western and southern Europe . . . . Throughout this whole area, which we could call a Mousterian culture area, there were a series of bands possessing related cultures between which similar cultural elements maintained circulation. Local differences in details of typology and technique of manufacture persisted, but all these subcultures possessed the same functional tool categories: scrapers, points, and knives.

Scrapers indicate a concern for the preparation of animal hides, which is reasonable for people living in a subarctic climate. It used to be thought that effective clothing was not developed until the ensuing Upper Paleolithic, with the invention and manufacture of bone needles, but there is no reason to deny the Neanderthals the use of skin clothing just because they had no needles; wrapped clothing bound on by thongs was utilized by the poorer peoples of Europe right up to historical times. Certainly the Neanderthals

must have been doing something with the skins they went to so much trouble to prepare, and it is reasonable to suppose that the manufacture of clothing was one such thing. Indeed, human survival in Europe during the last glaciation without clothing would have been impossible.

The Mousterian points, made on flakes of a variety of sorts, evidently were frequently hafted [or joined], and the inference can be made that spears were being so tipped. Whether they were thrusting spears or throwing spears we have no way of knowing, but they obviously played an important role in the Neanderthal way of life . . . .

Another cultural development at this time, also originally a part of the formal complex designated as Mousterian, was eventually to change the face of humanity in a literal sense. This was the beginning of culinary elaboration. We are not really attuned to thinking of gastronomy as a Neanderthal invention, and perhaps it would be stretching things a bit to make such a claim, but there is reason to believe that the Neanderthals were the ones who pioneered the use of cooking as a regular means of preparing food.

Source: C. Loring Brace, *The Stages of Human Evolution: Human and Cultural Origins*, third edition (Englewood Cliffs, New Jersey: Prentice-Hall, Inc., 1988), pp. 118–121.

# NEANDERTALS AND OURSELVES

Given the age of their fossils and their biological and cultural features, it is quite likely that the Neandertals were the direct descendants of the transitional forms. Almost all anthropologists agree on this point. What is still unclear is how the Neandertals are related to us.

## THE PRESAPIENS THEORY

Some scholars believe that there is no direct relationship between the Neandertals and modern humans. Those who support the **presapiens theory** are an example. In their view, *H. erectus* gave rise to two distinct human populations: a presapiens one that led from *H. erectus* to transitional forms such as Swanscombe and then on to *H. sapiens*, and another that led from *H. erectus* to transitional forms such as Steinheim and then on to **Homo neandertalensis** (*H. neandertalensis*), which supporters of the theory regard as a distinct species of humankind (see Figure 8.7). Proponents of the presapiens theory also maintain that before *H. neandertalensis* became extinct, Neandertals lived alongside early *H. sapiens* for a number of years (see QQ 8.9). Although once quite popular, the theory now has relatively few adherents. What makes the theory suspect is that the fossil record does not support the view that Steinheim and Swanscombe were evolving in different directions. On the whole the specimens are remarkably alike, and this contradicts one of the principal assertions of the theory.

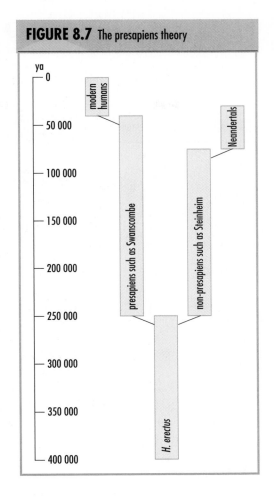

**FIGURE 8.7** The presapiens theory

## 8.9 THE PRESAPIENS THEORY

Henri Vallois summarizes the presapiens view.

• • •

The European *Homo sapiens* is not derived from the Neanderthal men who preceded him. His stock was long distinct from, and, under the name of Presapiens, had evolved in a parallel direction to theirs. Long debated, the Presapiens forms are thus not a myth. They did exist. The few remains of them we possess are the tangible evidence of the great antiquity of the phylum that culminates in modern man.

Source: Henri Vallois, "Neanderthals and presapiens." *Journal of the Royal Anthropological Institute*, Vol. 84, p. 128. Copyright 1954. Reprinted by permission of Wiley-Liss, a Division of John Wiley and Sons, Inc.

Another hypothesis, known as the **preneandertal theory**, likewise holds that *H. erectus* gave rise to the transitional forms. But this theory also states that the transitional forms then evolved directly into an early Neandertal population, called **third interglacial forms**. According to the theory, these third interglacial Neandertals then branched out in two directions: one branch became the so-called **progressive Neandertals** of Asia and North Africa who afterwards evolved into *H. sapiens*, and the other became what are known as the **classic Neandertals** of Western Europe, who were cut off from their counterparts elsewhere by a glacial advance and who ultimately became extinct.

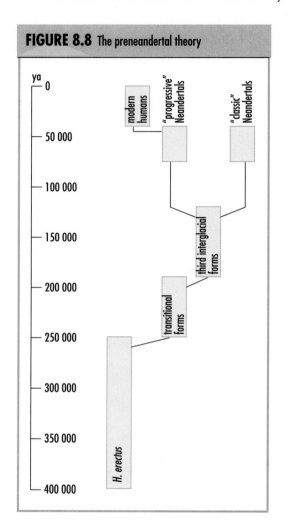

**FIGURE 8.8** The preneandertal theory

On the surface, the theory is appealing. Although the terms "classic" and "progressive" are no longer widely used, there are some noteworthy differences between the Neandertal fossils from Western Europe and those discovered in the Near East and in North Africa. The ones from Western Europe tended to have well-developed brow ridges, long low skulls with a protruding middle face, relatively large teeth, and a robust, stocky postcranial skeleton — features that some have suggested were specialized adaptations to cold weather. In contrast, the Neandertals who lived in the Near East and in North Africa, where the climate was milder, looked more like we do. It is because of this similarity that supporters of the preneandertal theory regard the so-called progressive Neandertals of the Near East and North Africa as *Homo sapiens neandertalensis* (*H. sapiens neandertalensis*) — a subspecies of our own kind. See QQ 8.10.

## 8.10 THE PRENEANDERTAL THEORY

F. Clark Howell describes the preneandertal view.

• • •

An examination of the early Neanderthal distribution reveals a definite morphological gradient extending from the Near East through central Europe as far as . . . Germany. The more anatomically modern of the Third Interglacial forms occur in the Near East area, there being a tendency further westward for characters more like those of the classic Neanderthals to occur. In this regard, the classic Neanderthals . . . represent a further extension westward of the gradient and a step later in time . . . .

It is . . . not inconceivable, and indeed appears highly probable, that isolation of western representatives of the early Neanderthal gradient for several tens of thousands of years in a rather limited area and subjection to the rigors of an extreme arctic environment, would account for the appearance of the classic Neanderthal population. Under this environment, selection would have been severe, chance for genetic drift at an optimum, and opportunities for migration reduced to a minimum. These multiple factors brought about a distinctive race of mankind occupying this area . . . [who later became extinct].

It is among the more eastern representatives of the early Neanderthal group that evolution was leading toward modern man.

Source: F. Clark Howell, "The Place of Neanderthal Man in Human Evolution." *American Journal of Physical Anthropology*, Vol. 9, No. 4 (1951), pp. 406–409.

## THE NEANDERTAL-PHASE THEORY

Perhaps the most important objection to the preneandertal theory has been raised by supporters of the **Neandertal-phase theory**. In their opinion, the physical differences between the Western European Neandertals and those discovered elsewhere do not indicate that there were two distinct Neandertal populations. Instead, these scholars insist that differences are exactly what should be expected in such a widely distributed group, and that Neandertals are consequently best regarded as a single population. They also maintain, as two supporters of the theory once put it, that if a Neandertal "could be reincarnated and placed in a New York subway — provided he were bathed, shaved and dressed in modern clothing — it is doubtful whether he would attract any more attention than some of its other denizens." In other words, they believe that Neandertals were the first *H. sapiens* — the evolutionary end product of a straight line that ran from *H. erectus* through the transitional forms to modern humans. See Figure 8.9 and QQ 8.11.

## 8.11 THE NEANDERTAL-PHASE THEORY

Aleš Hrdlička comments favourably on the Neandertal-phase theory.

• • •

My conviction that the Neanderthal type is merely one phase in the more or less gradual process of evolution of man to his present form, is steadily growing stronger. It isn't only the skulls we have, but also other

The abovementioned claim and the theory on which is its based are not without merit, especially considering that, like ourselves, Neandertals were cultural animals. Their toolkit and shelters attest to the fact, as does the discovery that some of them buried their dead, perhaps with an afterlife in mind. The burial of a young boy in a cave in Teshik-Tash, Russia, whose skull was apparently placed inside a circle of mountain goat horns may be an example. There are also several Neandertal bodies that appear to have been intentionally buried in trenches at La Ferrassie, France. And at Shanidar Cave, Iraq, one of the Neandertal corpses found in the cave appears to have been placed on a bed of pine boughs and then adorned with wild flowers.

But while the technology and social practices of the Neandertals indicate that they possessed culture, this does not prove that they were the first *H. sapiens* in a biological sense. Given the fact that they did not look exactly like us — especially those from Western Europe— it is possible that the Neandertals were a separate species, and this makes the Neandertal-phase hypothesis at least somewhat suspect.

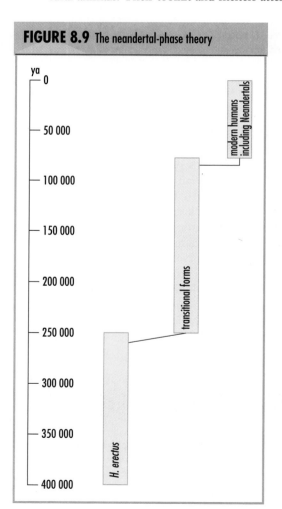

**FIGURE 8.9** The neandertal-phase theory

ya

0

50 000

100 000

150 000

200 000

250 000

300 000

350 000

400 000

modern humans including Neandertals

transitional forms

*H. erectus*

# THE REPLACEMENT THEORY

Recently, yet another theory has been advanced that addresses the relationship between the Neandertals and ourselves — the **replacement theory**. Like those who support the presapiens theory, advocates of the replacement theory also maintain that Neandertals played no direct role in our evolution and that they should be classified as a separate species. Supporters of the replacement theory, however, are also skeptical about the importance of transitional fossils such as Steinheim and Swanscombe. What they believe is that modern humans did not evolve from transitional forms in Europe, but rather from similar transitional forms in Africa. This, they say, took place about 200 kya, after which humans who evolved directly into *H. Sapiens* spread into the colder, northern areas where the Neandertals resided and then lived alongside them until the Neandertals became extinct. See Figure 8.10.

**FIGURE 8.10**   Map showing the regions that the Neandertals occupied

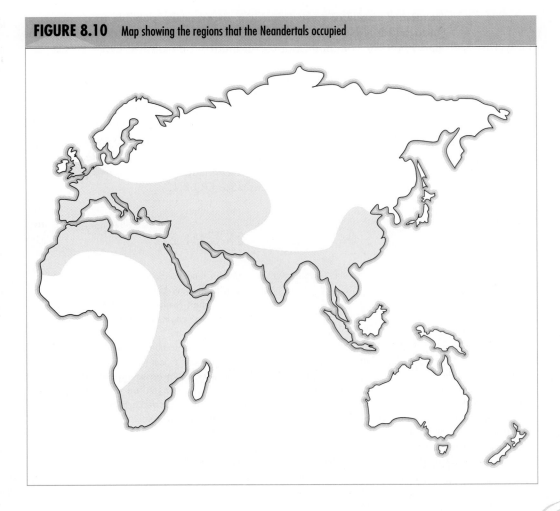

Whether the replacement theory becomes generally accepted remains to be seen. One point in its favour is that genetic research has shown that all living people may be the descendants of **Mitochondrial Eve**, a hypothetical representative of a small 200 000 year old African gene pool from which all modern human populations are thought to be derived. This view was brought to light by constructing a "fast ticking" molecular clock that uses mitochondrial DNA to estimate divergence times between modern human populations. As we discussed in Chapter 4, unlike nuclear DNA, mitochondrial DNA resides outside the nucleus of the cell in cellular objects called mitochondria. In the case of sperm, for example, mitochondrial DNA is located in the mid-section of the cell rather than in the head (nucleus), and since only the head of the sperm penetrates the egg, mitochondrial DNA is transmitted from one generation to the next strictly through the maternal line. Assuming that mitochondrial DNA mutates at a constant rate, it consequently stands to reason that it is theoretically possible to construct a mitochondrial-based molecular clock that can be used to determine when and where the common "mother" of modern humans had lived. This is precisely what Rebecca Cann and her colleagues did when they compared the mitochondrial DNA derived from the placentas of 147 newborn children from Africa, Europe, Asia, Australia, and Papua New Guinea. What they found is that the common thread that tied these children together was that they all carried at least some of Mitochondrial Eve's 200 000 year old African DNA, the implication being that Mitochondrial Eve was the "mother" of all modern humans. And given these findings the idea that the Neandertals were replaced by the descendants of a *H. sapiens* population that originated in Africa makes sense. See QQ 8.12.

## 8.12 THE STORY OF MITOCHONDRIAL EVE

Rebecca Cann recounts the story of Mitochondrial Eve, who is best regarded as a theoretical construct rather than as an individual.

• • •

In recent years, biochemists have learned to investigate human ancestry through living cells. By analyzing the DNA molecules of two animals and measuring the differences in the sequence of their components, they can gauge how long it has been since the two diverged from a common ancestor . . . .

Some ten years ago, biochemists discovered advantages to studying the separate set of DNA molecules found in mitochondria, outside the cell nuclei. Mitochondria, the tiny blob-shaped organelles that occur in all "higher" [multicellular] plants and animals . . . are the cells' engines, metabolizing food and water into energy . . . .

In 1979, I began gathering samples of mitochondrial DNA from the placentas of newborn children (which contain the same genes as the children themselves). By . . . [1986] Allan C. Wilson, Mark Stoneking, and I had collected a total of one hundred and forty-seven samples from children whose ancestors lived in five parts of the world: Africa, Asia, Europe, Australia, and New Guinea. Then . . . with a computer, we calculated the number of mutations that had taken place in each sample since it and the others evolved from a common ancestor. Fourteen of the samples had essentially the same base sequences as the others in the survey, leaving one hundred and thirty-three distinct types of mitochondrial DNA . . . .

The computer then constructed an evolutionary tree, placing the one hundred and thirty-three mitochondrial DNA types at the tips of the branches. Those with the fewest differences in their base sequences (the most closely related) are grouped together in clusters of tiny branches, and each of these clusters is, in turn, linked to others at the points (further in the past and, thus, closer to the tree trunk) at which they diverged from common ancestors. All but seven mitochondrial DNA types are on limbs that converge on a single large branch, and descendants of people from the five areas are mixed throughout this branch. A second major branch, much thinner than the first, contains the seven remaining mitochondrial DNA types, all of them from people of African descent. These seven are as different from one another, in the composition of their mitochondrial DNA, as are any of those on the more widely radiating, multiracial branch. Thus their common ancestor is just as old as the common ancestor of the one hundred and twenty-six on the larger branch. The base of the tree, the point at which the two branches split apart, is the position of the common mother — Eve. It was her children who diverged and spawned the two lines of descent . . . [and ultimately] modern man.

Source: Rebecca L. Cann, "In Search of Eve." *The Sciences*, Vol. 27, No. 5 (1987), pp. 30–35.

## THE QAFZEH CAVE REMAINS

Another point in favour of the replacement theory is that the most ancient *H. sapiens* remains in the Near East predate those of the Neandertals who also lived in the region. Recent *H. sapiens* finds in the **Qafzeh cave** in Israel have been discovered to be about 92 000 years old (see Figure 8.11). Meanwhile, the oldest Neandertal remains in the region, which come from nearby Kebara cave, are between 62 000 years old and 52 000 years old. This fact indicates not only that *H. sapiens* were in the Near East prior to the arrival of the Neandertals, but also that *H. sapiens* are more ancient than the vast majority of the Neandertals (*c.* 80 kya – 40 kya) and therefore could not possibly be their descendants. See QQ 8.13.

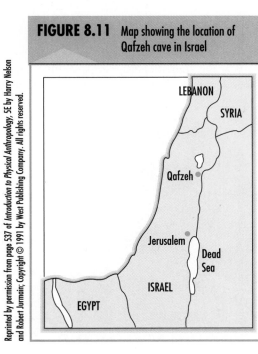

**FIGURE 8.11** Map showing the location of Qafzeh cave in Israel

LEBANON

SYRIA

Qafzeh

Jerusalem

Dead Sea

ISRAEL

EGYPT

Chris Stringer discusses the evolutionary implications of the 92 000 year old Qafzeh cave remains.

• • •

For the past half century, the relationship between Neanderthal and early modern human populations in south-west Asia has been a subject of considerable discussion and disagreement . . . . Helene Vallados and colleagues [now] report . . . dates of 92 000 years for the . . . Qafzeh cave (Israel), which suggest that early modern humans were living there twice as long ago as is generally believed. These dates indicate that some modern *Homo sapiens* preceded Neanderthals into the area, standing the conventional evolutionary sequence on its head . . . .

The palaeoanthropological implications of such an age are enormous. First, it appears to establish that some anatomically modern humans (admittedly with primitive features) did indeed precede Neanderthals in south-west Asia. Evolutionary models centred on a direct ancestor-descendant relationship between Neanderthals and modern *H. sapiens* must surely now be discarded, along with associated schemes designed to explain such a transition. Second, a minimum chronological overlap between modern *H. sapiens* and Neanderthals of about 60 000 years in Eurasia, coupled with only highly disputed evidence of gene flow between them, reinforce calls for a return to a greater taxonomic separation than that of a mere sub-species of *H. sapiens* . . . .

Another outcome of the new Qafzeh dates is that modern *H. sapiens* fossils are now known from Israel as early as they are known from southern Africa, and this may support claims for an early south African presence of modern humans . . . . Both palaeoanthro-pologist and geneticist proponents of the "Out of Africa" model, which proposes a late Pleistocene African origin for modern *H. sapiens*, have tended to favour a subSaharan home for the species . . . .

Source: Chris Stringer, "The Dates of Eden." *Nature*, Vol. 331, No. 6157, p. 565. Reprinted with permission from *Nature*. Copyright 1988 Macmillan Magazines Limited.

## PROTO-WORLD

There is also some linguistic evidence that has a bearing on the replacement theory. If the theory is correct, then it might be suspected that the four thou-

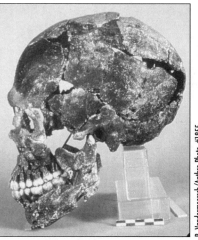

▼
One of the skulls discovered at Qafzeh
▲

B. Vandermeersch/Anthro-Photo, #1855

sand to eight thousand languages that are spoken in the world today should all be the evolutionary product of the language that was spoken by the *H. sapiens* population represented by Mitochondrial Eve. Historical linguists have shown that this may well be the case. By comparing the most stable words in the vocabularies of literally thousands of languages, some historical linguists have tentatively concluded that there was a parent language of all *H. sapiens*, which they call **proto-World**. Although long since extinct, proto-World has not only been traced to Africa, but has also been identified as the language that the *H. sapiens* who lived in the Qafzeh region spoke. This too supports the contention that our species originated in Africa, then spread to the north, and from there continued to venture farther afield.

Of course, it could be argued that the Neandertals who lived in the same region also spoke proto-World. But this is unlikely since research has shown that the tongue and vocal tract of the Neandertals were considerably different from those of their *H. sapiens* counterparts, so much so that some have insisted that the Neandertals were incapable of modern human speech. Although other experts believe that this conclusion is excessive, the fact that there were anatomical differences in the speech apparatus of Neandertals and *H. sapiens* also lends credence to the view that proto-world was the language that the *H. sapiens* (but not the Neandertals) who lived in the Qafzeh region spoke.

# What Happened to the Neandertals?

Given the abovementioned genetic, fossil, and linguistic information, the replacement theory may yet win the day. In the meantime, it is worthwhile to point out that even though the replacement theory can account for our origin, it does not explain why the Neandertals became extinct — assuming of course that they did become extinct, and are not modern *H. sapiens,* as supporters of the Neandertal-phase theory claim. Perhaps, as some of those who believe that the Neandertals became extinct have suggested, the Neandertals interbred with incoming humans who were more like we are or succumbed to a glacial advance or to disease. It may even be, as some have suggested, that the Neandertals were exterminated by incoming humans. Pending further research, the disappearance of the Neandertals also looms large as an unanswered theoretical problem in the study of our origins. See QQ 8.14.

**FIGURE 8.12** An artist's rendering of Neandertal

Neg. 4915 (Photo by D. Finnin/C. Chesek), Courtesy Department of Library Services, American Museum of Natural History

## 8.14 THE FATE OF THE NEANDERTALS

Erik Trinkaus describes how palaeoanthropologists are trying to determine the ultimate fate of the Neandertals.

• • •

Paleoanthropological research concerning the Neandertals, their neighbors, and the origins of modern humans has . . . experienced a renaissance in the past decade. New questions, new specimens, and new forms of data are providing both new answers to old persistent problems and new insights into this period of human evolution. We are beginning to reach a new consensus on the old "fate of the Neandertals" phylogenetic issue, with fewer and fewer paleontologists taking simplistic extreme positions and most recognizing the regionally complex nature of the transition.

Behavioral questions, oriented toward both phylogenetic and adaptational issues are being posed in more sophisticated forms, and researchers are developing analytical techniques that promise more precise answers to them. From this we should be increasingly close to understanding the proceses, both bilogical and cultural, that were involved in the origin of modern humans and to appreciating the role of the Neandertals in that process.

This should provide a window on the past, one that has implications for the evolutionary basis of modern human biology and culture and for the greater span of hominid evolution. It should also help us to comprehend our relatives, the Neandertals.

Source: Erik Trinkaus, "The Neandertals and Modern Human Origins," *Annual Review of Anthropology*, Vol. 15 (1986), p. 212.

# ANATOMICALLY MODERN HUMANS

In 1868, in the Dordogne region of southwestern France, in the village of Les Eyzies, an unusual collection of fossils was unearthed in a rock shelter by a group of workers who were constructing a roadbed for a railway. What they found were the partial remains of four or perhaps five individuals buried alongside flint tools and ancient animal bones. The human remains were those of a middle-aged man, one or two younger men, a young woman, and a two- or three-week-old child. These were called "Cro-Magnon," apparently to honour a hermit named Magnou who lived nearby. Today, for the sake of greater precision, these specimens and others like them are referred to as **anatomically modern humans**, the common designation for the most intelligent of the wise humans, ourselves.

## BIOLOGICAL FEATURES

Anatomically modern humans, who appeared at least 40 kya and possibly as long as 100 kya or more (see Table 8.3), are biologically indistinguishable from us. Their cranial capacity, which is estimated to have been about 1600 cm$^3$, is well within the range for modern humankind; as will be recalled, ours varies from about 900 cm$^3$ to about 2300 cm$^3$. In addition, their skulls were round; their foreheads were high; their brow ridges were weakly developed or absent; their faces, jaws, and teeth were relatively small; and their skeletons were composed of thin, light bones. And, like us, they occupied

**TABLE 8.3** The location and age of selected anatomically modern human remains discovered in Africa, Asia, Europe, Australia, North America, and South America

| CONTINENT | COUNTRY | AGE OF OLDEST REMAINS (kya) | |
|---|---|---|---|
| Africa | Ethiopia | 130 | (?)[1] |
| | Tanzania | 130 | (?) |
| | South Africa | 115 | (?) |
| | Sudan | 80 | (?) |
| | Morocco | 50 | |
| Asia | Israel | 90 | |
| | Borneo | 40 | (?) |
| | China | 20 | (?) |
| | Indonesia | 20 | (?) |
| Europe | Bulgaria | 43 | (?) |
| | Germany | 36 | |
| | (formerly) Yugoslavia | 34 | |
| | France | 30 | |
| | Czech Republic | 30 | |
| Australia | Australia | 40 | |
| North America | United States (California) | 30 | (?) |
| South America | Brazil | 30 | (?) |

[1](?) indicates a disputed age.

virtually every part of the globe. It is because of these features that anatomically modern humans are generally regarded as the first ***Homo sapiens sapiens*** (*H. sapiens sapiens*) — the subspecies of *H. sapiens* that is represented by contemporary humankind.

## ARTISTIC EXPRESSION

The intelligence of our immediate fore-runners was reflected in their art. Unlike earlier humans, anatomically modern humans recorded their impressions of the world around them in sculpture and painting. Their most outstanding sculptures were **Venus figurines** — small and delicate statues of pregnant women with exaggerated breasts and buttocks but

Neg. 326474, Courtesy Department Library Services, American Museum of Natural History

Venus figurine (Austria)

**FIGURE 8.13** Map showing the location of the cave galleries at Lascaux, France, and Altamira, Spain

FRANCE

Lascaux

Altamira

SPAIN

© 1975 Thames and Hudson Ltd., London

lacking facial features. Made from stone, bone, and ivory, the figurines may be some of the first examples of religious art, intended perhaps to represent a Mother Goddess who was responsible for the creation of life. Or, the figurines may represent something that we do not yet understand.

Anatomically modern humans also painted exceptionally beautiful pictures of the animals they hunted. These fabulous works of art have been found on the walls and ceilings of hidden caves, including the famous cave galleries at Lascaux, France, and at Altamira, Spain (see Figure 8.13). Although no one is certain, the paintings also may have had religious significance; perhaps they were intended to encourage supernatural powers to ensure a successful hunt. These paintings tell us that anatomically modern humans pursued mammoth, rhinoceros, bison, reindeer, and horse. On the other hand, the paintings may have been "art for art's sake" — a kind of intellectual play with concrete representations of familiar natural objects that our species seems uniquely equipped to produce and enjoy. See QQ 8.15.

## QQ 8.15 CAVE ART

John Halverson explains what he believes is the significance of Palaeolithic cave art.

• • •

It may be that the true significance of Paleolithic art lies in the history of consciousness. This art provides our earliest evidence of abstraction, the foundation of reflective thought. The images are abstracted from nature, yet concretely represent natural objects with their own independent existence, made, not given. Consciously created, they would invite a conscious response rather than the automatic or habituated reaction evoked by their natural counterparts. As external representations were disengaged from nature, so the percept [or mental product] was freed from the presence of its object and perhaps consciously differentiated, thus, like the figure on the wall, acquiring its own independent status. Knowing that they were making representations, the painters may have had an incipient awareness that it was a process of projection, a transferring of images in the mind to images on a wall. And this self-awareness of the mind in operation would be the first step in concept formation, the beginning of genuine thought . . . . If, as I suppose,

*294* CHAPTER EIGHT

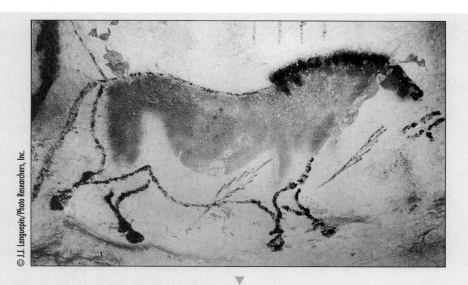

A painted horse from Lascaux, France

Paleolithic art had no practical function, it may be the first cultural work of mankind to be freed from praxis [or accepted practice] and therefore belongs to the general category of play and in the specific category of play with the mind. This is not the body play of animals or the mimetic [or imitative] play of higher primates, but a playing with signifiers. It was an activity undertaken for its own sake, with its own pleasures and rewards, but nevertheless pregnant with the cognitive future of humanity, for out of such activity would emerge conscious, reflective thought.

Source: John Halverson "Art for Art's Sake in the Paleolithic" *Current Anthropology*, Vol. 28, No. 1 (1987), pp. 70–71.

## REACHING AUSTRALIA AND THE NEW WORLD

No less impressive than their art were the far-ranging movements of *H. sapiens sapiens*. Unlike their predecessors, they were able to live in Australia, which they reached from Southeast Asia about 50 kya, when low sea levels exposed a land mass called **Sahul** — a continental shelf that united what are now Australia, Papua New Guinea, and Tasmania. At the time, the coastlines of Sahul and Asia were only about 100 kilometres apart. Presumably, the distance across the open sea was covered on bamboo rafts that *H. sapiens sapiens* launched from Asia. See Figure 8.14.

Anatomically modern humans were also the first people to reach the New World, where they arrived about 12 kya, and perhaps earlier. While there is doubt about which specific group of anatomically modern humans gave rise to the first Americans, there is far less uncertainty about the route they traveled.

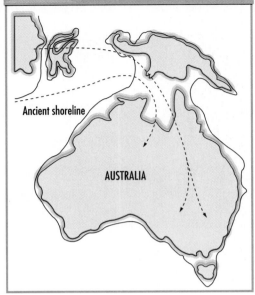

**FIGURE 8.14**    From southeast Asia to Sahul

Ancient shoreline

AUSTRALIA

The prevailing view is that they entered the New World via **Beringia**, a 1000 kilometre- to 2000 kilometre-wide land bridge between Siberia and Alaska (see Figure 8.15). Today, Beringia is under water, but during the last 40 000 years it was exposed frequently because more than 30 million square kilometres of the world's ocean water was locked up in glaciers in the high-altitude and high-latitude regions of Europe, Asia, and North America; this reduced the level of the world's oceans by about 100 metres and made it possible for anatomically modern humans to walk to the New World across land. Others may have come by boat, either by paddling across the Bering Strait, which is only about 90 kilometres wide, or possibly south of the strait across the open sea. It may even be that they came to the New World on foot across the ice since the Bering Strait is only about 45 metres deep and still occa-

sionally freezes over today. In any event, those who crossed over soon occupied both American continents and began to develop the distinctive American cultures that Europeans encountered thousands of years later. See QQ 8.16.

## QQ 8.16 ENTERING THE NEW WORLD

Knut Fladmark, Professor of Archaeology at Simon Fraser University, discusses the earliest human occupation of North America.

• • •

North American and the Old World approach each other most closely at the Bering Strait, and this is generally accepted as the main entry route for the early colonists. At different times in the past, people may have crossed the Strait by boats, on sea ice, or on dry land during glacial episodes when the water locked up in the great continental ice masses lowered world ocean levels by as much as 100 metres. Most of the interior of Alaska and the Yukon remained unglaciat-

ed at these times, so small bands of people following game, such as mammoths, into northeastern Siberia might have crossed the dry Strait into North America, unaware of any significant transition. Most archaeologists accept this theory, which implies an interior hunting way of life for the first North Americans. However, the Bering Land Bridge also had coastlines, and it is possible that some people, perhaps already adapted to a maritime existence on the shores of Asia, could have reached British Columbia around the North Pacific Rim, where rich stocks of fish and sea-mammals might have provided continuous resource "pathways" from Old World to New . . . .

Source: Knut R. Fladmark, *British Columbia Prehistory* (Ottawa: National Museums of Canada, 1986), p. 13.

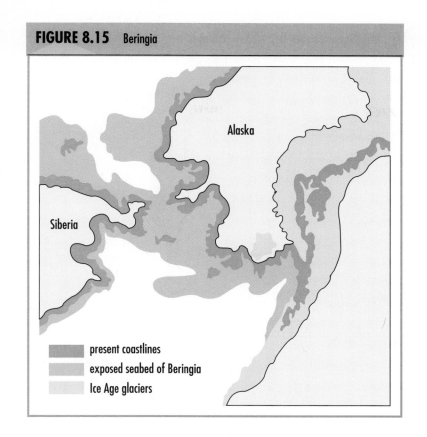

**FIGURE 8.15** Beringia

present coastlines
exposed seabed of Beringia
Ice Age glaciers

## TECHNOLOGICAL BREAKTHROUGHS

The skills required to produce art and occupy new territory also allowed anatomically modern humans to make important technological break-throughs. Although the first anatomically modern humans relied on Mousterian tools, within a short time they had mastered the **blade technique** — a method of making stone tools in which flakes at least twice as long as wide were struck from a core previously prepared for that purpose. This technique allowed them to make not only more tools from a single core, but also new implements including punches, needles, pins, and fasteners (see Figure 8.16 and QQ 8.17). Thereafter technology flourished. Along with a wide variety of stone, bone, and wooden tools, anatomically modern humans manufactured nets and snares, and these allowed them to incorporate more marine foods and small game into their diet. They also constructed larger and more efficient dwellings with improved hearths, and these were arranged in settlements that may have included upwards of 100 people. Thus, coupled with their other accomplishments, the technological skills of anatomically modern humans also mark them as the first undisputed *H. sapiens sapiens* — ourselves.

## FIGURE 8.16    Blade implements and their modern analogues

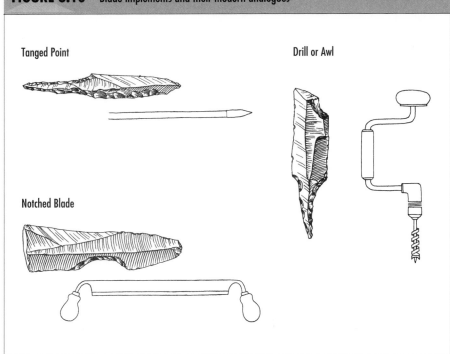

Tanged Point

Drill or Awl

Notched Blade

## 8.17 BLADE TOOLS

David Frayer discusses the nature of blade tools.

• • •

Probably the single most [outstanding technological] characteristic that is associated with the appearance of post-Mousterian groups is the wide usage of blade production techniques . . . . Although blade tools are not absent in earlier periods . . . [in the immediate post-Mousterian period] the prismatic [or many faceted] core forms the nearly universal step in stone tool production. Tools derived from these cores are a significant improvement over the earlier flake tools.

Blade tools have extremely thin cross-sections and sharp edges, providing more effective cutting and scraping surfaces. In some cases, the nonworking edge was even dulled before usage, presumably so that the implement could be used without damaging the hand. The significant feature of blade tools is that they can be modified into a variety of forms, designed for specific purposes.

Source: David W. Frayer, "Biological and Cultural Change in the European Late Pleistocene and Early Holocene." In Smith, Fred H. and Spencer, Frank, eds. *The Origins of Modern Humans: A World Survey of the Fossil Evidence* (New York: Alan R. Liss, Inc., 1984), pp. 213–214.

# PILTDOWN

As a postscript to the study of the fossil record it is worthwhile to recount the story of *Eoanthropus dawsoni*, Dawson's "dawn man", which is otherwise

known as **Piltdown**. In 1912 Charles Dawson, a British lawyer, reported that he had found fragments of a human skull at Piltdown, in Sussex, England. Later he discovered an ape-like jaw that he claimed belonged with the skull. It was a rare combination — a modern skull with the jaw of an ape — and, at the time of its discovery, it was hailed by many prominent scientists as the long-sought missing link that would close the gap between apes and humans. See QQ 8.18.

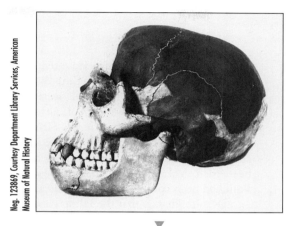

Neg. 123869, Courtesy Department Library Services, American Museum of Natural History

▼

Piltdown (black areas are the fragments that

Dawson claimed to have discovered)

▲

## 8.18 DAWSON'S "DAWN MAN"

**Charles Dawson describes his find.**

• • •

Several years ago I was walking along a farm-road close to Piltdown Common . . . when I noticed that the road had been mended with some peculiar brown flints not usual in the district. On enquiry I was astonished to learn that they were dug from a gravel-bed on . . . [a nearby] farm, and shortly afterwards I visited the place . . . . Upon one of my subsequent visits . . . one of the [work]men [there] handed to me a small portion of an unusually thick human parietal bone [from the top of the skull]. I immediately made a search, but could find nothing more . . . .

It was not until some years later, in the autumn of 1911, on a visit to the spot, that I picked up . . . anoth-

er and larger piece belonging to the frontal region of the same skull . . . . I . . . took it to Dr. A. Smith Woodward at the British Museum (Natural History) for comparison and determination. He was immediately impressed with the importance of the discovery, and we decided to employ labour and to make a systematic search . . . [during the spring of 1912].

Considering the amount of material excavated and sifted by us, the specimens discovered were numerically small and localized.

Apparently the whole or greater part of the human skull had been shattered . . . . [into] pieces . . . . Of these we recovered . . . as many fragments as possible. In a somewhat deeper depression of the undisturbed gravel I found the right half of a human mandible . . . [or lower jaw]. Dr. Woodward also dug up a small portion of the occipital bone of the skull

ARCHAIC *HOMO SAPIENS*, NEANDERTALS, AND OURSELVES

from within a yard of the point where the jaw was discovered, and at precisely the same level. The jaw appeared to have been broken . . . perhaps when it lay fixed in the gravel . . . .

Source: Charles Dawson, "On the Discovery of a Palaeolithic Human Skull and Mandible in a Flint-Bearing Gravel Overlying the Wealden (Hastings Beds) at Piltdown, Fletching (Sussex)." *Quarterly Journal of the Geological Society of London*, Vol. 69 (1913), pp. 117–121.

Although some scientists doubted the authenticity of Piltdown, it was accepted as genuine for many years. But in 1953, by using the fluorine test, it was determined that the fossils were a fraud. Most scholars now believe that the forgery was Dawson's idea, although no one is certain of his reasons. Others, however, claim that it was Sir Arthur Conan Doyle (the creator of Sherlock Holmes) who was to blame. In any event, the fossils were a hoax; although the skull was human the jaw had belonged to an ape. Its teeth had been filed down to make them appear human; then the jaws, teeth, and skull had been coloured with a chemical to make them appear old.

Despite the fact that the Piltdown story seems to show a careless attitude on the part of scientists, we are all ultimately indebted to the perpetrator of the hoax. More than anything else, the discovery of the fraud shows that the fossil record must be subject to constant scrutiny and revision, and that anthropologists' conclusions about our evolutionary history are not mere guesses, but, in the final analysis, carefully considered scientific opinions about the common heritage of each and every person who is alive today.

# KEY TERMS AND CONCEPTS

HOMO SAPIENS

REGIONAL CONTINUITY OR MULTIREGIONAL MODEL

SINGLE ORIGIN OR NOAH'S ARK MODEL

STEINHEIM (*shtyn'-hym*)

SWANSCOMBE (*swanz'-kohmb*)

ARCHAIC H. SAPIENS

TRANSITIONAL FORMS

LEVALLOIS METHOD (*li-val-wah'* + *method*)

NEANDERTALS (*nee-an'-der-tahlz*)

OLD MAN OF LA CHAPELLE (*old man of* + *lah* + *shah-pel'*)

MOUSTERIAN (*moo-steer'-ee-en*)

SCRAPERS

NOTCHES

BURINS (*bur'-inz*)

BORERS

DRILLS

DENTICULATES (*den-tik'-yoo-layts*)

PRESAPIENS THEORY

HOMO NEANDERTALENSIS (*homo* + *nee-an'-der-tahl-en-sis*)

PRENEANDERTAL THEORY

THIRD INTERGLACIAL FORMS

PROGRESSIVE NEANDERTALS

CLASSIC NEANDERTALS

HOMO SAPIENS NEANDERTALENSIS

NEANDERTAL-PHASE THEORY

REPLACEMENT THEORY

MITOCHONDRIAL EVE
(*my-toh-kawn'-dree-uul* + *Eve*)

QAFZEH CAVE (*kahf'-tzeh* + *cave*)

PROTO-WORLD

ANATOMICALLY MODERN HUMANS

HOMO SAPIENS SAPIENS

VENUS FIGURINES

SAHUL (*sah-ool'*)

BERINGIA (*beh-rin'-jee-uh*)

BLADE TECHNIQUE

EOANTHROPUS DAWSONI
(*ee-an'-throh-pus* + *daw'-soh-nee*)

PILTDOWN

# SELECTED READINGS

## INTRODUCTION

Leakey, R. and Lewin, R. *Origins Reconsidered: In Search of What Makes Us Human.* New York: Doubleday, 1992.

## HOMO ERECTUS AND OURSELVES

Howells, W. "Explaining Modern Man: Evolutionists Versus Migrationists." *Journal of Human Evolution*, 1976, pp. 477–495.

Mellars, P. and Stringer, C., eds. *The Origin and Dispersal of Modern Humans.* Edinburgh: Edinburgh University Press, 1988.

Stringer, C. "The Emergence of Modern Humans." *Scientific American,* Vol. 263 (1990), pp. 98–104.

Wolpoff, M., Wu, X., and Thorpe, A. "Modern *Homo sapiens* Origins: A General Theory of Hominid Evolution Involving Fossil Evidence from East Asia." In Smith, F. and Spencer, F., eds. *The Origins of Modern Humans: A World Survey of the Fossil Evidence.* New York: Alan Liss, 1984, pp. 411–484.

## ARCHAIC HOMO SAPIENS

Jacobs, K. "Human Origins." In Godfrey, L., ed. *What Darwin Began: Modern Darwinian and Non-Darwinian Perspectives on Evolution.* Boston: Allyn and Bacon, 1985, pp. 274–292.

Jones, J. and Rouhani, S. "How Small was the Bottleneck?" *Nature,* Vol. 319 (1986), pp. 449–450.

McBrearty, S. "The Origin of Modern Humans." *Man,* Vol. 25 (1990), pp. 129–143.

Oakley, K. "Swanscombe Man." *Proceedings of the Geological Association of London,* Vol. 63 (1952), pp. 271.

## NEANDERTALS

Brace, C. "Refocusing on the Neanderthal Problem." *American Anthropologist,* Vol. 64 (1962), pp. 729–741.

Trinkaus, E. and Howells, W. "The Neanderthals." *Scientific American,* Vol. 241 (1979), pp. 118–133.

## Myths about Neandertals

Hammond, M. "The Expulsion of the Neanderthals from Human Ancestry: Marcellin Boule and the Social Context of Scientific Research." *Social Studies of Science*, Vol. 12 (1982), pp. 1–36.

Kennedy, K. *Neanderthal Man.* Minneapolis: Burgess Press, 1975.

## The Appearance and Culture of the Neandertals

Rak, Y. "The Neanderthal: A New Look at an Old Face." *Journal of Human Evolution*, Vol. 15 (1986), pp. 151–164.

Solecki, R. Shanidar: *The First Flower People.* New York: Knopf, 1971.

Trinkaus, E. *The Shanidar Neandertals.* New York: Academic Press, 1983.

Trinkaus, E., ed. *The Mousterian Legacy.* Oxford: British Archaeological Reports, International Series, Number 151, 1983.

## Neandertals and Ourselves

Arsenburg, B., *et. al.* "A Reappraisal of the Anatomical Basis of Speech in Middle Paleolithic Hominids." *American Journal of Physical Anthropology*, Vol. 83 (1990), pp. 137–146.

Cann, R., Stoneking, M., and Wilson, A. "Mitochondrial DNA and Human Evolution." *Nature*, Vol. 325 (1987), pp. 31–36.

Gargett, R. "Grave Shortcomings: The Evidence for Neanderthal Burials." *Current Anthropology*, Vol. 30 (1989), pp. 157–190.

Lieberman, P. *Uniquely Human: The Evolution of Speech, Thought, and Selfless Behavior.* Cambridge: Harvard University Press, 1991.

Lieberman, P. and Crelin, E. "On the Speech of Neanderthals." *Linguistic Inquiry*, Vol. 2 (1971), pp. 203–222.

Smirnov, Y. "Intentional Human Burial: Middle Paleolithic (Last Glaciation) Beginnings." *Journal of World Prehistory*, Vol. 3 (1989), pp. 199–233.

Smith, F. "The Neandertals: Evolutionary Dead End or Ancestors of Modern People?" *Journal of Anthropological Research*, Vol. 47 (1991), pp. 219–238.

Spuhler, J. "Evolution of Mitochondrial DNA in Monkeys, Apes and Humans." *Yearbook of Physical Anthropology*, Vol. 31 (1988), pp. 15–48.

Straus, W. and Cave, A. "Pathology and the Posture of Neandertal Man." *Quarterly Review of Biology*, Vol. 32 (1957), pp. 348–363.

Thorne, A. and Wolpoff, M. "The Multiregional Evolution of Humans." *Scientific American*, Vol. 266 (1992), pp. 76–83.

Trinkaus, E. "The Neanderthals and Modern Human Origins." *Annual Review of Anthropology*, Vol. 15 (1986), pp. 193–218.

Wilson, A. and Cann, R. "The Recent African Genesis of Humans." *Scientific American*, Vol. 266 (1992), pp. 68–73.

## What Happened to the Neandertals?

Bower, B. "Neandertals' Disappearing Act." *Science News*, Vol. 139 (1991), pp. 360–363.

Brace, C. "The Fate of the "Classic" Neanderthals: A Consideration of Hominid Catastrophism." *Current Anthropology*, Vol. 5 (1964), pp. 3–34.

Kolata, G. "The Demise of the Neanderthals: Was Language a Factor?" *Science*, Vol. 186 (1974), pp. 618–619.

## Anatomically Modern Humans

Bräuer, G. "A Craniological Approach to the Origin of Anatomically Modern *Homo sapiens*." In Smith, F. and Spencer, F., eds. *The Origins of Modern Humans: A World Survey of the Fossil Evidence.* New York: Alan Liss, 1984, pp. 327–410.

Dillehay, T. and Meltzer, D., eds. *The First Americans: Search and Research.* Boca Raton: CRC Press, 1991.

Habgood, P. "The Origin of Anatomically Modern Humans in Australasia." In Mellars, P. and Stringer, C., eds. *The Human Revolution: Behavioral and Biological Perspectives in the Origins of Modern Humans.* Princeton: Princeton University Press, 1989, pp. 245–273.

Hoffecker, J., Powers, W., and Goebel, T. "The Colonization of Beringia and the Peopling of the New World." *Science*, Vol. 259 (1993), pp. 46–53.

Hopkins, D., Matthews, J., Schweger, C., and Young, B., eds. *Paleoecology of Beringia.* New York: Academic Press, 1982.

Kirk, R. and Szathmary, E., eds. *Out of Asia: Peopling of the Americas and the Pacific.* Canberra, Australia: The Journal of Pacific History Inc., 1985.

Laughlin, W. and Harper, A. "Peopling of the Continents: Australia and America." In Mascie-Taylor, C. and Lasker, G., eds. *Biological Aspects of Human Migration.* Cambridge: Cambridge University Press, 1988, pp. 14–40.

Leroi-Gourhan, A. "The Evolution of Paleolithic Art." *Scientific American*, Vol. 218 (1968), pp. 58–70.

Owen, R. "The Americas: The Case Against an Ice-Age Human Population." In Smith, F. and Spencer, F., eds. *The Origins of Modern Humans: A World Survey of the Fossil Evidence.* New York: Alan Liss, 1984, pp. 517–563.

Putnam, J. "In Search of Modern Humans." *National Geographic*, Vol. 174, No. 4 (1988), pp. 467–468.

Ruspoli, M. *The Cave of Lascaux: The Final Photographs.* New York: Abrams, 1986.

Stevens, A. "Animals in Palaeolithic Cave Art," *Antiquity*, Vol. 49 (1975), pp.54–57.

## Piltdown

Blinderman, C. *The Piltdown Inquest.* Buffalo, New York: Prometheus Books, 1986.

Millar, R. *The Piltdown Men: A Case of Archaeological Fraud.* London: Victor Gollanez, 1972.

Spencer, F. *The Piltdown Papers, 1908–1955: The Correspondence and other Documents Relating to the Piltdown Forgery.* New York: Oxford University Press, 1990.

Spencer, F. Piltdown: *A Scientific Forgery.* New York: Oxford University Press, 1990.

Weiner, J. *The Piltdown Hoax.* Oxford: Oxford University Press, 1955.

Winslow, J. and Meyer, A. "The Perpetrator at Piltdown." *Science '83,* September, 1983, pp. 33–43.

# CHAPTER NINE

## PALAEOLITHIC CULTURES

*contents at a glance*

# INTRODUCTION

One of the reasons that prehistoric humans such as *H. habilis* and *H. erectus* are regarded as our ancestors is because of their biological features. Another is that tools have been discovered alongside their fossils. Compared to the machines we use, the tools are relatively simple — mostly nothing more than roughly-shaped pieces of stone. However, there is no doubt that these implements played an important role in the evolution of our kind. Considering that our ancestors were competing with predators such as lions and scavengers such as hyenas, jackals, and vultures, it is hard to imagine how they could have survived without using tools. The first tools are also important in another respect — they signal the beginning of culture. See QQ 9.1.

Technology — one of the hallmarks of our species

Courtesy of National Aeronautics and Space Administration (STS-2 – Columbia)

## QQ 9.1 TOOLS AND THE ORIGIN OF CULTURE

Sherwood Washburn discusses the relationship between tools, evolution, and culture.

• • •

The earliest stone tools are chips or simple pebbles, usually from river gravels. Many of them have not been shaped at all, and they can be identified as tools only because they appear in . . . caves or other locations where no such stones occur. [However, t]he huge advantage that a stone tool gives to its user must be tried to be appreciated. Held in the hand it can be used for pounding, digging or scraping. Flesh and bone can be cut with a flaked chip, and what would be a mild blow with the fist becomes lethal with a rock in the hand. Stone tools can be employed, moreover, to make tools of other materials. Naturally occurring sticks are nearly all rotten, too large, or of inconvenient shape; some tool for fabrication is essential for the efficient use of wood. The utility of a mere pebble seems so limited to the user of modern tools that it is not easy to comprehend the vast difference that separates the tool-user from . . . [animals that do not use tools. But i]t was the success of the simplest tools that started the whole trend of human evolution and led to the civilizations of today.

Source: Sherwood Washburn, "Tools and Human Evolution," *Scientific American*, Vol. 203, No. 3, p. 63. Copyright © 1960 by Scientific American, Inc. All rights reserved.

Were it not for archaeologists, far less would be known about humankind's early experiments with culture. Their aim is to reconstruct these endeavours by examining prehistoric remains. Such remains can be divided into two general categories: ecofacts and artifacts. **Ecofacts** are what are left of prehistoric natural resources such as minerals, plants, and animals; **artifacts** are the remnants of manufactured prehistoric items ranging from weapons to art. Taken together, these materials comprise the **archaeological record**, which is the single most important source of evidence that archaeologists draw upon to reconstruct the cultures of the past. By looking closely at the contents of this ancient record, archaeologists have shed light not only on the cultural practices of the first members of our genus, but also on the cultures that bridged the gap between the first anatomically modern humans and ourselves.

# DISCOVERING THE PAST

## IN THE BEGINNING

Although no one is certain when humankind's fascination with archaeology began, from the fifteenth century onward the study of the material remains of prehistoric cultures became an increasingly important area of scholarly investigation. Before then, ancient objects were prized mainly for their beauty or for their commercial value. However, in the 1400s, researchers began to appreciate that these objects were the key to understanding how their own civilizations had developed. As a result, during the fifteenth century, Italian humanists began to study the ruins of ancient Rome, and soon afterward they and their intellectual counterparts in England, France, and Germany were studying the ancient civilizations of Greece, Egypt, Persia, and Assyria.

## DATING PREHISTORIC REMAINS

Since then our understanding of prehistory has been subject to constant improvement, in large part because of the techniques that archaeologists have embraced. Take the case of dating techniques. When Napoleon invaded Egypt in 1798, he brought 167 scientists and technicians with him to record information about Egyptian antiquities and to select, conserve, and transport artifacts from Egypt to France. Within three years, the members of Napoleon's Scientific and Artistic Commission had amassed an impressive collection of ancient objects, most of which were confiscated by the English in 1801, when Napoleon's army surrendered after it had been isolated for three years following the defeat of the French fleet by Admiral Nelson's forces at the Battle of Abukir Bay. Fortunately, the victors allowed the French scholars to sail home with their notes intact, and these were used to prepare a series of informative and stimulating books that paved the way for the future study of Egyptian prehistory. However, since neither the members of Napoleon's commission nor the antiquarians from the British Museum who acquired the collection could date the finds in an accurate way, the contents could not be arranged in proper chronological order.

Today the situation is different. In many cases ecofacts and artifacts can be assigned an absolute age — by using the potassium-argon method to date inorganic remains, and the carbon 14 method to date organic materials. In addition, depending upon the nature of what they unearth, archaeologists may bring other absolute dating techniques into play. As the contents of Table 1 indicate, one of these is **thermoluminescence**, which is used to estimate the absolute age of fired pottery and similar wares that are upwards of 10 000 years old. Another is **dendrochronology** or tree-ring dating, which is used to determine the absolute age of wooden beams and similar products, whose parent trees were felled as early as 7.4 kya. Still another is **obsidian hydration**, which is used to date volcanic glass (obsidian) upwards of 800 000 years old (obsidian was widely used in prehistoric times to manufacture cutting and scraping implements). It is largely because of the advent of these techniques that archaeologists can now arrange their finds in sequences that make chronological sense.

| TABLE 9.1 | Thermoluminescence, dendrochronology, and obsidian hydration |
|---|---|
| **TECHNIQUE** | **PURPOSE AND PROCEDURE** |
| Thermoluminescence | *Purpose:* to date fired pottery and similar wares that are upwards of 10 000 years old. *Procedure:* the age of the material is estimated by measuring the amount of light energy that has accumulated in the material since it was last superheated. |
| Dendrochronology (tree-ring dating) | *Purpose:* to date wooden beams and similar products, especially from the American Southwest, whose parent trees were felled as early as 7400 years ago. *Procedure:* the age of the material is determined by examining the "growth rings" displayed in the material and then matching these to a previously constructed master chronology. |
| Obsidian hydration | *Purpose:* to date obsidian — a volcanic glass that was widely used to manufacture prehistoric cutting and scraping implements — that may be upwards of 800 000 years old. *Procedure:* the age of the material is estimated by measuring the thickness of the weathering rind that forms in the material because of the absorption of water after the material is flaked. |

Equally important, even when ecofacts and artifacts cannot be assigned an absolute age, archaeologists can still rely on the principle of stratigraphy to determine the relative age of what they unearth. As will be recalled from Chapter 5, this principle is based on the assumption that items discovered at the bottom of an undisturbed collection of prehistoric remains are older than those found nearer the surface. Although not as precise as absolute dating, relative dating can also play a role in establishing the age of ancient remains. See QQ 9.2.

# 9.2 ABSOLUTE AND RELATIVE DATING IN ARCHAEOLOGY

David Thomas discusses the importance of absolute and relative dating techniques in archaeology.

• • •

The Fourth Egyptian Dynasty lasted from 2680 to 2565 B.C. The Roman Colosseum was constructed between A.D. 70 and 82. In the seventeenth century, Dr. John Lightfoot proclaimed that God created the entire Earth in 4004 B.C., at precisely 9 A.M. on October 23. Such dates are expressed as specific units of scientific measurement: days, years, centuries, or millennia, but no matter what measure — Lightfoot computed his estimate by projecting biblical lifespans — all absolute determinations attempt to pinpoint a discrete, known interval in time.

Archaeologists also measure time in a second, more imprecise manner by establishing the *relative date*. As the name implies, relative duration is monitored not through specific segments of absolute time but, rather, through relativistic relationships: earlier, later, more recent, after Noah's flood, prehistoric, and so forth. Although not as precise as absolute dating, relative estimates are sometimes the only dates available. Taken together, both forms of dating, absolute and relative, give archaeologists a way of controlling the critical dimension of time.

Source: David Hurst Thomas, *Archaeology*, second edition (New York: Holt, Rinehart and Winston, Inc., 1989), p. 282.

## LOCATING SITES

Great strides have also been made in locating **sites** — places where prehistoric materials are buried. Unlike Napoleon's technicians, archaeologists no longer depend strictly on monuments such as pyramids to guide them to sites; they also rely on **surface surveys** — inspecting the ground for signs of human activity, usually by walking or riding over the landscape in search of ancient material that is scattered on the surface. In other cases potential sites are identified from the air, by looking for discoloured patches of earth or stones that indicate where prehistoric peoples may have lived or worked. Quite frequently, archaeologists also ask local residents for information about where sites can be found. However, a site is excavated only if there is a strong suspicion, based on comprehensive prior research, that its contents will contribute to our understanding of human prehistory; because, once excavated, an archaeological site is, in effect, destroyed. See QQ 9.3 on page 310.

## ARCHAEOLOGY AND CULTURE

Over the years the goals of archaeology have expanded, and this too has contributed to our understanding of the past. Today archaeologists are concerned not only with the ancient empires of Europe and the Near East, but also with cultures that developed in other parts of the world. Moreover, instead of simply describing and classifying artifacts as the first archaeolo-

Martha Joukowsky examines the nature of archaeological sites and explains what makes them worthy of study.

• • •

Archaeology covers a wide diversity in time, space, subject matter, and approach. What is the nature of the evidence, or how does the archaeologist know when a site is a site? Some people think that archaeologists excavate because they think they will find something in one special spot. This was sometimes the case in the past, as when Heinrich Schliemann planned his excavations at Troy. But today when archaeologists excavate in one spot, it is almost always because they *know* after prolific research, surveys, and study that a site exists and is important, and that it will provide a relevant key to a gap in our understanding of a past culture . . . .

Therefore, the selection of a site for excavation depends upon what the excavator is attempting to find, what he or she is attempting to answer by the excavation.

▼

Archaeological site in southern Ontario

▲

Source: Martha Joukowsky, *A Complete Manual of Field Archaeology: Tools and Techniques of Field Work for Archaeologists* (Englewood Cliffs, New Jersey: Prentice-Hall, Inc., 1980), p. 38.

gists did, modern researchers focus their attention on the cultural significance of what they unearth.

When an archaeologist discovers ecofacts and artifacts that are indicative of a specific activity that a particular group of prehistoric people undertook in the past — say, for instance, discarded caribou bones and broken arrows found together in an ancient garbage pit — the collection is called a **subassemblage**. And when the subassemblages of an archaeological site are grouped together to reveal how the members of an entire community lived, the remains are called an **assemblage**. Prehistoric cultures are reconstructed by carefully examining the contents of multiple assemblages. If the remains of these assemblages are from the same general area, were deposited at about the same time, and if the artifacts they contain are similar enough in terms of their form and their function to be grouped together, then the assemblages are said to constitute an **archaeological culture**. The goal of archaeology is to observe, record, and interpret such cultures. In this sense, archaeologists are first and foremost cultural anthropologists. See QQ 9.4.

## 9.4 ARCHAEOLOGY

Bruce Trigger, Professor of Anthropology at McGill University, discusses the scope and nature of archaeology.

• • •

If the social sciences possessed an almost total understanding of socio-cultural process and were able to control for the effects of a sufficient number of parameters, it might be possible to predict past or future on the basis of the present and to demonstrate that no other past or future was possible. Until such feats can be done, the study of prehistory must be essentially an explanation of the archaeological record. Insofar as an understanding of past development is an important aspect of understanding current variations in behaviour and culture, such a study, while idiographic [or concerned with specific cases], is a vital part of a scientific study of man. Moreover, since the formulation and testing of generalizations about human behaviour can be based on contemporary, as well as archaeological, data, while the past can be explained only with the help of archaeological data, the explanation of past events must be regarded as the most important goal of archaeological research.

Source: Bruce G. Trigger, *Time and Traditions: Essays in Archaeological Interpretation* (New York: Columbia University Press, 1978), pp. 51–52.

# CLASSIFYING PREHISTORIC CULTURES

## THE THREE AGE SYSTEM

In order to make sense of the cultures they study, archaeologists arrange them in groups. Between 1812 and 1859, the most popular method of achieving this goal was based on the **Three Age System** that was developed by the Danish scholar Christian Thomsen, who divided prehistoric cultures into three main groups. According to Thomsen's scheme, the earliest prehistoric cultures were **stone age cultures**, which are now known to have arisen in Africa about 2.5 mya and featured implements made from stone. These, he maintained, gave rise to **bronze age cultures**, which arose about 5 kya in the Near East and featured implements made from bronze. Then came **iron age cultures**, which arose about 4 kya in Asia, and featured tools that were fashioned from iron.

However, by the middle of the nineteenth century it had become clear that the stone age included much more cultural material than Thomsen had originally suspected, and that it needed to be expanded in order to make better sense of the past. As a result, Thomsen's classification of ancient societies was replaced by one that was developed by Sir John Lubbock. Based on Lubbock's system, archaeologists now divide stone age cultures into three subsidiary groups: old stone age or **palaeolithic cultures**, which appeared towards the end of the Pliocene epoch (*c.* 2.5 mya); middle stone age or **mesolithic cultures**, which originated at the beginning of the Holocene epoch (*c.* 11 kya); and new stone age or **neolithic cultures**, which appeared shortly thereafter (*c.* 8 kya). See Table 9.2 and QQ 9.5.

**TABLE 9.2**  A simplified chronology of prehistoric cultures

| AGE | DIVISIONS | SUBDIVISIONS | FIRST INAUGURATED | ASSOCIATED CULTURAL REMAINS | ASSOCIATED FOSSIL REMAINS |
|---|---|---|---|---|---|
| stone | palaeolithic | lower palaeolithic | 2.5 mya | Oldowan and Acheulian | *Australopithecus,* early *Homo,* and *H. erectus* |
| | | middle palaeolithic | 75 kya | Mousterian | Neandertals |
| | | upper palaeolithic | 40 kya | Various | Anatomically modern humans |
| | mesolithic | | 11 kya | Various | Anatomically modern humans |
| | neolithic | | 8 kya | Various | Anatomically modern humans |
| bronze[1] | | | 5 kya | Various | Anatomically modern humans |
| iron[1] | | | 4 kya | Various | Anatomically modern humans |

[1] These cultures are often referred to as civilized cultures.

# 9.5 THOMSEN AND LUBBOCK

V. Gordon Childe discusses the classification systems developed by Christian Thomsen and John Lubbock.

• • •

The classification into the Three Ages was originally devised for archaeological objects, relics and monuments, in order to show which of them belonged together. The terms Stone, Bronze, and Iron Ages go back to a Dane, Thomsen, who about 1812 used them for the arrangement and classification of exhibits in the newly-founded Museum of Northern Antiquities at Copenhagen. Thomsen had decided to group together objects made and used in the same period of time. No written records were available to show when the . . . inhabitants of Denmark made and used the objects to be catalogued and exhibited. But Thomsen knew that bronze had been used for cutting-tools before iron, and stone before bronze. All objects current before bronze came into use were therefore classified in the first division and labelled "Stone Age." All objects, of whatever material, found in graves or otherwise associated with bronze swords, spears, or axes, were labelled "Bronze Age," and so on.

This classification was adopted in other European countries, for it was found that in Great Britain, France, Switzerland, Italy, and Germany also, stone was used for weapons and tools before bronze, and bronze before iron. By the crucial year 1859, the division of prehistoric European antiquities among the three ages was generally accepted. But by that year it appeared that the first division was of unwieldy size . . . . So the Stone Age had to be subdivided. Lubbock proposed the principle of division that in the sequel won universal acceptance. He termed "Palaeolithic," or Old Stone Age, those implements found in association with remains of extinct and always wild animals, and sharpened by chipping but never by grinding. On the other hand, those found accompanied exclusively by the bones of recent animals — including domesticated species — and sometimes sharpened by grinding or polishing, he labelled "Neolithic," or New Stone Age . . . . By 1900 cultures were known belonging to the geological Recent, but still without domesticated animals or cultivated plants and polished stone implements. To accommodate them a "Mesolithic," or Middle Stone Age, was ultimately created.

Source: V. Gordon Childe, *Social Evolution.* Cleveland (Ohio: World Publishing Company, 1951), pp. 28–30.

## SUBDIVIDING THE PALAEOLITHIC

Since palaeolithic (or old stone age) cultures originated during the late Pliocene and dominated the entire Pleistocene or Ice Age, which is a span of almost 2.5 million years, these cultures are usually subdivided into three additional groups: **lower palaeolithic cultures**, which arose about 2.5 mya and persisted until about 75 kya; **middle palaeolithic cultures**, which developed about 75 kya and persisted until about 40 kya; and **upper palaeolithic cultures**, which originated about 40 kya and persisted until about 11 kya (see Table 9.2). Although our Plio-Pleistocene and Pleistocene ancestors have been examined in preceding chapters, in this chapter we will concentrate on what archaeologists have learned about lower and middle palaeolithic cultures and explain more fully what they have discovered about upper palaeolithic ones. In the next chapter we will turn our attention to mesolithic, neolithic, and civilized cultures, which were ultimately based on palaeolithic achievements.

# LOWER PALAEOLITHIC CULTURES

## OLDOWAN CULTURE

Based on the ecofacts and artifacts that date back to the Plio-Pleistocene, most archaeologists believe that the first recognizable cultural practices arose in Africa in conjunction with the appearance of Oldowan tools, about 2.5 mya. Whether their makers were australopithecines, habilines, or both, is unknown. As will be recalled, fossils belonging to both groups have been found in association with Oldowan implements. Nor is it clear how Oldowan tools were used. Some archaeologists believe that the Oldowan toolkit was used primarily to hunt and butcher small game such as lizards, tortoises, rodents, rabbits, young pigs, and young antelope; while others maintain that Oldowan tools were used, at least in part, to process the scavenged remains of large game animals that were killed by non-human predators. The same implements may also have been used to transform bones and branches into digging sticks in order to remove edible foods, such as roots, tubers, bulbs, and insect larvae, from the ground. Whatever the functions of the Oldowan tools, since the sites where they are found do not indicate that their makers built shelters, it is almost certain that these early people were nomadic. They likely lived in small groups that roamed the grasslands of Africa in search of plant and animal foods that were eaten at open-air sites. It is also likely that it was at such sites that the young were cared for and where the immature learned how to survive by observing and imitating their elders. See Figure 9.1.

## ACHEULIAN CULTURE

Acheulian culture, which superseded Oldowan culture, takes its name from Acheulian tools, which were first made about 1.5 mya. Among others, such tools included Acheulian hand axes, which were likely used for a wide variety of

**FIGURE 9.1** An artist's rendering of a group of Oldowan tool users

The Natural History Museum, London

purposes including hunting, butchering, wood-working, cutting, scraping, digging, and pounding. Although there is absolutely no doubt that *H. erectus* manufactured and used Acheulian tools, whether they invented these implements is unclear; perhaps the habilines invented Acheulian tools. The subsistence skills of *H. erectus* are likewise in dispute — whether they were capable of dispatching large game animals, such as elephants, remains to be determined — as are the ideas that they used spoken language to communicate, and that they established seasonal camps that featured huts equipped with hearths. All that is certain is that *H. erectus* knew how to control fire and moved from Africa into Asia and possibly into Europe, where they adapted to local conditions by harvesting the plants and animals that were available in their new homelands. While these were formidable cultural feats that made it possible for *H. erectus* to live in larger groups than their immediate ancestors, it may be that in most other respects their cultural lives had not changed dramatically. Whatever the case, taken together, the Oldowan and Acheulian traditions represent humankind's first undisputed experiments with technology — the hallmark of lower palaeolithic cultures. See QQ 9.6, and Figure 9.2 on page 316.

## QQ 9.6 LOWER PALAEOLITHIC CULTURES

J. Desmond Clark expresses his opinion about the nature of lower palaeolithic cultures.

Some 2.5–2.0 million years ago, contemporary with the period of global cooling heralding the onset of the

Pleistocene glacial/interglacial epoch, some . . . hominids turned to including meat in their diet and the earliest stone tools and bone food waste date to this time. These hominids are likely to have lived in small groups, to have ranged widely across the landscape and to have carried back the results of this foraging to a central place where food was shared among the members of the group . . . .

By 1.5 million years ago the first full-statured hominid, *Homo erectus*, appears. He is best known from the crania found at Koobi Fora and the virtually complete skeleton of a youth from West Turkana. These were large, powerful hominids and it was possible competition for resources with *Homo erectus* that led to the extinction of the robust Australopithecines about 1.0 million years ago. Two patterns of tool-kits are now present — the new, large bifaces of the Acheulian techno-complex and a continuation of the earlier Oldowan core and flake complex. This suggests to me two sets of equipment with different functional associations which may also have social implications, as, at many sites, these two traditions intermingle. Camping places were now larger, the range of tools broader and the sites contain more tools. By later Acheulian times, groups had moved out from the tropical savannas and now occupied a wide range of ecological niches throughout the continent only excepting the tropical lowland and montane [or mountain] forests and true desert. This was also the time of the diaspora into Eurasia and adaptation to a way of life very different from that in the African tropics. But the tool-kits were basically the same throughout the whole of the Old World occupied by Acheulian people and this suggests that basic life ways were also much the same throughout, though different animal and plant resources were being used.

Source: J. Desmond Clark, *Blood From Stones* (Indiana: Department of Anthropology, Indiana University), 1987, pp. [12–13].

# MIDDLE PALAEOLITHIC CULTURES

## MOUSTERIAN CULTURE

Middle palaeolithic cultures were based on the technological achievements of Neandertals, who invented Mousterian tools about 75 kya. Like Oldowan and Acheulian implements, Mousterian tools were fashioned mainly from stone that could be easily and predictably worked. Mousterian implements, however, were smaller, more varied, and more efficient than Oldowan and Acheulian ones. As already mentioned, the Neandertal toolkit included scrapers, notches, burins, borers, drills, denticulates, and improved hand axes, and these gave their makers much greater control over the environment than did earlier implements. Among other things, some Mousterian tools were used to prepare hides for clothing, blankets, and other protective coverings which, coupled with fire, allowed Neandertals to live in caves, in rock shelters, and in skin tents in northern Europe where winter temperatures were extremely cold. There, as in Asia and Africa, Neandertals also learned how to attach a stone point to the tip of a wooden pole. The result was a spear — the weapon that Neandertals used to hunt the large mammals (mammoth, woolly rhinoceros, bison, cave bear, reindeer, and wolf) that provided them with the bulk of their food. Neandertals may also have been the first to intentionally bury their dead, and this has encouraged some

The Natural History Museum, London

**FIGURE 9.2** An artist's rendering of a group of Acheulian tool users

anthropologists to speculate that they, like modern humans, may have believed in an afterlife. See QQ 9.7.

## BEAR SKULLS IN ALPINE CAVES

There is some evidence that Neandertals may even have had a bear cult, which is to say, they may have worshipped bears — if the ethnographic anal-

## 9.7 MOUSTERIAN CULTURE

F. Clark Howell comments on the culture of the Neandertals.

• • •

Neanderthal . . . had sure control of fire. He used it regularly, presumably could create it when he needed it and had progressed to the point of digging hearths in the floors of his caves. He was also a home builder as well as a cave dweller. There are several sites in Russia that appear to have served as dwellings of Neanderthal . . . . One such is marked by a rough ring of hearths; outside that is a large circle of heavy

elephant bones and tusks that may have served — along with wood, which is no longer preserved — as a framework to support animal skins . . . .

In keeping with the growing complexity of his life and the greater variety of his possessions and talents, Neanderthal . . . also apparently stood on the edge of becoming both an esthete and a mystic. For the first time in human experience, faint signs of decoration and artistic appreciation appear. He began scratching designs on bones. Some interesting objects have emerged from a dig at Tata in Hungary, none of them with any apparent utilitarian purpose. Rather, they seem to have been fashioned for esthetic or ceremonial reasons . . . .

Neanderthal [may also have] . . . buried his dead, which suggests an awareness of the transitoriness of life . . . [and a] concern about the future . . . .

Beyond these suppositions — and it must be remembered that they are only suppositions — it is impossible to go. However, some kind of intellectual and fantasy life was not only possible for Neanderthal . . . considering his intellectual potential, but is also overwhelmingly logical.

Source: From *Life Nature Library: Early Man*, by F. Clark Howell and the Editors of Time-Life Books © 1973, 1980 Time-Life Books Inc.

ogy is correct — in much the same way as some northern hunting and gathering peoples did until recent times. The Ainu of northern Japan are an example. Towards the end of winter, the Ainu captured a bear cub that they treated with the utmost respect until it was sacrificed in an elaborate religious ceremony that was held the following year. This religious observance, the Ainu believed, encouraged the spirit of the cub to return to the forest and tell the animals that lived there that they should willingly give themselves up to the Japanese hunters who pursued bear and other game in order to meet their subsistence needs.

In the case of the Neandertals, the species that they may have worshipped is a now extinct giant cave bear weighing between 400 kilograms and 450 kilograms (*Ursus spelaeus*) that lived in northern Europe during the Pleistocene. Some archaeologists believe that the Neandertals venerated these huge beasts, because the skulls of *Ursus spelaeus* appear to have been intentionally deposited in caves that are associated with other Neandertal remains (see Figure 9.3). One such cave in the Swiss Alps, for example, yielded thirteen bear skulls; seven of these were found in what appeared to be a roughly constructed stone chest that was covered with a massive stone slab, and the other six were set into the walls of the cave. Another cave in southern France housed a similar collection; it contained twenty

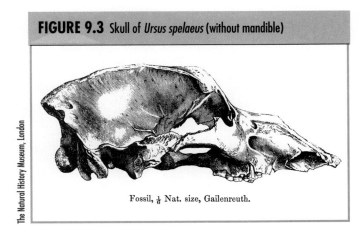

**FIGURE 9.3** Skull of *Ursus spelaeus* (without mandible)

The Natural History Museum, London

Fossil, ⅛ Nat. size, Gailenreuth.

bear skulls in a rectangular pit that was likewise covered by a massive stone slab (see QQ 9.8). Although a recent taphonomic study has suggested that the arrangement of the bear skulls in the two caves might well have been due to natural causes, the possibility that a bear cult existed is certainly an idea worth pursuing.

## 9.8 BEAR SKULLS

Anthony Wallace describes the scene as reported by those who originally discovered the skulls of cave bears in Alpine caves.

• • •

In the interior chambers of several Alpine caves occupied by Neanderthals there have been found, carefully arranged and protected, the skulls and occasional other bones of numerous cave bears. In some cases,

the skulls were carefully packed in stone chests, constructed of limestone slabs; in other cases, the skulls were mounted on niches in the wall, sometimes protected by flat stones. The skulls were often oriented east to west, and sometimes were intact, indicating that they had not been used for food; in some cases, long bones, unsplit for marrow, accompanied the skull.

Source: Anthony F. C. Wallace, *Religion: An Anthropological View* (New York: Random House, Inc., 1966), pp. 226–227.

## THE ALTRUISTIC HYPOTHESIS

Several students of Mousterian culture have also suggested that Neandertals were altruistic, in other words, that their behaviour was sometimes governed, as is our own, by an unselfish concern for the welfare of others of their kind.

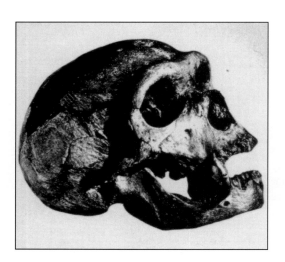

▼

The skull of the Old Man of La Chapelle

▲

The Old Man of La Chapelle-aux-Saints, whose remains were discovered at the beginning of the present century in a cave in southern France, was once regarded as an outstanding example of this supposed aspect of Neandertal behaviour; it was alleged that he owed his longevity to the kindness of his fellow human beings.

When Marcellin Boule examined the Old Man's remains in 1913 he estimated that the specimen had died between 50 years and 55 years of age. Later, when it was determined that the Old Man had suffered from arthritis (a fact that had escaped Boule's attention), many concluded that the only way such a disabled person could have lived so long was to have been fed and cared for by those he had lived among. Recent studies, however, have indicated not only that the Old Man's arthritis was less debilitating than originally suspected, but also that he likely was in his mid-30s when he died. And the Old Man was not the only Neandertal who died when he was relatively young; the longest-lived specimens from Shanidar Cave in Iraq also apparently died quite young, between 40 years and 45 years of age. The altruistic hypothesis, therefore, remains to be corroborated, and is not, as was once thought, all but confirmed. See QQ 9.9.

 **9.9 NEANDERTAL LONGEVITY AND THE ALTRUISTIC HYPOTHESIS**

Erik Trinkaus and David Thompson comment on Neandertal longevity and make the case for the appearance of altruism with the emergence of anatomically modern humans.

• • •

Boule . . . suggested that La Chappelle-aux-Saints . . . had an age of between 50 and 55 years (hence frequent reference to it as a "viellard"). However, Vallois . . . correctly noted that the cranial sutures are incompletely fused and suggested an age probably less than 40. Such a lower age estimate would correspond with the rather modest wear on the surviving premolars . . . . [Others have] suggested an age in the 20s or the low 30s . . . . It therefore appears unlikely that this individual was much older than the mid-30s (and may have been much younger) . . . [when he died].

This . . . in conjunction with the new age determinations of the Shanidar partial skeletons, indicates the extreme rarity and possible absence of Neandertals greater than 40 to 45 years in the fossil record . . . . If it is a reflection of Neandertal demography, it has interesting implications.

An absence or extreme rarity of postreproductive individuals would imply . . . that postreproductive survival was absent among the Neandertals. This would be particularly interesting, since a significant separation of senescence [or death] from the cessation of reproduction appears to be unique to humans among the primates . . . . Such a conclusion would therefore indicate that this uniquely human life cycle pattern appeared only at or subsequent to the appearance of anatomically modern humans . . . . Current interpretations of Upper Pleistocene human functional morphology and Paleolithic archaeology suggest that there was a major increase in the levels of organizational complexity and information processing among early modern humans relative to those of their Neandertal predecessors . . . and it appears likely that it was only with the emergence of modern humans that many aspects of a modern hunting and gathering adaptive system appeared. A shift in human life cycle parameters at this time in human evolution, leading to greater potential for transgenerational communication, would therefore fit with the apparent contemporaneous increased reliance on information and enhanced ability to support less productive individuals.

Source: Erik Trinkaus and David D. Thompson, "Femoral Diaphyseal Histomorphometric Age Determinations for the Shanidar 3, 4, 5, and 6 Neandertals and Neandertal Longevity" *American Journal of Physical Anthropology*, Vol. 72, No. 1 (1987), pp. 127–128.

# AN EXTRAORDINARY SEXUAL DIVISION OF LABOUR?

An interesting hypothesis has recently been advanced in connection with another aspect of Neandertal behaviour, namely, the relationship between the sexes. Ethnographic research has shown that cultures are characterized by a **sexual division of labour**, in which different socio-economic tasks are assigned to men and women on the basis of their sex. Among contemporary foragers and those in the recent past, for example, hunting is usually done by men and gathering by women, and since male and female foragers freely share what they harvest with everyone in the group, both sexes are directly involved in promoting the economic well-being of the unit.

According to Lewis Binford, the division of labour among the Neandertals may have been much more extreme than that practised by the foragers. Based on the distribution of ecofacts and artifacts at **Combe-Grenal**, a series of rock shelters in a box canyon in southwestern France (see Figure 9.4), Binford has suggested that while the Neandertal women and children who lived there spent most of their time in the immediate vicinity of the rock shelters, the men ventured farther afield, frequently leaving the women and children behind for weeks at a time. Additionally, and most importantly, Binford believes that while there was certainly interaction between the sexes, Neandertal men and women tended to fend for themselves.

The archaeological evidence, Binford says, supports these conclusions. He has identified two types of sites at Combe-Grenal — nest sites and scraper sites — and what makes these different, he says, is that they contain different types of debris. Whereas nest sites contain readily available materials that Binford believes were procured by women from the immediate vicinity of Combe-Grenal, scraper sites contain materials that he says men brought back to the rock shelters from the countryside. And, since the food wastes and implements in the sites are different, it would seem that, unlike modern hunters and gatherers and those of the recent past, the Neandertal men and women who lived at Combe-Grenal rarely shared what they produced with the opposite sex. Although Binford admits that Combe-Grenal may be the exception rather than

**FIGURE 9.4** Location of Combe-Grenal

FRANCE

Atlantic Ocean

Combe-Grenal

SPAIN

© 1975 Thames and Hudson Ltd., London

the rule, even if he is mistaken, his hypothesis does force scholars, as he puts it, "to say we damn well better ask some questions," including the question of how a sexual division of labour might have originated. See QQ 9.10.

## QQ 9.10 NEANDERTALS AND THE DIVISION OF LABOUR

Joshua Fischman comments on Binford's hypothesis.

• • •

What really jumped out at Binford was that most of the material in the nests and the scraper sites came from different places. The stone used to make the flakes in the nests was local. "In the nest, the raw material came from the cave itself or at most a hundred sixty feet [or about 52.5 metres] away," he says. "But in the locations with the scrapers, it came from more distant sites. The uplands plateau, about two miles [or about 3.2 kilometres] away, was a common source . . . ." "Now this," Binford says, "looks like a system . . . ."

"The raw material in the nests suggests that you're not very mobile if you live there — all the material comes from the immediate area," Binford says. "But with the scrapers, you're moving over substantial areas, and you're bringing in animal parts." That difference in ranges points to a dividing line in Combe-Grenal life, and Binford thinks the line fell between Neanderthal men and women . . . .

"I can't find — and I have looked — anything like this in any archeological remains of fully modern man," Binford says. "This looks like we've got a situation in which females are essentially taking care of

themselves much of the time. Fully modern man obtains food and brings it back. Then it's prepared and eaten by reproductive units. I don't think Neanderthals did that . . . ."

Instead, Binford says, there's a duplication of effort at Combe-Grenal, with the men at their own fires, cracking open their animal skulls. There was some sharing of the skull contents, testified to by the bone splinters found in the nests, but for the most part women were on their own, foraging for plants and fruits and cooking them on their own slow-burning fires. There was cattail pollen associated with the flakes, indicating these plants were a regular part of the females' diet.

"I'm not arguing that males and females weren't interacting," Binford says. "Rather, that it's an interaction we don't commonly see in modern humans. There's independent food preparation, different land-use patterns, different uses of technology. In modern humans the relationships are more integrated. But the Neanderthals are separate, yet they're interacting. That's the important point. That's what makes them different."

Source: Joshua Fischman, "Hard Evidence," *Discover*, Vol. 13, No. 2 (1992), pp. 49–50.

# UPPER PALAEOLITHIC CULTURES IN EUROPE AND ASIA

## ATLANTIC EUROPE TO INNER ASIA

About 40 kya, perhaps 50 000 years or more after the appearance of *H. sapiens sapiens*, blade tools were invented. Such tools signal the beginning of upper palaeolithic cultures. One of the most remarkable things about these cultures is their rate of technological change. Although important technological breakthroughs were made by lower palaeolithic and middle palaeolithic peoples, their breakthroughs were separated by hundreds of thousands of years. Upper palaeolithic peoples expanded their toolkits much faster, in bursts of inventiveness that are measured in thousands rather than in hundreds of thousands of years.

Also, unlike Mousterian culture, which was widespread but almost everywhere the same, upper palaeolithic cultures were regionally distinct. And while one upper palaeolithic culture did not necessarily evolve into another in a unilinear fashion, they did sometimes arise in sequence. The cultures that appeared between Atlantic Europe and Inner Asia from about 34 kya until about 11 kya are an example. As the contents of Table 9.3 indicate, these included **Aurignacian culture** (*c.* 34 kya – 30 kya), whose craftspeople specialized in manufacturing stone and bone points; **Châtelperrronian culture** (*c.* 32 kya – 28 kya), whose toolmakers manufactured extremely sharp flint knives; **Gravettian culture** (*c.* 30 kya – 20 kya), whose artisans fashioned finely worked stone chisels and ivory tools; **Solutrean culture** (*c.* 20 kya – 18 kya), whose archaeological marker consists of beautiful, laurel-leaf blades; and **Magdalenian culture** (*c.* 18 kya – 11 kya), whose members manufactured serrated stone knives, stone borers, and bone harpoons. Equipped with such tools, the *H. sapiens sapiens* who lived between Atlantic Europe and Inner Asia became better hunters than the Neandertals. Moreover, judging by their art and their burials, supernatural powers and forces played important roles in their lives. See QQ 9.11.

## 9.11 UPPER PALAEOLITHIC CULTURES BETWEEN ATLANTIC EUROPE AND INNER ASIA

Grahame Clark describes the technological achievements of upper palaeolithic peoples living between Atlantic Europe and Inner Asia.

• • •

The technology of . . . [Upper] Palaeolithic peoples was founded on lithic [or stone] industries based on the production, by some kind of punch, of blades or flakes relatively narrow in proportion to their length and having more or less regular, parallel flake-scars. From such blades were made a variety of types adapted to different functions: knife blades and projectile points and barbs were made by . . . retouching which . . . facilitated mounting in a handle or shaft; scrapers for dressing skins and probably for shaping other materials were formed by trimming blades to a convex edge; and burins, vitally important for working antler and bone as well as for engraving, were made by striking off flakes more or less counter to the main axis of the blade. The much greater use of antler and bone, facilitated by the flint burin, was another leading feature of . . . [Upper] Palaeolithic industry. Here again many different forms were produced, such as various patterns of lance- and harpoon-head, spear-throwers, perforated batons, leather-working tools, awls, and delicately eyed needles. All in all, the . . . [Upper] Palaeolithic industries give the impression of being far more diversified and highly specialized than those practised by . . . [earlier] Palaeolithic peoples.

Source: Grahame Clark, *World Prehistory: An Outline* (Cambridge, England: Cambridge University Press, 1961), pp. 50–51. Reprinted with the permission of Cambridge University Press.

**TABLE 9.3**  Sequential appearance of upper palaeolithic cultures between Atlantic Europe and Inner Asia

| CULTURE | DURATION (kya) | LOCATION OF SITES | EXAMPLES OF ARTIFACTS |
|---|---|---|---|
| Aurignacian | c. 34–30 | | bone point |
| Châtelperronian[1] | c. 32–28 | | flint knife |
| Gravettian[2] | c. 30–20 | | stone chisel |
| Solutrean | c. 20–18 | | laurel-leaf blade |
| Magdalenian | c. 18–11 | | bone harpoon |

[1] The Châtelperronian is also sometimes referred to as the Lower Perigordian.
[2] The Gravettian is also sometimes referred to as the Upper Perigordian.

Unfortunately, far less is known about upper palaeolithic cultures elsewhere in Asia. The picture in the Indian subcontinent and in Southeast Asia, for example, is still largely incomplete, and what transpired in Vietnam and China is only beginning to be understood. Japan and northeastern Siberia are the only exceptions. In Japan, which was joined to mainland China during most of the Ice Age, hundreds of late Pleistocene sites have been excavated, and these have made it clear not only that the use of blades was widespread, but also that the first residents of Japan were excellent hunters and fishers. They hunted a wide variety of small game; captured large amounts of shellfish; and traveled by dugout canoe to deep- and shallow-water fishing locations. The upper palaeolithic peoples of Japan also frequently camped near the sea.

Meanwhile, in northeastern Siberia, the **Dyukhtai Tradition** had emerged (see Figure 9.5). Persisting from about 18 kya until about 14 kya, the tradition featured wedge-shape cores that were used to kill large migratory

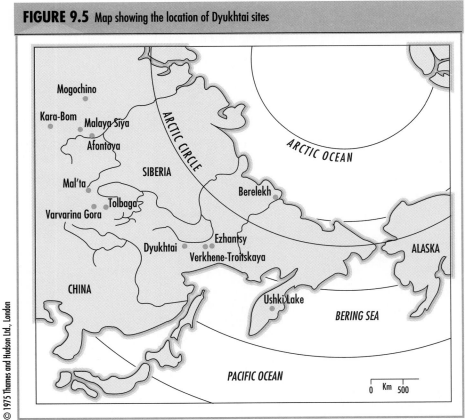

**FIGURE 9.5** Map showing the location of Dyukhtai sites

© 1975 Thames and Hudson Ltd., London

game, such as reindeer, and an assortment of smaller microblades that were used for butchering as well as for other purposes (see Figure 9.6). Older analogues of Dyukhtai implements have also been recovered from sites in northern China, and this has led some archaeologists to conclude that this is where the tradition originated. Some also believe that those who manufactured Dyukhtai tools were the ancestors of the Palaeo-Eskimo hunters who entered the New World via Beringia, and then spread rapidly across the far north from west to east, from Alaska into Arctic Canada and then on to Greenland. This, they say, took place about 8 kya, after the ancestors of contemporary North American Indian people already had arrived. See QQ 9.12.

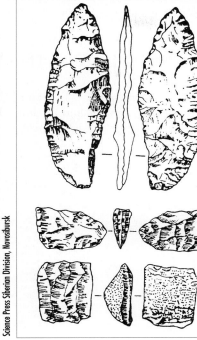

**FIGURE 9.6**  Dyukhtai tools recovered from Verkhene-Troitskaya, Siberia — a biface (top) and two wedge-shaped microblades (bottom)

Based on drawing in The Most Ancient Stages of Human Settlement of Northeastern Asia by Mochanov, 1977, Science Press Siberian Division, Novosibursk

## QQ 9.12 FROM SIBERIA TO ALASKA AND BEYOND

Clifford Hickey, Professor of Anthropology at the University of Alberta, explains why and how the first substantial population of Palaeo-Eskimo hunters may have migrated from Siberia to Alaska.

· · ·

[The contents of the archaeological record] suggest that a rapid migration took place from west to east, into previously unoccupied Arctic regions [about 8 kya]. A major question is, of course, why did this movement occur at this time? This migration can be correlated with a period of somewhat warmer climate and this fact may

provide a clue. It seems likely that the . . . [Asian ancestors of the first Eskimos] were adapted to a range of environments, including large rivers, the Boreal Forest and adjacent tundra barrens. Almost certainly they moved in a seasonal round sequentially exploiting various regions. They probably pursued small game, birds and fish as well as reindeer (caribou) in the forested regions during the winter. They may have followed reindeer out onto the tundra during the summer as historic Siberian hunters did, and if summers became longer and milder some people may have stayed out on the tundra for increasing lengths of time. In this way they may have developed expertise in Arctic conditions, including the ability to exploit seals along the frozen coasts between Siberia and Alaska. This may have led

# THE SITUATION IN AUSTRALIA

While much remains to be learned about upper palaeolithic cultures in Asia, considerably more is known about the cultures of the anatomically modern humans who migrated from Asia to Australia and the New World. In

The remains of a cremation at Lake Mungo

Photo by Prof. D.J. Mulvaney from *Ascent to Civilization: The Archaeology of Early Man* by John Gowlett, 1984, © 1984 by Roxby Archaeology Limited

Australia, for example, the remains of one such group have been unearthed at a 25 000-year-old to 30 000-year-old site on the shore of **Lake Mungo**, in the southeastern corner of the continent (see Figure 9.7). The dozen or so ancient Australians who originally camped at the site specialized in manufacturing heavy, horseshoe-shaped stone cores and more delicate stone scrapers. Archaeologists believe that these implements were used to kill and process the small burrowing animals that the Lake Mungo people hunted in winter, and the shellfish and perch that they took from the lake in the spring and summer. The charred remains of a young woman found in a shallow pit at the site also suggest that the Lake Mungo people cremated and buried their dead. See QQ 9.13.

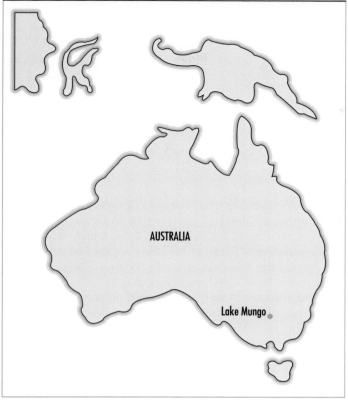

**FIGURE 9.7** Map showing the location of Lake Mungo

## 9.13 LAKE MUNGO

J. Peter White and James O'Connell describe the discovery of the Lake Mungo site and comment on its significance.

• • •

On 5 July 1968, geomorphologist Jim Bowler was walking along the southern end of the sand dune which curved around the eastern shore of the now dry Lake Mungo in western New South Wales . . . . He knew already that the dune had been built up in several stages during later Pleistocene times and had noted stone artefacts lying on the eroded surfaces. But

on this morning he encountered something different — a small cluster of broken bones embedded in a part of the dune he knew to be more than 20 000 years old. He suspected that these bones were food bones remaining from human activities, and they appeared to have been buried during the time the dune was forming . . . .

[W]hen Bowler recognized that his find was within dated sediments and asked [archaeologist Harry] Allen and other prehistorians to investigate it, he opened up a research field of some potential. Not only were the bones human, but it now seemed possible to order prehistoric material in sequence, give it a firm chronology and look at it in relation to changing local environments . . . . The potential of Lake Mungo and similar sites in the area is only just beginning to be realized, even after more than ten years of work. In part, this is because of the nature of the data and the scale on which research needs to proceed . . . .

Source: J. Peter White and James F. O'Connell, *A Prehistory of Australia, New Guinea and Sahul* (New York: Academic Press Inc., 1982), p. 33.

# UPPER PALAEOLITHIC CULTURAL DEVELOPMENTS IN THE AMERICAS

## THE ICE-FREE CORRIDOR

Thousands of miles away from Australia, perhaps as long as 20 kya or more, and certainly no later than by about 12 kya, the first *H. sapiens sapiens* who migrated from Asia to the Americas also began to fashion their implements to meet their needs in their new homeland. These were the first Americans, who presumably entered the Americas either in watercraft or on foot via Beringia, and once in the New World made their way south either by paddling southward along the coast of British Columbia, or, as most archaeologists believe, by traveling on foot through an **ice-free corridor** that stretched through the surrounding ice fields from the northwestern corner of the con-

**FIGURE 9.8** The ice-free corridor

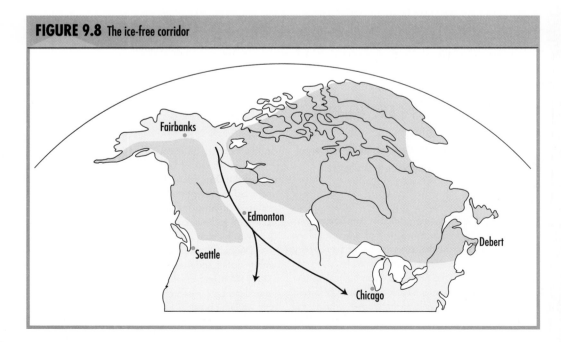

tinent down through what are now British Columbia and Alberta into Montana (see Figure 9.8). In either event, once free of the ice, the first Americans moved into the southwestern United States and into South America where the climate was comparatively warm. Later, when the ice sheets had melted, some moved back into the northern United States and then into Canada. Many archaeologists believe that those who returned to the north did so because they were following the movements of the large game they then hunted; these animals apparently moved north in conjunction with global warming, which allowed them to subsist more easily in the north. See QQ 9.14.

## 9.14 THE ICE-FREE CORRIDOR

Gail Helgason, an archaeologist who works for the Archaeological Survey of Alberta, describes the route that most archaeologists believe the first Americans followed to enter the heart of the continent.

• • •

Many believe . . . [that an] ice-free corridor funnelled the first people into the heart of North America at various times during the last surge of glaciation, called the Wisconsinan [or Würm]. That period began about 75 000 years ago and lasted until about 10 000 years ago . . . .

We know people lived south of the ice sheets, in what is now the United States, Mexico and South America, at a time when much of Canada, including Alberta, was still partly gripped in ice. But how did they get there?

The ice-free corridor theory offers an appealing explanation. It holds that the first prehistoric passage from the region around the Bering Strait (Beringia) into the heart of North America opened between lobes of ice sheets along the Eastern slopes of the Rocky Mountains. This provided a "highway" which stretched from the Yukon and Alaska into present-day Montana. The corridor, the theory goes, was open most of the time during the Late Wisconsinan [or Würm]. It extended 4000 kilometres from the Richardson Mountains in the Yukon to the Lethbridge-Waterton Lakes region of southern Alberta, probably cutting through the present-day townsites of Edson, Nordegg, Rocky Mountain House and Calgary.

The answer to an enduring mystery — the peopling of the Americas — may be enshrouded within a mountainous ribbon of land which includes a lengthy portion in Alberta. A majority of archaeologists believe the interior of North America was first populated when ancestral Native Americans moved down that as yet elusive passage. The highly advanced civilizations of Peru and Mexico, the Pueblo dwellers of the American southwest and the mound-building societies of the southeastern U.S. all trace their origins to the people who made their way along the first great migration route, a passageway in which Alberta may have figured prominently.

Source: Gail Helgason, *The First Albertans: An Archaeological Search* (Edmonton: Lone Pine Publishing, 1987), pp. 35–36. Copyright © 1987. Used by permission of Lone Pine Publishing, Edmonton, Alberta.

## PALAEO-INDIANS

Archaeologists have learned much of what they know about the cultural practices of the first Americans by excavating campsites and kill sites; the results indicate that by about 12 kya the first clearly discernible North American upper palaeolithic hunting culture had made its debut. Its practitioners are called **Palaeo-Indians**, which literally means old Indians, and towards the end of the Pleistocene they established the **Big Game Hunting Tradition** on the American Great Plains, which is now dry but was then a lush savanna dotted with wooded valleys.

The tradition was based on the exploitation of large game, especially mammoths, which were hunted and butchered by people working in concert. Spreading from the American Southwest into other parts of the United States, and also into Canada where some Palaeo-Indians specialized in hunting bison and caribou, the Big-Game Hunting Tradition featured the **Clovis point** — a lance-shaped stone projectile between 1.5 centimetres and 4 centimetres wide and about 13 centimetres long with flutes or grooves extending about halfway up from the base to the tip on opposite faces (see Figure 9.9). The flutes were removed to make it easier to join the point to a wooden spearshaft; it was this weapon that the Palaeo-Indians used to kill the large game that they depended upon for their survival. See Figure 9.10 and QQ 9.15.

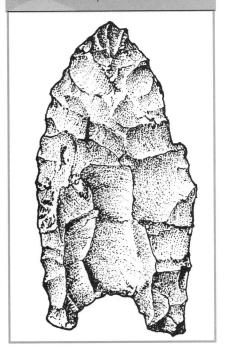

**FIGURE 9.9** Fluted Palaeo-Indian point from the Debert site, Nova Scotia

For roughly 2000 years, from about 12 kya until about 10 kya, the Palaeo-Indians were able to cope with their late Pleistocene environment in what was apparently a highly successful manner. The distribution of their sites throughout North America indicates as much. Nevertheless, early in the Holocene (*c.*10 kya), when the large game animals that the Palaeo-Indians subsisted upon started to become extinct, they inaugurated a wide range of cultural experiments that allowed them to replace their traditional way of life with a number of alternative adaptive systems. These experiments paved the way for the development of the Indian cultures that Europeans subsequently encountered on their voyages to the New World. It was during their expeditions to the Maritimes, for example, that the first contact between Europeans and the descendants of the Palaeo-Indians who had settled in what is now Canada took place.

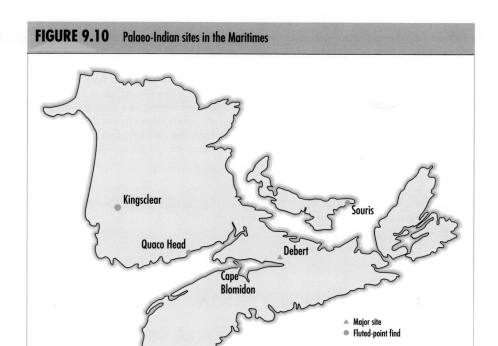

**FIGURE 9.10** Palaeo-Indian sites in the Maritimes

Kingsclear

Souris

Quaco Head

Debert

Cape
Blomidon

▲ Major site
● Fluted-point find

0    Kilometres    100

## 9.15 PALAEO-INDIAN CULTURE

James Tuck, Professor of Anthropology at Memorial University, looks at the first Americans and describes a Palaeo-Indian site in the Maritimes — the Debert site in Nova Scotia.

• • •

The Palaeo-Indian Period is the oldest generally recognized period of human occupation in North America. Sites are characterized by a complex of chipped-stone tools and weapons, the most distinctive being a type of spear or dart point known as a fluted point. It takes its name from two shallow channels, or flutes, formed by removing long flakes from opposite faces of the stone. Archaeologists believe that these channels facilitated hafting the point to a spear or dart shaft. Evidence of

Palaeo-Indian cultures was found as early as the late 1920s in the southwestern United States and shortly thereafter in Canada and the eastern states . . . .

Although Palaeo-Indian artifacts are now known from all of the Maritime Provinces, the Debert site remains one of the most important locations in northeastern and perhaps all of the eastern North America for understanding the Palaeo-Indian Period . . . .

[Excavations at this site indicate that] the people of Debert probably built simple structures framed with wood (or perhaps even bone) and covered with skins. In view of the strong possibility that the Debert site was a winter encampment, these dwellings were probably quite substantial, at least in terms of their ability to insulate the occupants from the cold. [In addition, each had two or more hearths] . . . .

Perhaps the most easily recognizable implements found at Debert are the fluted spear or dart points that doubtless tipped the weapons the hunters used to bring down game, their principal food; at Debert this was probably caribou.

Once secured, game was butchered with a variety of implements. Thin, sharp flakes of stone and specially made stone knives . . . were used to cut the skin, meat and tendons. Carcasses were probably dismembered with large stone choppers or cleavers, and simple cobble hammerstones may have been used to break open bones in order to extract the marrow.

The Debert people used hides, antler, bone, wood, bark and other materials to make clothing, weapons, domestic articles, and a variety of other objects, including tools themselves. Small scrapers flaked from stone in such a way as to produce a thick but sharp edge removed the fat and flesh from skins and probably also shaped wood, bone and antler. Stone awls and perforators were probably used to sew clothing, the covers for tent frames, hide containers, and perhaps even bark baskets. Wood, too, may have been carved and cut to make handles for stone tools.

Source: James A. Tuck, *Maritimes Provinces Prehistory* (Ottawa: National Museums of Canada, 1984), pp. 5–9.

# THE PICTURE IN AFRICA

Meanwhile, in Africa, cultural developments had followed a similar course. There too blade tools mark the transition from the middle to the upper palaeolithic, and there too blade implements were first manufactured by anatomically modern humans, who had built on the cultural legacies of their ancestors. In northeastern Africa, blade tools date back to about 25 kya, and in northwestern Africa and south of the Sahara, as far back as 40 kya in some locations. As was the case elsewhere in the world at the time, those who fashioned these tools were hunters and gatherers who focused their attention on developing the technology and skills that were necessary to capture and process the game that they exploited. The anatomically modern humans who manufactured Dabban tools are an example. Best known from archaeological deposits in Libya that are between 40 000 years old and 13 000 years old, the **Dabban industry** featured a wide variety of blades, scrapers, and chisels that archaeologists suspect their manufacturers used to dispatch and process the animals that they hunted in the coastal ranges of the Mediterranean (see Figure 9.11). Plant foods also likely figured prominently in the Dabban diet, although the evidence for their having gathered such foods is less obvious in the archaeological record. It is unfortunate that more is not known about the Dabban and other late stone age industries in Africa, for there is little doubt that these figured prominently in the subsequent cultural developments of the original homeland of our species. Once considered to be culturally backward and irrelevant to prehistory, most archaeologists now believe that the anatomically modern humans who inaugurated upper palaeolithic cultural experiments in Africa were the direct ancestors of the historic hunting and gathering cultures of central, eastern, and northwestern Africa. See QQ 9.16.

## FIGURE 9.11 Dabban blades

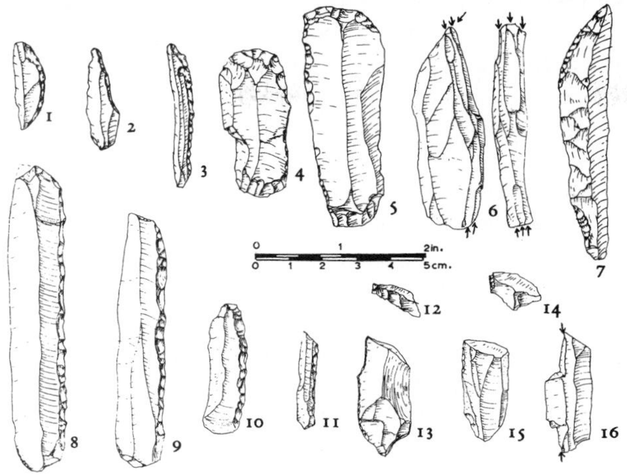

# QQ 9.16 THE COURSE OF AFRICAN PREHISTORY

**J. Desmond Clark discusses the significance of African prehistory.**

• • •

Since the focus of African history has now shifted from that of the colonizing power to the African peoples themselves, it is also time for a more objective approach to the *prehistory* of the continent, unencumbered by preconceptions inherited from an early emphasis on the cultural sequence of western Europe . . . . Although still very far from comprehensive, the picture that is emerging is very different from the previously painted one of isolation and stagnation interspersed with spasmodic stimulation from foreign migrants. While the peoples of Africa have not been slow to take advantage of the opportunities for cultural exchange with non-African peoples, this exchange operated not merely in one direction but in both and the world in general, as well as Africa, is thereby the richer . . . .

Even though there are, from time to time, some who think they can get along without it, a feeling for

history lies deep in us all. This is hardly surprising since every society has its behaviour determined by the traditions that have grown out of the circumstances of its past. There is no part of the world where this is appreciated more than in Africa. It is precisely here that prehistory is making one of its most significant contributions, since it alone can provide time-depth and knowledge of all except the more recent past from which present-day culture derives. New patterns are developing, new goals are being set and new leaders are rising up to effect the changes, but every year the increasing awareness of the importance of the past becomes more apparent in the greater emphasis that is being placed upon the traditional inheritance and the traditional way of life. Continuity through the African past is the clue to the changes taking place in the new Africa today.

Source: J. Desmond Clark,. *The Prehistory of Africa* (Great Britain: Thames and Hudson, 1970), pp. 222–223.

# THE HUNTING AND GATHERING WAY OF LIFE

Finally, it is worthwhile to point out that even though the people who lived in the various regions of the world during the latter part of the Pleistocene manufactured tools that were best suited to local conditions, those tools were used for similar purposes everywhere. The terminal phase of the Pleistocene was the heyday of the hunting and gathering way of life, a period when people depended exclusively on wild plants and animals for their livelihoods, presumably moving from one place to another in an annual round.

Judging by what has been learned about hunting and gathering peoples at the time of contact with Europeans (if the analogy is appropriate), it was also the heyday of **band societies** — small, politically autonomous groups that likely contained between 50 to 75 persons who were related through blood and through marriage, and who occupied comparatively large territories where they were generally self-sufficient with respect to food, clothing, and shelter. Within such groups — the late sixteenth century and early seventeenth century Northern Ojibwa Indians of Northern Ontario are an example — there may have been a charismatic individual whom others acknowledged as their leader because he or she possessed superior cultural skills (see QQ 9.17). On

## 9.17 BAND SOCIETIES

The late Edward Rogers, who was one of the foremost experts on the Northern Ojibwa, and J. Garth Taylor, Arctic Curator, Canadian Ethnology Service, Canadian Museum of Civilization, describe Northern Ojibwa bands at the time of contact with Europeans.

• • •

The . . . Northern Ojibwa during the late 1700s and early 1800s . . . appear to have been organized in bands, the core of which consisted of an elder and his married sons with male affines [related by marriage] often included. This extended family may have numbered between 20 and 40 people. With the addition of other nuclear families, most likely relatives, the total aggregate may have been anywhere from 50 to possibly 75 individuals . . . .

Each band was led by a senior male of the core family. Frequently, if not always, these individuals acted as the "trade chiefs" . . . . The position of leader appears to have been based on his ability to secure for his followers abundant trade goods, to excel as a hunter, and to command superior religious knowledge. Nevertheless, such individuals were charismatic, not autocratic leaders. Their followers could sever their allegiance at will . . . .

Each band tended to habitually exploit a particular territory, a "hunting range" or "hunting area" . . . . Such territories were not rigidly bounded and trespass was not resented. Indications are that when big game was abundant, the band remained together throughout the year.

Source: Reprinted from *Handbook of North American Indians, Volume 6: Subarctic*, ed. June Helm, gen. ed. William C. Sturtevant (Washington, DC: Smithsonian Institution Press), 233, by permission of the publisher. Copyright 1981 Smithsonian Institution. All rights reserved.

the other hand, decisions affecting the group as a whole were almost certainly made by consensus, which means that all of the adults in the band were involved. Under the circumstances, it is not surprising that bands are frequently described as egalitarian. Later, with the advent of agriculture and animal husbandry, non-egalitarian forms of political organization (including the state) would come to the fore, but more about this in the following chapter.

# KEY TERMS AND CONCEPTS

ECOFACTS

ARTIFACTS

ARCHAEOLOGICAL RECORD

THERMOLUMINESCENCE
(*thur'-moh-loom-in-es'-ens*)

DENDROCHRONOLOGY
(*den'-droh-krawn-ahl'-uh-jee*)

OBSIDIAN HYDRATION
(*awb-si'-dee-en + hydration*)

SITES

SURFACE SURVEYS

SUBASSEMBLAGE

ASSEMBLAGE

ARCHAEOLOGICAL CULTURE

THREE AGE SYSTEM

STONE AGE CULTURES

BRONZE AGE CULTURES

IRON AGE CULTURES

PALAEOLITHIC CULTURES
(*pay'-lee-oh-lith'-ik + cultures*)

MESOLITHIC CULTURES
(*meh'-soh-lith'-ik + cultures*)

NEOLITHIC CULTURES

LOWER PALAEOLITHIC CULTURES

MIDDLE PALAEOLITHIC CULTURES

UPPER PALAEOLITHIC CULTURES

BEAR CULT

ALTRUISTIC HYPOTHESIS

SEXUAL DIVISION OF LABOUR

COMBE-GRENAL (*kom-bu' gre-nal*)

AURIGNACIAN CULTURE
(*aw-rig-nay'-shen + culture*)

CHATELPERRONIAN CULTURE
(*sha-tel-pur-oh'-nee-en + culture*)

GRAVETTIAN CULTURE (*gra-ve'-tee-en + culture*)

SOLUTREAN CULTURE
(*suh-loo'-tree-en + culture*)

MAGDALENIAN CULTURE
(*mag-duh-lin'-ee-en + culture*)

LAKE MUNGO

DYUKHTAI TRADITION
(*dy-yuhk'-ty + tradition*)

ICE-FREE CORRIDOR

PALAEO-INDIANS

BIG-GAME HUNTING TRADITION

CLOVIS POINT (*kloh'-vis + point*)

DABBAN INDUSTRY (*dah-bawn + industry*)

BAND SOCIETIES

# Selected Readings

## Introduction

Fagan, B. *Quest for the Past: Great Discoveries in Archaeology.* Prospect Heights, Illinois: Waveland Press, 1988.

Fagan, B. *In the Beginning: An Introduction to Archaeology* (seventh edition). New York, HarperCollins 1991.

Sharer, R. and Ashmore, W. *Archaeology: Discovering Our Past.* Palo Alto, California: Mayfield Publishing, 1987.

## Discovering The Past

Baillie, M. *Tree-Ring Dating.* Chicago: University of Chicago Press, 1982.

Kelly, J. and Hanen, M. *Archaeology and the Methodology of Science.* Albuquerque, New Mexico: University of New Mexico Press, 1988.

Longacre, W. "Archaeology as Anthropology." *Science*, Vol. 144 (1964), pp. 1454–1455.

Michael, H. and Ralph, E., eds. *Dating Techniques for the Archaeologist.* Cambridge: MIT Press, 1971.

Taylor, R. *Radiocarbon Dating: An Archaeological Perspective.* New York: Academic Press, 1987.

Trigger, B. "Distinguished Lecture in Archeology: Constraint and Freedom — a New Synthesis for Archeological Explanation." *American Anthropologist*, Vol. 93 (1991), pp. 551–569.

Trigger, B. *A History of Archaeological Thought.* Cambridge: Cambridge University Press, 1990.

Wiley, G. and Sabloff, J. *A History of American Archaeology* (second edition). San Francisco: W. H. Freeman, 1980.

## Classifying Prehistoric Cultures

Daniel, G. *The Three Ages.* Cambridge: Cambridge University Press, 1942.

Lubbock, J. *Prehistoric Times: As Illustrated by Ancient Remains and the Manners and Customs of Modern Savages.* London: Williams and Norgate, 1865.

## Lower Palaeolithic Cultures

Bartstra, G. "*Homo erectus*: The Search for his Artifacts." *Current Anthropology*, Vol. 23 (1982), pp. 318–320.

Bordes, F. *The Old Stone Age.* London: Weidenfeld and Nicholson, 1968.

Gowlett, J. "Culture and Conceptualization: The Oldowan-Acheulian Gradient." In Bailey, G. and Callow, P., eds. *Stone Age Prehistory.* Cambridge: Cambridge University Press, 1986, pp. 243–260.

Isaac, G. "The Archeology of Human Origins: Studies of the Lower Pleistocene in East Africa," 1971–1981. *Advances in World Archaeology*, Vol. 3 (1984), pp. 1–87.

Jelinek, A. "The Lower Paleolithic: Current Evidence and Interpretations."

*Annual Review of Anthropology*, Vol. 6 (1977), pp. 11–32.

O'Connel, J., Hawkes, K., and Jones, N. "Hadza Scavenging: Implications for Plio/Pleistocene Hominid Subsistence." *Current Anthropology*, Vol. 29 (1988), pp. 356–363.

Toth, N. "The Oldowan Reconsidered: A Close Look at Early Stone Artifacts." *Journal of Archaeological Science*, Vol. 12 (1985), pp. 101–120.

## MIDDLE PALAEOLITHIC CULTURES

Binford, L. "Human Ancestors: Changing Views of their Behavior." *Journal of Anthropological Archaeology*, Vol. 4 (1985), pp. 292–327.

Binford, L. "The Hunting Hypothesis, Archaeological Methods, and the Past." *Yearbook of Physical Anthropology*, Vol. 30 (1987), pp. 1–9.

Bordes, F. *A Tale of Two Caves.* New York: Harper and Row, 1972.

Bordes, F. and de Sonneville-Bordes, D. "The Significance of Variability in Paleolithic Assemblages." *World Archaeology*, Vol. 2 (1970), pp. 61–73.

Chase, P. and Dibble, H. "Middle Paleolithic Symbolism: A Review of Current Evidence and Interpretations." *Journal of Anthropological Archaeology*, Vol. 6 (1987), pp. 263–296.

Mellars, P. "A New Chronology for the French Mousterian Period." *Nature*, Vol. 322 (1986), pp. 410–411.

Movius, H. "The Mousterian Cave of Teshik-Tash, South-Central Uzbekistan, Central Asia." *Bulletin of the American School of Prehistorical Research*, Vol. 17 (1953), pp. 11–71.

Tappen, N. "The Dentition of the 'Old Man' of La Chapelle-aux-Saints and Inferences Concerning Neanderthal Behavior." *American Journal of Physical Anthropology*, Vol. 67 (1985), pp. 43–50.

Trinkaus, E., ed. *The Mousterian Legacy.* Oxford: British Archaeological Reports, International Series, 164, 1983.

## UPPER PALAEOLITHIC CULTURES IN EUROPE AND ASIA

Akazawa, T. and Aikens, C., eds. *Prehistoric Hunter-Gatherers in Japan.* Tokyo: University of Tokyo Press, 1986.

Bailey, G., ed. *Hunter-Gatherer Economy in Prehistory: A European Perspective.* Cambridge: Cambridge University Press, 1983.

Dickson, D. *The Dawn of Belief: Religion in the Upper Paleolithic of Southwestern Europe.* Tucson: University of Arizona Press, 1990.

Jacobson, J. "Recent Developments in South Asian Prehistory and Protohistory." *Annual Review of Anthropology*, Vol. 8 (1979), pp. 467–502.

McGhee, R. *Canadian Arctic Prehistory.* Ottawa: National Museum of Man, National Museums of Canada, 1978.

Yi, S. and Clark, G. "The "Dyuktai" Culture and New World Origins." *Current Anthropology*, Vol. 26 (1985), pp. 1–20.

Wu, R. and Olsen, J., eds. *Paleoanthropology and Paleolithic Archaeology in the People's Republic of China.* New York: Academic Press, 1985.

## The Situation In Australia

Allen, J., Golson, J., and Jones, R., eds. *Sunda and Sahul.* New York: Academic Press, 1977.

Flood, J. *Archaeology of the Dreamtime.* Sydney: Collins, 1983.

## Upper Palaeolithic Cultural Developments In The Americas

Agenbroad, L., Mean, J., and Nelson, L., eds. *Megafauna and Man: Discovery of America's Heartland.* Flagstaff, Arizona: Northern Arizona University Press, 1990.

Dincauze, D. "An Archaeological Evaluation of the Case for the Pre-Clovis Occupations." *Advances in World Archaeology,* Vol. 3 (1984), pp. 275–323.

Bonnichsen, R. and Young, D. "Early Technological Repertoires: Bone to Stone." *Canadian Journal of Anthropology,* Vol. 1 (1980), pp. 123–128.

Haynes, C. "The Clovis Culture." *Canadian Journal of Anthropology,* Vol. 1 (1980), pp. 115–121.

Kelly, R. and Todd, L. "Coming into the Country: Early Paleoindian Hunting and Mobility." *American Antiquity,* Vol. 53 (1988), pp. 231–244.

MacDonald, G. *Debert: A Palaeo-Indian Site in Central Nova Scotia.* Ottawa: Anthropological Papers, Number 16, National Museums of Canada, 1968.

Wiley, G. *An Introduction to American Archaeology* (two volumes). Englewood Cliffs, New Jersey: Prentice-Hall, 1966.

## The Picture In Africa

Phillipson, D. *African Archaeology.* Cambridge: Cambridge University Press, 1985.

Wendorf, F. and Schild, R., eds. *Prehistory of the Eastern Sahara.* New York: Academic Press, 1980.

## The Hunting And Gathering Way Of Life

Price, T. and Brown, J. *Prehistoric Hunter-Gatherers: The Emergence of Cultural Complexity.* San Diego: Academic Press, 1985.

Steegman, A., ed. *Boreal Forest Adaptations.* New York: Plenum, 1983.

Winterhalder, B. and Smith, E., eds. *Hunter-gatherer Foraging Strategies: Ethnographic and Archeological Analyses.* Chicago: University of Chicago Press, 1981.

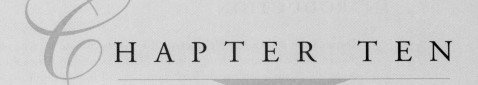

# CHAPTER TEN

# MESOLITHIC, NEOLITHIC, AND CIVILIZED CULTURES

*contents at a glance*

# INTRODUCTION

Towards the end of the Pleistocene, the earth's climate became warmer and the landscape began to change. In the Near East and in Africa the grasslands began to expand, and in the high-latitude and high-altitude regions of the world the glaciers began to melt. As a result, the barren lands of Europe, Asia, and North America were transformed into forests and meadowlands. Meanwhile, many of the large and medium-size herbivores that upper palaeolithic peoples had hunted had either become extinct or more circumscribed in their movements, partly as a result of human predation and partly because of environmental change. It was in this new post-Pleistocene environment that mesolithic or middle stone age cultures made their initial appearance, about 11 kya, at the beginning of the Holocene epoch. Although once thought to be an uninteresting interval in world prehistory, the mesolithic was characterized not only by a host of technological innovations (see Table 10.1), but also by comparatively elaborate and stable settlements. See QQ 10.1.

**TABLE 10.1** Mesolithic Innovations and their Uses

| INNOVATION | USE |
|---|---|
| awls | basketry |
| baskets, some made watertight and fire-resistant with clay | collecting seeds, fruits, insects, water; parboiling; boiling bones to make soup |
| fire-cracked rock | boiling or steaming plants; steaming open shellfish and snails; making soups from crushed bones; extracting oils from crushed nuts |
| grinding stones, mortars and pestles | processing grass seeds, toxic nuts (e.g., acorns), small animal bones and cartilage, nuts etc. |
| nets | capturing fish, small animals (e.g., rabbits), birds |
| weirs | capturing fish |
| fishhooks, harpoons, leisters | catching fish |
| basket traps | capturing fish and small animals |
| elaborate snares and traps | capturing small and large animals, birds, and fish |
| bows and arrows | hunting |
| microblades or microliths (sickles, knives, arrowheads, and spearheads) | all activities requiring cutting and piercing, i.e., hunting, harvesting, butchering |
| domesticated dogs | hunting, transport |
| sleds and canoes | transport over long distances for exploitation of distant resources |
| storage pits, storage scaffolds, drying/smoking racks, and blades to cut thin filets for drying | long term storage of meat, fish, and plants |

Source: Brian Hayden, *Archaeology: The Science of Once and Future Things* (New York: W. H. Freeman and Company, 1993), p. 196.

## 10.1 MESOLITHIC CULTURES

T. Douglas Price looks at the accomplishments of mesolithic peoples.

• • •

Several interrelated themes dominate . . . the dynamic character of human adaptation during the . . . Mesolithic . . . .

Technology develops toward greater efficiency in transport, in tools and in subsistence procurement. Subsistence equipment, both implements and facilities, becomes more diverse in form, more specialized in function, and more abundant in number. An incredible range of fishing gear, including nets, weirs [or fish traps], leisters [or pronged fish spears], hooks, and harpoons, is known. Ground stone artifacts appear for the first time as axes, celts [or axe-like implements], plant-processing equipment, and other tools. Large canoes and snow sleds . . . are known from this period in northern Europe. Projectile weapons are armed with a vast array of specialized tips, made of bone, wood, antler, and stone . . . . Other chipped stone implements begin to take a secondary role in many activities, becoming tools to make tools of wood, bone, and antler that are used directly in food procurement.

Settlements are larger, of longer duration, and more differentiated, both in terms of the internal organization of the settlement and in the number and variety of sites and locales in use. Large co-resident groups and permanent occupation are hallmarks of Mesolithic settlement in several areas . . . .

Subsistence activities appear to be greatly intensified during the Mesolithic. Resource procurement becomes more specialized and diversified — more specialized in terms of the nature, technology, and organization of foraging and more diversified in terms of the numbers and kinds of species and habitats exploited. The number of species incorporated into the diet is much greater than that documented for the Upper Paleolithic in almost all areas . . . .

Source: T. Douglas Price, "The European Mesolithic," *American Antiquity*, Vol. 48, No. 4 (1983), p. 770. Reproduced by permission of the Society for American Archaeology.

# MESOLITHIC CULTURE IN THE AMERICAS

## ARCHAIC CULTURE

One mesolithic culture that has long fascinated New World archaeologists is **Archaic culture**, which arose about 8 kya in the southwestern and eastern regions of North America, where it had evolved from a Palaeo-Indian base. Archaic culture then spread into neighbouring regions, and eventually into South America and northern North America, where it flourished until it gave rise to subsequent prehistoric American Indian cultures about 3 kya. Although much remains to be learned about archaic cultures in South America, their North American counterparts have been studied extensively.

## LAURENTIAN ARCHAIC CULTURE

Like other mesolithic cultures, Archaic culture was characterized by a number of subcultures that appeared over the years in different ecological zones. Among these was **Laurentian Archaic culture**, a mesolithic culture that persisted from about 6 kya until about 3 kya in the Lower Great Lakes region of southern Ontario and adjacent areas. It was during this period that Laurentian people developed one of the hallmarks of their culture. This was a human burial cult, whose archaeological marker consists of stone, bone, and copper implements and ornaments that were buried with certain deceased individuals (see Figure 10.1). Since the majority of the abovementioned grave goods are associated with the skeletal remains of adult males, archaeologists surmise that such people were accorded a relatively high status in Laurentian society. Considering that adult males likely were primarily responsible for pursuing deer, elk, and bear, and that sharing the meat of these big game animals with kinsmen and friends doubtlessly enhanced their prestige, the argument makes all the more sense. Archaeologists also believe that Laurentian people pursued big game mainly in winter, when hunting units composed of one or several families resided on their hunting grounds in the wilderness. As

**FIGURE 10.1**  Laurentian Archaic artifacts

Ground slate bayonet

Stone plummet

Copper dart head

Slate spear-thrower weight

Stone gouge

Slate dart head

Dart heads

Chipped stone drill

Copper adze

From *Ontario Prehistory*, by J.V. Wright. Copyright © 1972, National Museums of Canada.

warm weather approached, these areas were abandoned in favour of sites near open water where several hunting units would congregate. This resulted in the formation of temporary summer encampments where the residents earned the bulk of their livelihoods by fishing and gathering plant foods.

# SHIELD ARCHAIC CULTURE

While Laurentian Archaic people were establishing a foothold in southern Ontario, a related mesolithic people were establishing themselves in the northern part of the province and elsewhere in northern Canada (see Figure 10.2).

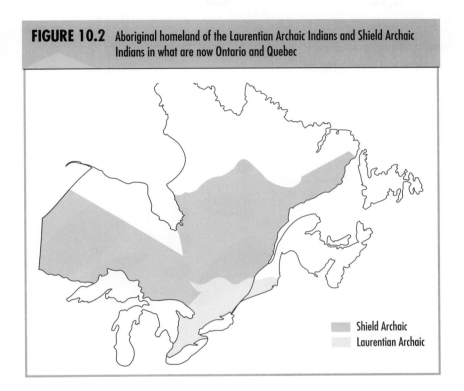

**FIGURE 10.2** Aboriginal homeland of the Laurentian Archaic Indians and Shield Archaic Indians in what are now Ontario and Quebec

Shield Archaic
Laurentian Archaic

These were the Shield Archaic Indians, whose **Shield Archaic culture** persisted from about 6500 ya until about 2500 ya, in the region stretching from the Keewatin District in the Northwest Territories to Cape Breton in Nova Scotia.

Like Laurentian culture, Shield Archaic culture had also evolved from a Palaeo-Indian base. During the course of its development, its members created an increasingly sophisticated toolkit in order to better satisfy their socio-cultural wants and needs in their new northern homeland. The implements they manufactured attest to their subsistence skills. Foremost among these were **composite tools** — implements consisting of several parts that were joined together into a single unit. The bow and arrow is an outstanding example. Equipped with this weapon, Shield Archaic Indians were able to kill big game such as caribou from a distance. This made them all the more productive, especially in northern Ontario where caribou and fish were the mainstays of their diet. Some archaeologists also believe that the immediate cultural descendents of the Shield Archaic Indians (the first major Indian population to reside in northern Ontario) gave rise to the Cree, Ojibwa, and Algonkin Indians who lived in the northern part of the province in historic times. See QQ 10.2.

## QQ 10.2 SHIELD ARCHAIC CULTURE

K. C. A. Dawson, Professor of Anthropology at Lakehead University (retired), summarizes what is known about the Shield Archaic peoples who inhabited northern Ontario.

• • •

The Northern Shield Archaic peoples . . . were the first substantial population in Northern Ontario. The early beginnings of the period [of their tenure in the northern part of the province] are still poorly defined . . . . However, it is possible from present information to reconstruct a sizeable segment of the people's lifeways, which appear to have been similar if not identical to that of the Northern Algonkian speakers of the early historic period. In Northern Ontario, stone tools . . . while manifesting some local variation, reflect a common technology different from the tool kits of adjacent contemporaneous populations. The varying degrees of cultural continuity seen in the tool kit, hunting practices and social organization suggest that these Archaic peoples are the ancestors of the historic Cree, Ojibwa, and Algonkin peoples . . . .

Throughout the Precambrian Shield country, the peoples relied on big game and fish for their food. Therefore sites were located at lake and river narrows which are natural caribou crossings and fishing areas. Bears, beaver, hare and water fowl supplemented the diet and edible plant resources . . . although comparatively sparse, would also have been collected in season. There would probably have been a sexual division of labour with men concentrating on hunting and women on fishing and gathering plants. The resources available would have had a direct bearing on the patterns of labour; for example, a region rich in fish would place greater emphasis on the role of women . . . .

Over the years, there are marked changes in the tool kit. There is a general decrease in size . . . . Large cutting and scraping tools, core-scraper tools (large rough scraper-like tools made from cores), large

▼

Shield Archaic artifacts

▲

scrapers, bifaces and unifaces disappear, while there is a significant increase in the number of small scrapers. Projectile points are reduced in size and the shape changes to side-notched and trianguloid forms . . . . This radical shift in missile technology probably indicates the introduction of the bow and arrow . . . .

Source: K. C. A. Dawson, *Prehistory of Northern Ontario* (Thunder Bay, Ontario: Thunder Bay Historical Museum Society, 1983), pp. 9–11.

# MESOLITHIC CULTURE IN EUROPE AND THE NEAR EAST

## THE MAGLEMOSIANS

**Maglemosian culture**, a mesolithic culture that flourished from about 9500 ya until about 7700 ya in the peat-swamp-forest country of northern Europe, stretching from England to Denmark, was based on much the same underlying principles as Archaic culture. Like Archaic peoples, the Maglemosians made use of a wide variety of materials that they fashioned into a formidable array of implements and utensils. Among other things, the Maglemosian toolkit included polished stone axes with handles that were used to fell trees. These were then fashioned into rough planks that were used to construct wooden huts, and into dugout canoes up to three metres long that were propelled with wooden paddles. The Maglemosians were also adept at manufacturing **microliths** — small stone blades one centimetre to five centimetres long that were set into pieces of wood, bone, and antler in order to fabricate composite tools such as barbed arrowheads, barbed spearpoints, and small woodworking implements (see Figure 10.3). Although microliths were first employed by upper palaeolithic peoples, mesolithic peoples such as the Maglemosians manufactured many more of these tiny blades than their technological forerunners had done, and put them to a wider variety of uses.

One of the most impressive things about Maglemosian implements is that they indicate that the old nomadic way of life that had dominated the Ice Age was coming to an end. Equipped with their new tools, the Maglemosians and mesolithic peoples like them elsewhere in the world were no longer forced to follow their food as earlier

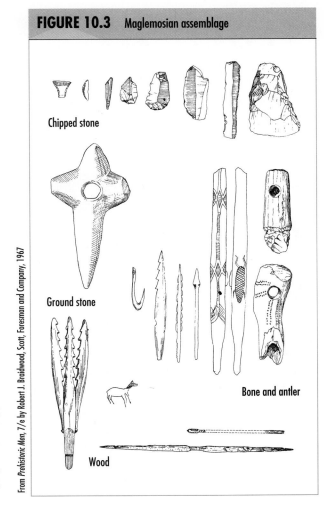

**FIGURE 10.3**   Maglemosian assemblage

Chipped stone

Ground stone

Bone and antler

Wood

From *Prehistoric Men*, 7/e by Robert J. Braidwood, Scott, Foresman and Company, 1967

palaeolithic peoples had done. Instead, they used their technological expertise to produce what they needed to survive from their immediate post-Pleistocene surroundings, often dividing their time between inland hunting camps in winter and coastal fishing camps in summer.

## THE NATUFIANS

The trend towards a more settled way of life during the middle stone age was even more pronounced in the Near East. There, the **Natufians**, a mesolithic people who lived in what are now Israel, Jordan, and Syria between about 12 500 ya and about 10 500 ya, established the first long-term residential base camps, and perhaps even the first year-round settlements. The size of their sites, the substantial structures that the Natufians built, and the ecofacts and artifacts found in Natufian sites attest to their sedentary lifestyle. These demonstrate that while the Natufians hunted and fished with tools that were similar to those employed by other mesolithic peoples (see Figure 10.4), they also added an important new food to their diet — wild grain, which they harvested with microlith-tipped sickles and ground into flour with milling stones. The Natufians also dug pits in front of the caves and rock shelters in which they lived in order to store the grain, and, at Eynan, in Israel, they stored their harvest beneath circular pit houses that were arranged in small villages. Excavations at Eynan have revealed the remains of three Natufian villages, one on top of another, each consisting of about 50 dwellings that were connected with stone walkways (see QQ 10.3). With the advent of these villages, the **neolithic revolution** was about to begin — a food-producing revolution in which people began to depend on domesticated rather than on wild species for their livelihoods.

**FIGURE 10.4** Natufian assemblage

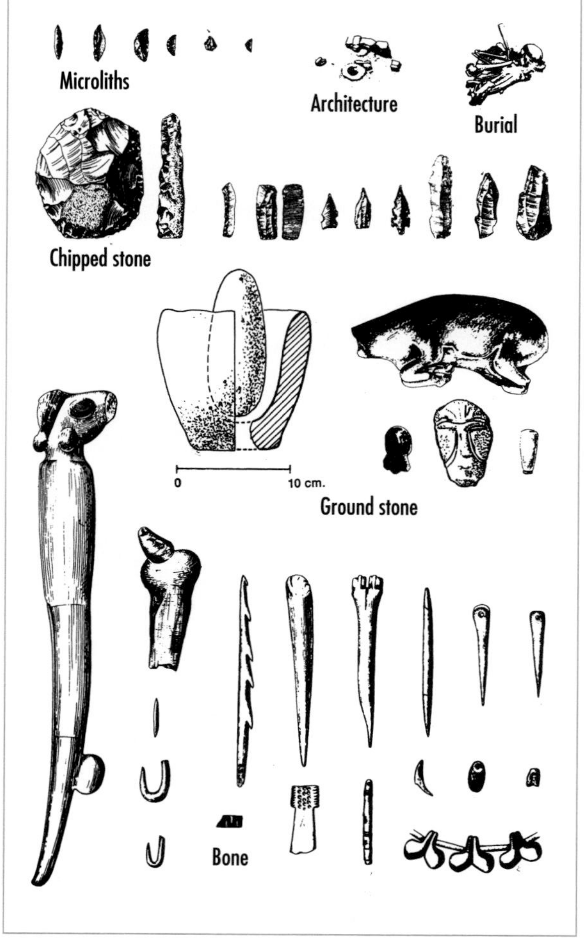

Microliths

Architecture

Burial

Chipped stone

Ground stone

0      10 cm.

Bone

From *Prehistoric Men*, 7/e by Robert J. Braidwood, Scott, Foresman and Company, 1967

## 10.3 NATUFIAN SETTLEMENT PATTERNS

Ofer Bar-Yosef and Anna Belfer-Cohen discuss the evidence that indicates that the Natufians had adopted a sedentary way of life and comment on the consequences of this new way of life.

• • •

The size of Natufian sites and the presence of built structures are interpreted as reflecting somewhat larger bands than in preceding periods. One of the largest Natufian sites, Mallaha (Eynan) in the Hula Valley, has been classified as a village . . . . The investment of leveling slopes in order to build houses on terraces, the preparation of plaster, the transportation of heavy undressed stones into open-air and cave sites . . . and the digging of underground storage pits . . . indicate energy expenditure expected in base camps but not in ephemeral, season occupations. However, it is the faunal and not the architectural evidence which, in our view, testifies to very long seasonal or permanent residence in these larger sites. For the first time human commensals [or animals that live with humans largely without injury to themselves] (house mouse, rat, and house sparrow) are found, and in large numbers . . . .

The house mouse in Natufian base camps seems to be morphologically different from the ordinary *Mus musculus* [or house mouse] in pre-Natufian deposits . . . thus indicating a morphological change which resulted from the prolonged duration of occupation by humans . . . .

Natufian sedentism, whether year-round or only partial, marked both the planning depth of anticipated logistical mobility, with well-built houses and installations, and a departure from the Paleolithic mode of residential mobility. The process of "abandoning mobility" was deepened in Early Neolithic times, with increasing investments in building dwellings, storage facilities, and cultivating fields. These developments, beginning about 10 000 [years] B.P. [or before the present], occurred during a period of improved climate, which a few millennia earlier would have led . . . [their ancestors] to retreat to a traditional hunting-gathering way of life. Thus, the emergence of the Natufian was "a point of no return" that became consolidated in the Early Neolithic.

Source: Ofer Bar-Yosef and Anna Belfer-Cohen, "The Origins of Sedentism and Farming Communities in the Levant." *Journal of World Prehistory*, Vol. 3, No. 4 (1989), pp. 473–490.

# NEOLITHIC CULTURES

Beginning in the Near East about 8 kya or earlier, and shortly thereafter in other parts of the world, neolithic cultures appeared. At one time, archaeologists identified these cultures by the presence of pottery, polished stone tools, and the remnants of woven articles in archaeological sites. Today, however, the term *neolithic* is used to refer to the cultures of those who built upon the food-producing revolution that originated in the Near East, planting crops and rearing animals not necessarily for long-term economic gains, but almost certainly with short-term adaptive goals in mind.

## ÇATAL HÜYÜK

The residents of **Çatal Hüyük**, a neolithic town in southern Turkey (see Figure 10.5) that was occupied between about 8250 ya and about 7400 ya,

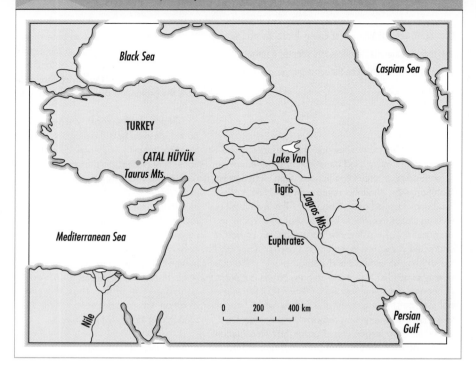

**FIGURE 10.5** Location of Çatal Hüyük

are among those who built upon the food-producing revolution that originated in the Near East. Excavations at the thirteen hectare site straddling one of the branches of the Carsamba River, where Çatal Hüyük once stood, indicate that the 4000 to 6000 neolithic people who lived there resided in side-by-side single and double storey sun-dried brick houses that were equipped with storerooms and verandas that opened onto courtyards (see Figure 10.6). Their material remains also reveal that they worshipped in more than forty similarly constructed, richly decorated shrines whose walls were adorned with paintings, reliefs, and engravings. These works of religious art frequently featured women and bulls — subjects that some believe represent the opposing themes of fertility and death.

Although those who initially lived in the region depended mainly on wild cattle for their food supply, by about 8 kya their descendants had domesticated these animals as well as sheep, goats, and dogs. In addition, they grew barley and three kinds of wheat. The residents of Çatal Hüyük also made contact with and traded with their neighbours; the obsidian that they quarried from the nearby Taurus Mountains has been found in sites distributed throughout Anatolia (the plateau between the Black and Mediterranean seas), the Levant (the lands bordering the shores of the

From *Çatal Hüyük* by James Mellaart (Published by McGraw-Hill Book Company, 1967 and Thames and Hudson, 1967)

**FIGURE 10.6** Schematic reconstruction of several houses and shrines at Çatal Hüyük

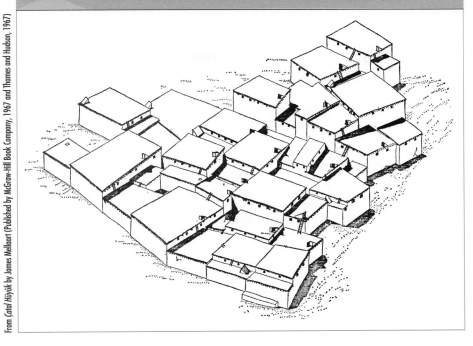

Mediterranean and Aegean seas), and Cyprus. Equally important, although the mechanics of prehistoric cultural diffusion from the Near East into Europe are still being debated, most archaeologists believe that it was due to the movement of people and the dissemination of ideas and materials from towns such as Çatal Hüyük that the neolithic way of life spread from the Near East into Europe. See QQ 10.4.

## 10.4 A NEOLITHIC TOWN IN TURKEY

James Mellaart comments on the strategic importance of neolithic towns such as Çatal Hüyük.

• • •

Excavations on the Anatolian plateau of Turkey a few years ago provided an answer to an archaeological question of long standing about Neolithic culture. The Neolithic is the stage of civilization at which men began to cultivate crops and domesticate animals and as a result of these activities to dwell in permanent settlements; in the Near East this state occurred roughly between 7000 B.C. and 5000 B.C. The question was how Neolithic culture had moved from the Near East into Europe. The answer was that the movement was overland, by way of the Anatolian plateau. Such a route had long seemed to archaeologists a logical supposition, but until Neolithic communities were excavated on the plateau there had been no direct evidence to support the supposition . . . .

[The] people who lived in these Neolithic communities] may not have been the first to learn the arts of cereal cultivation and animal husbandry, but . . . [b]y the seventh millennium they had created the first Mediterranean civilization of which Çatal Hüyük is such an impressive representative. In time the offshoots

of that civilization reached the Aegean shore, and by the sixth millennium Anatolian colonists were laying the foundations for the ultimate development of civilization in Europe.

Source: James Mellaart, "A Neolithic City in Turkey," *Scientific American*, Vol. 210, No. 4, pp. 94–104. Copyright © 1964 by Scientific American, Inc. All rights reserved.

## THE NEOLITHIC IN THE OLD AND NEW WORLDS

Given that the neolithic revolution likely spread into Europe from the Near East, it might be expected that neolithic peoples elsewhere in the world were also influenced by those in the Near East. Although this is probably true with respect to the neolithic cultures that appeared in Saharan and sub-Saharan Africa and in the Indian subcontinent, most archaeologists believe that neolithic cultures in the Far East and in the Americas developed entirely on their own. Under the circumstances, it is not surprising that different species of plants and animals were domesticated in different parts of the world (see Table 10.2). The contrast between the species that were domesticated in the Near East and in the Far East is an example. In the Near East, a wider range of animals yielding more meat and dairy products was domesticated, whereas in the Far East soya beans largely provided the nourishment that was acquired in the Near East from meat and dairy products. There were also differences in the species that were domesticated in the Old World and in the New. Working with what was available, neolithic peoples in the Old World domesticated more animals than plants, while those in the New World did the opposite. See QQ 10.5.

 ## 10.5 DOMESTICATED PLANTS AND ANIMALS

Philip Smith, Professor of Anthropology at the University of Montreal, describes the species that were domesticated in the Old World and in the New.

• • •

New World food production was based on combinations of starch and protein cultigens [or domesticated plants], especially maize [or corn], beans, squash,

pumpkin, manioc [or cassava], white potatoes, sweet potatoes, chili pepper, chocolate, and tomatoes. Domesticated animals (the llama, alpaca, turkey, guinea pig, Muscovy duck, and dog) were relatively unimportant for food or other purposes. In the Old World, food production was based on other combinations of starch and protein cultigens and on a wide range of domesticated animals, with the proportions varying considerably from one continent to the other. Cultigens included such cereals as wheat, barley, rice, millet [a cereal grass with a small grain], and sorghum [a cereal grass with dense grain]; root crops such as yams and taro [a starchy tuber], vegetables, legumes and many tree crops of fruits and nuts. The principal animals were sheep, goats, pigs, cattle, water buffalo, camels, yaks, and reindeer. In the Far East, however, especially in monsoon Asia, domesticated animals were restricted mainly to pigs, fowl, and draught animals: fish and plant proteins, especially soya beans, largely took the place of meat and dairy products.

Source: Philip E. L. Smith, *Food Production and its Consequences*, second edition (Menlo Park, California: Cummings Publishing Company, 1976), p. 13.

**TABLE 10.2** The first appearance of selected domesticated species in the archaeological record and the location where these were brought under human control

| SPECIES | FIRST APPEARANCE (ya) | LOCATION |
|---|---|---|
| *Plants* | | |
| wheat | 9 500 | Near East |
| barley | 9 500 | Near East |
| common bean | 9 500 | Mesoamerica |
| squash | 9 200 | Mesoamerica |
| rice | 8 000 | Far East |
| lima bean | 7 900 | South America |
| gourd | 7 200 | South America |
| chili pepper | 7 200 | Mesoamerica |
| maize (corn) | 7 000 | Mesoamerica |
| cotton | 6 500 | South America |
| sunflower | 3 500 | North America |
| potato | 2 400 | South America |
| | | |
| *Animals* | | |
| dog | 11 000 | Near East and North America |
| sheep | 11 000 | Near East |
| goat | 9 500 | Near East |
| cattle | 9 500 | Near East |
| guinea pig | 8 200 | South America |
| pig | 6 000 | Far East |
| ass | 6 000 | Africa |
| llama | 6 000 | South America |
| alpaca | 6 000 | South America |
| horse | 5 000 | Europe |
| camel | 5 000 | Near East |
| chicken | 4 200 | Far East |
| cat | 3 500 | Africa |

# EXPLAINING THE NEOLITHIC

What remains to be determined is why neolithic peoples began to rely on domesticated as opposed to wild plants and animals for their survival, for this required a major a shift in traditional hunting and gathering subsistence techniques. Although none is universally accepted, many explanations for this cultural transformation have been proposed.

## THE OASIS HYPOTHESIS

Some of the explanations are based on the idea that the neolithic revolution was produced by a specific cause. The **oasis hypothesis** is an example. According to its original proponents, the cultivation of plant foods arose in the Near East as a result of the new hot and dry climate that prevailed there at the end of the Pleistocene — a climatic change that was allegedly caused by the contraction of high pressure areas over the diminishing European glaciers, which deflected the rainfall that had formerly fallen in North Africa and the Near East over Europe. These climatic conditions, supporters of the

Water buffalo pulling a plow much like the plows used in the Near East about 5000 years ago

DeVore/Anthro-Photo

oasis hypothesis said, not only reduced the number of wild animals in the region, but also led to the formation of oases that attracted migratory hunters and gatherers because the oases contained a relatively abundant supply of wild plants that were capable of providing the hunter-gatherers with the nourishment that they needed to survive. Those who originally proposed the oasis hypothesis also maintained that the people who settled at these resource-rich locations were soon forced to learn how to cultivate crops in order to feed their expanding population. Others subsequently pointed out that herbivores would also have been attracted to the oases, making it possible for the residents to domesticate these animals by protecting them from predators such as lions, leopards, and wolves that were also attracted to the oases. The residents of the oases reportedly did so by allowing the herbivores that were once their prey to feed at the oases, and by studying herbivore behaviour in order to determine how best to make these animals less aggressive and bring them under human control. Thus, according to the oasis hypothesis, climatic change was ultimately responsible for the neolithic revolution. See QQ 10.6.

## 10.6 THE OASIS HYPOTHESIS

Geologist R. Pumpelly, an early supporter of the oasis hypothesis, explains how he believes the neolithic revolution began.

• • •

With the gradual shrinking in dimensions of habitable areas and the disappearance of herds of wild animals [as a result of post-Pleistocene climatic change], man, concentrating on the oases and forced to conquer new means of support, began to utilize the native plants;

and from among these he learned to use seeds of different grasses growing on the dry land and in the marshes at the mouths of larger streams on the desert. With the increase of population and its necessities, he learned to plant the seeds, thus making by conscious or unconscious selection, the first step in the evolution of the whole series of cereals.

Source: R. Pumpelly, *Explorations in Turkey, the Expedition of 1904: Prehistoric Civilization of Anau*, Vol. 1 (Washington, D. C.: Publications of the Carnegie Institution, No. 73, 1908), pp. 65–66.

## THE READINESS HYPOTHESIS

Although popular for many years, by the 1950s, research had shown that the oasis hypothesis suffered from a serious flaw. The oasis hypothesis was based on the idea that there were prolonged droughts in the Near East when the Pleistocene came to an end. However, by the 1950s climatologists had determined that post-Pleistocene climatic change in the Near East was not nearly as extreme as proponents of the oasis hypothesis had alleged, and that the oases that are currently located in the region did not appear until after the neolithic revolution had begun.

Armed with this knowledge, and with the results of his own extensive archaeological fieldwork and experience in the Near East, Robert Braidwood proposed an alternate explanation — the **readiness hypothesis**, which attributes the origin of agriculture and of animal husbandry to humankind's cumulative acquisition of botanical and zoological knowledge in specific seminal or *nuclear zones*. According to Braidwood, by about 10 kya, hunters and gatherers in the resource-rich **fertile crescent** that extended from the Levant to Iraq (see Figure 10.7) had become familiar enough with the plants and animals in their immediate environment to begin to consciously manipulate their patterns of growth. Those who did so, he said, became the first farmers and herders.

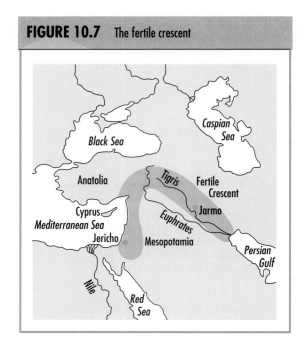

FIGURE 10.7  The fertile crescent

Braidwood also went on to say that, somewhat later, due to the fact that they were then ready to domesticate plants and animals, certain hunting and gathering peoples in nuclear zones in the Americas and in Asia followed a similar course, as eventually did other hunter-gatherers elsewhere in the world, who learned about agriculture and animal husbandry via cultural diffusion from the nuclear zones where the food-producing revolution had begun.

However, by the 1960s, it had become clear that the readiness hypothesis was also flawed. As many archaeologists were quick to point out, Braidwood's idea that cumulative knowledge amassed in nuclear zones was the driving force behind the neolithic revolution cannot be tested in a scientific way. There is simply no material in the archaeological record that indicates why certain people were "ready" to domesticate plants and animals only at the end of the Pleistocene and not before. The readiness hypothesis also seems improbable because hunters and gatherers presumably already possessed detailed knowledge about their sources of food long before the Pleistocene had come to an end.

## THE MARGINAL HABITAT HYPOTHESIS

In view of the shortcomings of the abovementioned hypotheses, these are no longer in vogue. Instead, archaeologists are currently debating the merits of other more recent and more sophisticated explanations. One is Lewis Binford's **marginal habitat hypothesis**, which points to population pressure as the ultimate source of the neolithic revolution. In Binford's view, because agriculture and animal husbandry require a much greater investment of time and energy than hunting and gathering, there must have been a good reason why some nomadic hunters and gatherers abandoned their traditional subsistence pursuits in favour of domesticating their food.

The scenario that Binford has proposed to account for this seemingly unorthodox turn of events begins at the end of the Pleistocene, when a general rise in sea levels attributable to global warming created new coastal plains that were richer in food resources than bordering inland areas farther removed from the sea. Although this encouraged people to occupy the new resource-rich coastal plains, where they subsisted by fishing and foraging and tended to become more sedentary, it also resulted in a marked increase in population, so much so that the residents soon overshot the carrying capacity of their new homelands. This, in turn, encouraged some of these people to migrate to the less productive, marginal interior areas nearby. However, as well as being less productive, some of these marginal areas were also already inhabited, and when the migrants from the coastal plains arrived competition for food between them and the original inhabitants once again caused the carrying capacity of the land to be overshot. Binford maintains that it was in such interior marginal regions — especially in the Near East, Asia, Mesoamerica, and South America, where plants and animals that were capable of being domesticated were present — that both the newcomers and the original inhabitants invented agriculture and animal husbandry in order to survive. See QQ 10.7.

 **10.7 THE MARGINAL HABITAT HYPOTHESIS**

Lewis Binford explains how he believes the neolithic revolution originated.

. . .

Change in the demographic structure of a region which brings about the impingement of one group on the territory of another would . . . upset an established equilibrium system, and might serve to increase the population density of a region beyond the carrying capacity of the natural environment . . . . [In fact, p]opulation growth . . . might well be so great that daughter communities would frequently be forced to reside [elsewhere] . . . .

From the standpoint of the populations already in the recipient zone, the intrusion of immigrant groups would disturb the population density equilibri-

um system and might raise the population density to the level at which we would expect diminishing food resources. This situation would serve to increase markedly for the recipient groups the pressures favoring means for increasing productivity. The intrusive group, on the other hand, would be forced to make adaptive adjustments to their new environment . . . . There would be strong selective pressures favoring the development of more efficient subsistence techniques by both groups . . . .

It is proposed here that it was in the selective context outlined above that initial practices of cultivation occurred.

Source: Lewis R. Binford, "Post-Pleistocene Adaptations." In Binford, Sally R. and Binford, Lewis R., eds. *New Perspectives in Archeology* (Chicago: Aldine Publishing Company, 1968), pp. 328–332.

## THE CYBERNETICS MODEL

Kent Flannery's **cybernetics model** portrays the neolithic revolution in a different way. Instead of regarding domestication as a discovery or an invention that was prompted by climatic change, population pressure, or any other prime mover or cause, the cybernetics model focuses on how early neolithic peoples slowly developed the resource procurement systems that transformed them from nomadic hunters and gatherers into sedentary agriculturalists who also raised livestock where feasible.

Assuming that early neolithic peoples preferred to exploit some food resources over others and already knew that certain foods were available only on a seasonal basis, Flannery holds that such peoples relied on the positive and negative feedback they acquired from manipulating their food resources to make decisions about how their resource procurement systems could be altered to better satisfy their nutritional needs. Increased productivity would encourage them to continue to manipulate a resource procurement system and lack of response would curtail further experimentation. Based on this cumulative feedback, Flannery says, which was generated by countless deliberate and intentional minor choices that people made about their subsistence endeavours from time to time in various locations, resource procurement systems based on agriculture and animal husbandry eventually came to the fore. The way in which wild maize was transformed from an incidental foodstuff in Mesoamerica into a major domestic crop may be a case in point. In Flannery's opinion, the positive feedback that was generated by accidentally introducing small amounts of wild maize into new environments, where it flourished because of gene flow from related grasses such as teosinte, encouraged people to domesticate the crop and to consciously and intentionally improve the yield until domestic maize had become a staple of the Mesoamerican food supply. See QQ 10.8.

## QQ 10.8 THE CYBERNETICS MODEL

Kent Flannery discusses the advantages of the cybernetics model over other competing explanations for the origin of domesticated species.

• • •

The use of a cybernetics model to explain prehistoric cultural development . . . has certain advantages. For one thing, it does not attribute cultural evolution to "discoveries," "inventions," "experiments," or "genius," but instead enables us to treat prehistoric cultures as systems. It stimulates inquiry into the mechanisms that counteract change or amplify it, which ultimately tells us something about the nature of adaptation. Most importantly, it allows us to view change not as something arising *de novo*, but in terms of quite minor deviations in one small part of a previously existing system, which, once set in motion, can expand greatly because of positive feedback.

The implications of this approach for the prehistorian are clear: it is vain to hope for the discovery of the first domestic corn cob, the first pottery vessel, the first hieroglyphic, or the first site where some other major breakthrough occurred. Such deviations from the pre-existing pattern almost certainly took place in such a minor and accidental way that their traces are not recoverable. More worthwhile would be an investigation of the mutual causal processes that amplify these tiny deviations into major changes in prehistoric culture.

Source: Kent V. Flannery, "Archeological Systems Theory and Early Mesoamerica." In Meggers, Betty J., ed. *Anthropological Archeology in the Americas* (Washington, D. C.: Anthropological Society of Washington, 1968), p. 87.

# THE COEVOLUTIONARY MODEL

David Rindos' **coevolutionary model** of the origin of the neolithic revolution also merits consideration. Like the cybernetics model, the coevolutionary model is based on the assumption that the neolithic revolution was a process that took many years to unfold. Rindos' focus, however, is exclusively on the origin of agriculture. Moreover, while Flannery holds that the domestication of plants was intentional, Rindos maintains the opposite. He argues that, at various times and places, people and plants established mutually beneficial coevolutionary relationships in which humans, acting as the unintentional agents of natural selection rather than as the intentional agents of artificial selection, unconsciously encouraged the evolution of domesticated plants by sowing and harvesting crops that they believed would best satisfy their immediate nutritional needs. Rindos also maintains that because those nutritional needs were immediate, it would be wrong to conclude that fledgling farmers had any long term goals in mind when they planted and harvested crops, including the goal of inventing agriculture for the purposes of subsistence. That prehistoric peoples had such long-term economic goals in mind, Rindos says, is a modern cultural myth based on mistaken popular notions about the rise of large-scale food-producing societies such as our own, and on the equally mistaken idea that modern humans are independent of nature when it comes to the production of food. See QQ 10.9.

 **10.9 THE COEVOLUTIONARY MODEL**

David Rindos looks at the relationship between human behaviour and the origin of domesticated plants.

• • •

Man selects, but his selection is similar to nature's — he selects the best, the most useful, the most desirable, the most vigorous, the most successful (in other words, the most "fit") plant present in the immediate environment at a given time. Splendid adaptations to man's desires occur, but they evolve over spans of time that preclude man's ever knowing what the fruits of his selection will be.

Man, like nature, is an unconscious agent selecting only for immediate benefit. Yet, most models proposed for the origin of agriculture have not stressed the unconscious aspects of man's interaction with plants. This in large part is because of a bias on the part of many who see in the origin of agriculture the beginnings of modern civilization. Forgetting that the concept of methodical selection has to do with the pursuit of goals by means of a sophisticated understanding of breeding systems, we apply it to man's earliest interactions with plants. We assume that modern man is in control of his destiny and forget that his options when dealing with the biological world are in large part dictated by processes over which he has no control. Man may indeed select, but he cannot direct the variation from which he must select.

Source: David Rindos, *The Origins of Agriculture: An Evolutionary Perspective.* (Orlando, Florida: Academic Press, Inc., 1984), p. 4.

Still others have attributed the neolithic revolution to the practice of keeping animals for sacrificial purposes; the need to raise more food to feed the world's expanding population; and to experiments undertaken by a small number of individuals who learned how to domesticate plants and animals via their endeavours. There are also scholars who maintain that the neolithic revolution was the logical outcome of the decision made by some mesolithic hunters and gatherers to exploit small game and plants, instead of big game animals, in areas where big game proved to be an unreliable resource because it was susceptible to overexploitation. Each of these explanations, and each of those cited above, has its adherents as well as its critics.

# CIVILIZED CULTURES

At the end of the neolithic, following hard on the heels of the food-producing revolution, prehistory blends into history. Starting about 5500 ya, and

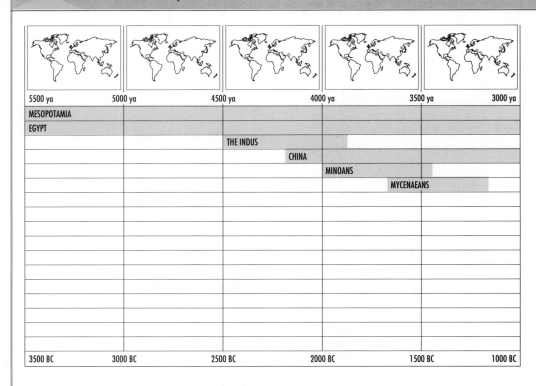

**FIGURE 10.8** Timechart of early civilizations

based on the accomplishments of neolithic peoples in various parts of the world, **civilized cultures** began to appear: first in the Near East, next in the Far East, and finally in the Americas (see Figure 10.8). Although the term "civilized" is often taken to mean "better" or "superior," this is not what anthropologists have in mind when they speak about civilized cultures. Conceptual ethnocentrism, like all forms of ethnocentrism, is simply unacceptable in the discipline. Instead, anthropologists employ the modifier "civilized" to identify cultures that are characterized by a specific constellation of social and institutional traits. Their social features, which are known as their **primary characteristics**, include the development of urban centres, the rise of occupational specialization, the taxation of surplus wealth, the formation of distinct social classes, and the advent of centralized government. Their institutional features, which are known as their **secondary characteristics**, and which were generated by the effective operation of the abovementioned social system, include large-scale public works, long-distance trade, standardized monumental artwork, writing, and scientific achievements. See QQ 10.10.

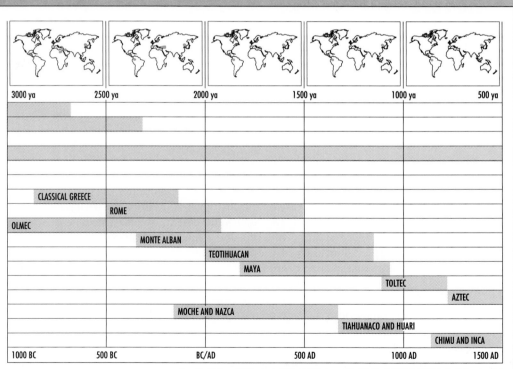

From *The Making of Civilization* by Wilkins & Whitehorse. Alfred A. Knopf, 1986 © 1986 Roxby Archaeology Ltd.

## 10.10 CIVILIZED CULTURES

Charles Redman identifies the social and institutional features of civilized cultures.

• • •

*Primary Characteristics*

1  Size and density of cities: the great enlargement of an organized population meant a much wider level of social integration.
2  Full-time specialization of labor: specialization of production among workers was institutionalized, as were systems of distribution and exchange.
3  Concentration of surplus: there were social means for the collection and management of the surplus production of farmers and artisans.
4  Class structured society: a privileged ruling class of religious, political, and military functionaries organized and directed the society.

5  State organization: there was a well structured political organization with membership based on residence . . . .

*Secondary Characteristics*

6  Monumental public works: there were collective enterprises in the form of temples, palaces, storehouses, and irrigation systems.
7  Long-distance trade: specialization and exchange were expanded beyond the city in the development of trade.
8  Standardized, monumental artwork: highly developed art forms gave expression to symbolic identification and aesthetic enjoyment.
9  Writing: the art of writing facilitated the processes of organization and management.
10  Arithmetic, geometry, and astronomy: exact, predictive science and engineering were initiated.

Source: From *The Rise of Civilization* by Charles L. Redman. Copyright © 1978 by W.H. Freeman and Company. Reprinted with permission.

# EGYPTIAN CIVILIZATION

## CHRONOLOGY

Ancient **Egyptian civilization**, which flourished along the banks of the Nile River between about 5 kya and about 2 kya, exemplifies the pattern of a civilized culture. The history of ancient Egypt is traditionally divided into four major periods of growth and high achievement, two intermediate periods of political upheaval and civil strife, and a final period of political decline. For greater precision, each of these periods is then subdivided into numbered dynasties that include a series of monarchs in the same line of descent.

The first period of growth and high achievement was the protodynastic period, during which warring neolithic villages along the Nile River were consolidated into a single political unit. This set the stage for the development of three united kingdoms: the Old Kingdom, which is noted above all else for the construction of pyramids; the Middle Kingdom, which arose after the First Intermediate Period of political turmoil when the prestige and power of the monarchy were restored and Egyptian art and literature flourished; and the New Kingdom, which arose after the Second Intermediate Period of political upheaval when Egypt was once again reunited and the country's military power and influence in the outside world reached their peak. The New Kingdom was followed by the Final Period of Decline during which Egyptian civilization collapsed. See QQ 10.11 and Table 10.3.

## 10.11 EGYPTIAN CHRONOLOGY

Jean Yoyotte discusses the chronology of ancient Egypt.

• • •

Egyptologists have superimposed on the [ancient Egyptian's own] method of arranging Egyptian kings in dynasties . . . a division of the dynasties into groups corresponding with epochs or periods. This double classification, which is easy to remember, has the advantage of retaining the traditional arrangement [in which each dynasty is known by a number and its town of origin] and well describes the whole sequence of political events in pharaonic history.

Between the obscure prehistoric period and the Late Period . . . there were three 'Kingdoms' during which Upper and Lower Egypt were united, prosper-

ous and ruled over by a strong monarchy. Each in turn succumbed to centrifugal forces (the 'feudal' ambitions of nobles and local self-interests) and gave way to an Intermediate Period during which the country was divided almost into independent principalities.

The Old, Middle and New Kingdoms were periods of internal security and a strong foreign policy; eras when great buildings were constructed and artists produced their finest works. The Intermediate Periods were times when the economy was impoverished, foreign invasion or infiltration and civil war were taking place and there was little artistic achievement.

Source: Jean Yoyotte, "History." In Posener, Georges, ed. *Dictionary of Egyptian Civilization,* trans. by Alix Macfarlane (New York: Tudor Publishing Company, 1959), pp. 126–127.

**TABLE 10.3** A Simplified Chronology of Ancient Egyptian Civilization

| PERIODS | APPROXIMATE DATE (B.C.) | DYNASTYS |
|---|---|---|
| Protodynastic | 3168 – 2705 | 1 – 2 |
| Old Kingdom | 2705 – 2250 | 3 – 6 |
| First Intermediate Period | 2250 – 2035 | 7 – 10 |
| Middle Kingdom | 2035 – 1668 | 11 – 13 |
| Second Intermediate Period | 1720 – 1550 | 15 – 17 |
| New Kingdom | 1550 – 1070 | 18 – 20 |
| Final Period of Decline | 1070 – 332 | 21 – 31 |

Source: Cyril Aldred, *The Egyptians,* revised and enlarged edition (New York: Thames and Hudson Inc., 1984), p. 9.

## SOCIAL FOUNDATIONS

Although many of the accomplishments of ancient Egypt were made during the course of the Middle and the New Kingdoms — when great strides were made in art, literature, and commerce and Egypt became increasingly powerful — it was during the heyday of the Old Kingdom that the social innovations on which Egyptian civilization was based were established. These included the rise of urban centres such as Memphis, which then served as Egypt's administrative capital; the emergence of a wide array of occupational specialists ranging from peasant farmers and labourers through a host of public officials to members of the royal family; the inauguration of a system of taxation that caused much of the wealth that was produced by peasant farmers and labourers to be transferred into the hands of a ruling elite; and the entrenchment of a rigid social hierarchy that featured a divine king (later called a *pharaoh*) at the top, a privileged aristocratic class composed of scribes, architects, medical practitioners, priests, military officers, and political advisors in the middle, and, at the bottom, a huge illiterate underclass of peasant farmers and labourers, whose agricultural productivity and economic contributions provided the wealth that supported the kingdom (see QQ 10.12). The Old Kingdom was also characterized by the appearance of one of the first large centralized governments in the world — a government whose monarchs made laws that were administered by public officials and enforced by soldiers and police, and which enabled a succession of kings and their aristocratic advisors to preside over the public and private affairs of upwards of one million people. At its zenith, the Old Kingdom empire stretched from the southern reaches of the Nile River north along the banks of the river for almost 1300 kilometres to the headwaters of the Nile at the Mediterranean Sea. See Figure 10.9.

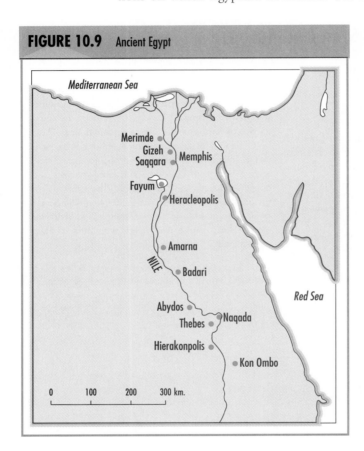

**FIGURE 10.9**   Ancient Egypt

# 10.12 ANCIENT EGYPTIAN SOCIETY

Cyril Aldred discusses Egyptian society during the Old Kingdom.

• • •

Under the divine authority of the Pharaoh, Egypt during the Old Kingdom achieved a vigorous, characteristic and self-assured culture, untroubled by doubts and unfaltering in its belief that material success depended upon completing a practical education, doing right for the king, respecting superiors, and exercising moderation in all things. The ideal of the golden mean is as much as evidence in the calm and disciplined art as in the books of precepts which the sages wrote for their posterity. Such a civilization is essentially aristocratic. At first only the king mattered, but as his divinity came to be shared to some degree by his children and descendants, his exclusive powers, like the centralized authority of the state, began to disperse among a ramified privileged class who boasted of their acquaintance with the king and partook in some degree in his immortality. It was for them that all the economic and artistic activities were created. It was they whose wishes of eternity were satisfied with tombs and endowments. They formed, however, no idle court nobility. Included in this elite were the architects, designers, writers, thinkers, theologians and master craftsmen of the day . . . . All this achievement was the exotic flower upon a plant whose root was the eternal Egyptian peasant, forever toiling in the fields, living with his animals for the moment only, never far from sudden plague and famine, hedged around by grosser superstitions than those of his masters, but able unlike them to escape the inhibitions of polite society, and preserving intact the same sardonic gusto, the same manners and customs that have brought him virtually unscathed through five thousand years of changing history.

Source: Cyril Aldred, "The First Flowering in Ancient Egypt." In Piggott, Stuart, ed. *The Dawn of Civilization: The First World Survey of Human Cultures in Early Times* (New York: McGraw-Hill Book Co. Inc., 1961), p. 132.

## INSTITUTIONAL FRAMEWORK

The institutional framework that appeared in conjunction with the operation of the abovementioned social system was equally impressive. It was during the Old Kingdom, for example, that irrigation projects such as building dikes to entrap water for subsequent distribution were first undertaken in order to increase agricultural production along the banks of the Nile. Although modest by contemporary standards, given adequate rainfall, these irrigation projects, which were capable of distributing seasonal floodwaters from natural flood basins into dry areas that otherwise would have been unproductive, were technological marvels at the time. The Old Kingdom was likewise the period during which massive shrines and tombs were constructed to celebrate the achievements of Egypt's rulers and nobles. Among others, these included the three great pyramids at Gizeh, which were erected during the reigns of kings Cheops (*c.* 2650 B.C.), Chephren (*c.* 2620 B.C.), and Mycerinus (*c.* 2600 B.C.). Situated on the outskirts of Cairo, the pyramids, whose thresholds are protected by the Great Sphinx, once formed part of a royal cemetery that has been described as "possibly the most celebrated group of monuments in the world." The massive scale of the pyramids bears witness to the technological expertise that the ancient Egyptians possessed. See QQ 10.13.

# 10.13 THE GREAT PYRAMID OF CHEOPS AT GIZEH

Thomas Patterson describes the Great Pyramid of Cheops and comments on its construction.

• • •

The largest of . . . [the three great pyramids], the Great Pyramid of Cheops, is more than 755 feet [or about 230 metres] on a side at the base and covers an area of more than 13 acres [or about 5 hectares]. It rises 481 feet [or about 146 metres] above the desert today

nearby quarries; however, the fine-grained ones were brought by barge from quarries across the river.

Workmen's barracks were located next to the construction site; they housed between 2500 and 4000 men who were engaged full time in the construction work at the building site. This is 0.25 to 0.40 percent of the total population of Egypt at the time. In other words, 1 to 2 percent of the adult male working population was involved in full-time work at the building site. This does not include gangs of men who worked at the quarries, those who transported the

The Pyramids of Gizeh; the Great Pyramid is in the centre

The Bettmann Archive

and was even taller in ancient times, since some of the fine-grained limestone blocks that covered it have been stripped away. There are more that 2 million blocks in the pyramid weighing up to 15 tons [or about 13.6 tonnes] each; the nine slabs that form the roof of the king's burial chamber each weigh about 50 tons [or about 45.3 tonnes]. Many of these blocks came from

heavy blocks to the sorting area at the foot of the pyramid, or the people who supplied the workmen with food, or the people who were engaged in other projects.

Source: Thomas C. Patterson, *Archaeology: The Evolution of Ancient Societies* (Englewood Cliffs, New Jersey: Prentice-Hall, Inc., 1981), pp. 201–202.

It was also during the Old Kingdom that trading networks were established with Middle Eastern and African neighbours; that sciences such as mathematics and astronomy came to the fore; and that hieroglyphs were adopted as the nation's official script.

## EGYPTIAN HIEROGLYPHS

Otherwise known as "sacred carvings" or "sacred engravings," **Egyptian hieroglyphs** occupy a special place in the study of ancient Egyptian society. Although once thought to be a form of picture writing, Egyptian hieroglyphs are actually a combination of **semograms** and **phonograms**: while the former convey meaning through the use of symbols (⌐♪, for example, means pharaoh), the latter represent specific sounds in Egyptian speech, much like the letters in the English alphabet (□, for example, represents the *p* sound). Carved in stone, written with a reed pen on papyrus and on leather, and painted on wood and on clay tablets, these signs and symbols were commonly used to commemorate the political and military achievements of ancient Egypt's powerful ruling class. Although the Egyptians later developed two other forms of script that were both written with connected letters — **hieratic**, which was used mainly for religious purposes, and **demotic**, which was used primarily to record secular events — hieroglyphs (of which there are more than 700) were employed from the beginning of the Old Kingdom until just after the birth of Christ.

## CHAMPOLLION'S ACHIEVEMENT

The key to deciphering the meaning of Egyptian hieroglyphs was an artifact known as the **Rosetta Stone**, a massive black basalt slab that bears a trilingual inscription written in hieroglyphic, demotic, and Greek characters that honours Ptolemy V (203 – 180 B.C.). His name, like the names of all pharoahs written in heiroglyphics, appears on the stone inside a special ring with a knot at the base called a **cartouche**. Weighing about 762 kilograms and measuring 118 centimetres long, 77 centimetres wide, and 30 centimetres thick, the stone was discovered in July 1799 near the ancient town of Rosetta (Rashid in modern Egypt) by a group of Napoleon's soldiers who were digging the foundations for a fort. The hieroglyphs inscribed on the commemorative slab, which date back to the ninth year of Ptolemy's reign, remained a mystery until they were finally deciphered in 1922 by Jean-François Champollion, a French scholar. Champollion broke the code by comparing the hieroglyphic message on the top of the slab with the demotic message in the middle and the Greek message at the bottom, and by comparing the signs and symbols enclosed in Ptolemy's cartouche with those enclosed in the cartouches of other Egyptian monarchs such as Cleopatra VII (69 – 30 B.C.). See QQ 10.14 and Figure 10.10.

## QQ 10.14 BREAKING THE CODE

John Wilson tells how Jean-François Champollion used the Rosetta Stone to decipher Egyptian hieroglyphs.

• • •

Jean-François Champollion . . . was eight years old when the Rosetta Stone was found. He also was gifted in languages and was barely of age when he was named Professor of Ancient History at Grenoble. The young man was something of a hothead. He was a democrat, anticlerical, and anti-imperial, which was dangerous in Napoleon's day. For some years, until he settled down to unravel the hieroglyphs, Champollion's political writings and acts lost him jobs and threatened his personal freedom. He had spent a year in Paris following the false lead that hieroglyphic was merely a symbolic language. [The] . . . young man then went off on his own analysis. He seems to have been aware of the work of others, but essentially he worked it out for himself — playing the demotic against the Greek, the hieroglyphic against the demotic, the hieroglyphic against the Greek . . . [and] com-

paring the positions of the *p-o-l* in the cartouche of Ptolemy with the positions of the *l-o-p* in the cartouche of Cleopatra. With a genius which was chiefly methodical, he broke the system open. He compared the hieroglyphic and hieratic versions of the Book of the Dead. His famous "Lettre à M. Cacier relative à l'alphabet des hiéroglyphes phonétiques," read before the French Academy in 1822, was the unlocking of the door. From that time on, with increasing facility, the words of the ancient Egyptians could be read.

Source: John A. Wilson, *Signs & Wonders Upon Pharaoh: A History of American Egyptology* (Chicago: University of Chicago Press, Inc., 1964), p. 19.

**FIGURE 10.10** The Rosetta Stone

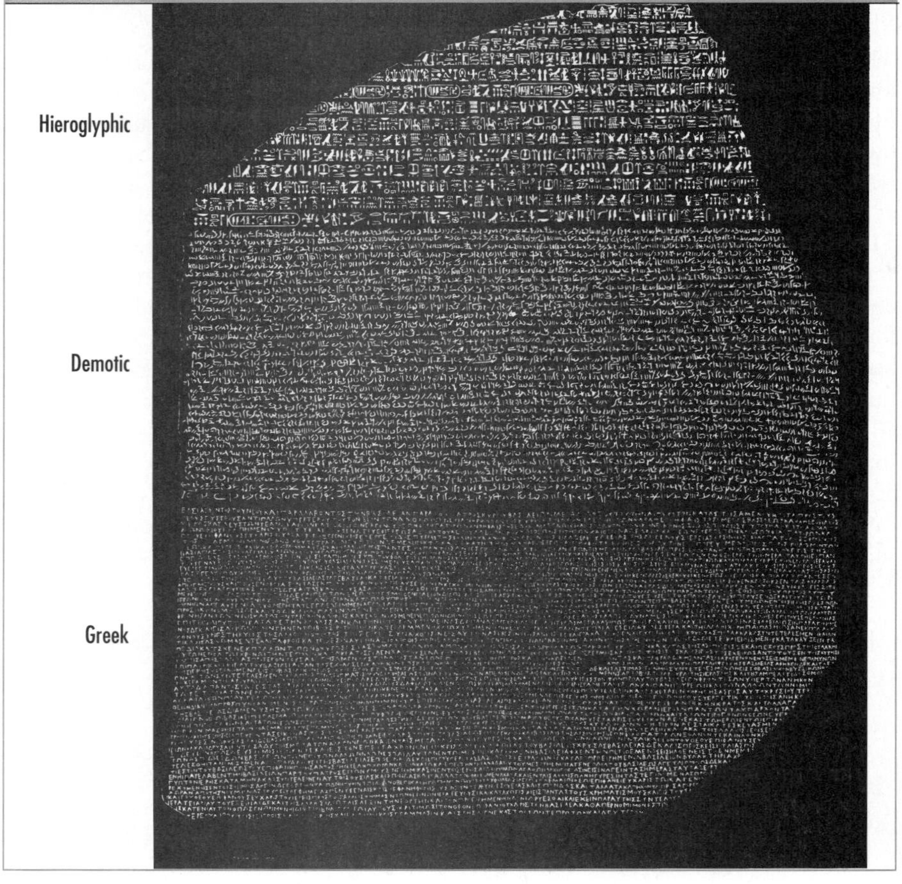

Hieroglyphic

Demotic

Greek

The Bettmann Archive

## FINAL DECLINE

The written word has provided archaeologists with considerable insight into the history of Egyptian civilization, including its final decline. For roughly 4000 years after the Old Kingdom ended, despite sometimes verging on the brink of collapse, ancient Egypt was able to maintain its cultural vigor. The nation's strength was ultimately sapped during the Final Period of Decline because of internal political and economic chaos and external military pressure. In about 525 B.C., Cambyses, a Persian king, invaded Egypt and made the country a vassal state. Later, when Alexander the Great (356 – 323 B.C.) occupied Egypt in 332 B.C., it was absorbed into the Macedonian Empire.

# THE RISE AND FALL OF CIVILIZED CULTURES

Archaeological research has also shown that, no matter where they appeared, virtually all ancient civilized cultures developed in a similar way. The course of that development can be divided into five sequential stages. During the first stage, which occurred shortly after the food-producing revolution had begun, neolithic peoples organized themselves into increasingly large village communities. Next they settled in and around valleys associated with major waterways where the natural resources that they depended upon for their livelihoods were relatively abundant. Then they began to produce surplus food, either plant or animal or both. Shortly thereafter, with the advent of occupational specialization and the rise of centralized government, the surplus came to be used to support a large occupationally specialized urban population composed of artisans and craftspeople and religious, military, and political specialists. Finally, in conjunction with these events, militarism developed, and, over the years, greatness was achieved and lost.

# THE AZTECS

## THE VALLEY OF MEXICO

The rise and fall of **Aztec civilization** illustrates the process. The Aztecs, or Mexica as they called themselves, flourished in what is now Mexico from about 1325 A.D. until their civilization was destroyed by Spanish invaders some 200 years later. Although the Aztecs ultimately extended their political dominion over a territory that encompassed more than 250 000 square kilometres of land, the heartland of their civilization was centred in the Valley of Mexico (see Figure 10.11), whose sixteenth century soil and water resources made it one of the most fertile ecological zones in the Americas. It was in the Valley of Mexico that Aztec farmers grew the corn, beans, squash, tomatoes, and other crops that the vast majority of the Mexica subsisted upon. The crops frequently were grown on **chinampas** — small raised fields surrounded by water that the Aztecs

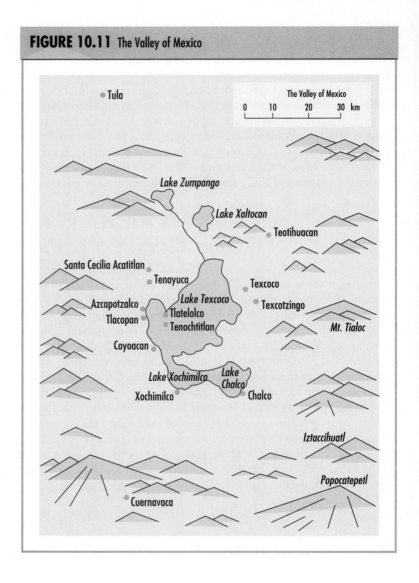

**FIGURE 10.11** The Valley of Mexico

made by dragging floating mats of vegetation from open water to marshy locations where these were anchored in place with stakes and cypress roots and then dressed with fertile muck from below. During the heyday of the Aztec empire thousands of chinampas or "floating gardens" dotted the water in their homeland; some can still be seen in modern Mexico today. See QQ 10.15.

## SOCIAL ORGANIZATION

The crops that were grown by Aztec farmers proved sufficient to support a decidedly stratified population, as many as 250 000 of whom may have lived

# 10.15 CHINAMPAS

**Richard Adams describes chinampa gardening.**

• • •

[T]he famous *chinampa* or floating garden technique . . . is essentially one of land reclamation or swamp drainage . . . . These so-called floating gardens never actually floated, but were created by making efficient use of the standing water, rich alluvium, and the marsh grasses and other vegetation in the swamps. Mats of floating water plants growing on the deeper and open ponds in the basin were used to build up the marshy areas for chinampas. These mats, long, rectangular strips of vegetation, were towed to suitable sites in the marshes and dragged onto selected spots for the new chinampa. The mats were then anchored by stakes from the native cypress, which would take root and eventually become trees, like those which one sees at Xochimilco today. Successive layers of vegetation were dragged into place until the chinampa site was raised above water . . . . The next stage of construction was to pile on lake mud from the [surrounding] canal bottom. This mud was periodically renewed and was also rejuvenated by the use of night soil [or human excrement]. Special canoe-latrines collected human waste . . . for this purpose.

One of the most efficient aspects of the chinampa system was in the use of the farm land thus created. A small section might be set aside in a corner for germination purposes and all crops planted there first. Tomatoes, corn, squashes, and other vegetables would be planted in small squares of mud and the resultant plants transplanted to crop plots elsewhere on the chinampa. Thus, germination losses were taken on a relatively small amount of ground and the transplanted

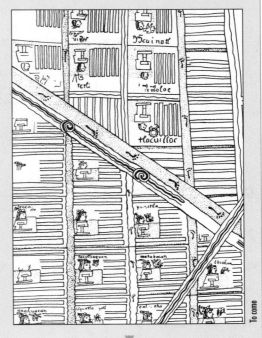

Part of an early sixteenth-century map of chinampas in Tenochtitlan

crops were made up of those plants which already had a good start, presumably being the hardiest. The main part of the chinampa was devoted to either a variety of truck crops or a main food crop such as corn.

Source: Richard E. W. Adams, *Prehistoric America* (Boston: Little, Brown and Company (Inc.), 1977), p. 27.

in the island city of **Tenochtitlán**, the Aztec capital, whose 60 000 dwellings and twenty-five pyramid temples now lie buried beneath Mexico City. The Spanish documents, Aztec records, and archaeological investigations that have helped to reveal the splendour of what was Tenochtitlán also indicate that the most important and prestigious resident was the Aztec god-king who was drawn from the highest ranking social class. This was the nobility, which included rulers, chiefs, and nobles. Also drawn from their ranks were the

MESOLITHIC, NEOLITHIC, AND CIVILIZED CULTURES

administrators, theologians, and military officials who governed Aztec society on a day-to-day basis. Below them was an intermediate class composed of skilled artisans such as painters, featherworkers, goldsmiths, silversmiths, and lapidaries (gem-workers), and merchants who bought and sold utilitarian items as well as the luxury goods that the artisans produced. Then came the commoners, who comprised the bulk of the population, most of whom were farmers who worked either their own land or land that was owned by the nobility. Slaves were at the bottom rung of the social ladder, not entitled to be citizens or to receive pay for their manual labour and sacrificed in relatively large numbers to various gods in the Aztec pantheon. Although the system allowed for social mobility, movement from one social class to another was the exception rather than the rule. See Table 10.4 and QQ 10.16.

## TABLE 10.4  Major divisions of Aztec social class structure

| NOBILITY | VERNACULAR SINGULAR | PLURAL | REPRESENTATIVES |
|---|---|---|---|
| Rulers | *tlatoani* | *tlatoque* | Supreme rulers of major political bodies (empires, cities, towns) |
| "Chiefs" | *tecutli* | *tetecutin* | Controlled a more restricted area than tlatoque; usually occupied high military and government positions |
| Nobles, "sons of nobles" | *pilli* | *pipiltin* | Children of rulers and "chiefs"; occupied governmental, religious, and military positions |
| **INTERMEDIATE POSITIONS** | | | |
| Merchants | *pochtecatl* | *pochteca* | Merchants organized into guilds and trading over long distances; often agents of the state |
| Luxury artisans | *tolteccatl* | *tolecca* | Artisans of crafts such as gold- and featherworking; some were apparently guild organized; others worked for the state |
| **COMMONERS** | | | |
| "Free commoners" | *macehualli* | *macehualtin* | Organized into calpulli, these persons were agriculturalists, fishers, and producers of utilitarian crafts |
| Rural tenants | *mayeque* | *mayeque* | Commoners who worked on the private lands of the nobility |
| Slaves | *tlacotli* | *tlacotin* | Slaves provided much urban labor for the nobility, and attained their status through gambling, economic necessity, or a criminal act (usually theft) |

Source: Frances F. Berdan, *The Aztecs of Central Mexico: An Imperial Society* (New York: Holt, Rinehart and Winston, 1982), p. 46.

# 10.16 AZTEC SOCIAL ORGANIZATION

Frances Berdan discusses the social class structure of the Aztecs.

• • •

As in other complex societies, Mexica social organization was characterized by rules which provided for the differential allocation of power, privilege, prestige, and property. The rules identified highly valued positions in the society in conformity with cultural emphases, defined avenues of status achievement, and indicated how rewards were to be distributed to the successful occupants of social positions. The Mexica system of social stratification combined ascription and achievement: Although access to most positions was controlled by birthright, the positions also required validation and allowed some mobility through achievement.

The fundamental social division was between nobility and commoners, with intermediate positions occupied by certain specialists, largely merchants and artisans of luxury goods . . . . Persons in each of these strata had specific rights and obligations, and carried out particular activities . . . . [M]any of these activities were associated with the religious, military, governmental, or commercial hierarchies.

Source: Frances, F. Berdan, *The Aztecs of Central Mexico: An Imperial Society* (New York: Holt, Rinehart and Winston, 1982), p. 45.

## MILITARISM

Convinced that they were "chosen people" selected for greatness by their gods, the Aztecs relied on their political prowess and military might to extend their sphere of influence. When they arrived in the Valley of Mexico in the latter part of the twelfth century, the region was already inhabited by sedentary farmers and nomads. For the next one hundred years, the Aztecs lived as wanderers among these people, sometimes acting as mercenaries on behalf of the original inhabitants and surviving as best they could by combining fishing with chinampa agriculture and mercantilism. Then, in about 1325 A.D., the Aztecs established two settlements in a swampy inlet of Lake Texcoco: Tlatelolco and Tenochtitlán (see Figure 10.12). The residents of Tenochtitlán eventually subdued those from their sister city and then proceeded to expand, bringing ever more people in the region under their political control. In the

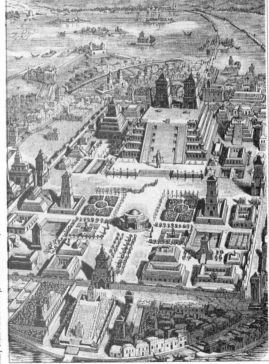

**FIGURE 10.12** An artist's rendering of the Great Temple Precinct in the centre of Tenochtitlán

Mary Evans Picture Library/Photo Researchers, Inc.

early part of the fifteenth century the Mexica concluded an advantageous political alliance with two rival Mexican states: Texcoco and Tlacopa, and this allowed them to extend their rule to the west and to the east under the auspices of what was known as the **Triple Alliance**. It was, however, an alliance in name only; the Mexica were in control. The Aztecs' dominion over Mexico ended in 1521, when, after a 90 day siege, Spanish conquistadors under Cortés ransacked the great city of Tenochtitlán and leveled it to the ground. See QQ 10.17.

## 10.17 TENOCHTITLÁN

Bernal Díaz del Castillo, a Spanish conquistador who was with Cortés in Mexico, describes the Aztec capital.

• • •

So we stood looking about us, for that huge and cursed temple stood so high that from it one could see everything very well . . . . [W]e saw the . . . [lake] which supplies the city [with fresh water], and we saw the bridges on the three causeways which were built at certain distances apart through which the water of the lake flowed in and out from one side to the other, and we beheld on that great lake a great multitude of canoes, some coming with supplies of food and others returning loaded with cargos of merchandise; and we saw that from every house of that great city . . . that it was impossible to pass from house to house, except by drawbridges which were made of wood or in canoes;

and we saw [temples] and oratories like towers and fortresses and all gleaming . . . a wonderful thing to behold . . . .

After having examined and considered all that we had seen we turned to look at the great market place and the crowds of people that were in it, some buying and others selling, so that the murmur and hum of their voices and words that they used could be heard more than a league off. Some of the soldiers among us who had been in many parts of the world, in Constantinople, and all over Italy, and in Rome, said that so large a market place and so full of people, and so well regulated and arranged, they had never beheld before.

Source: *The Discovery and Conquest of Mexico* by Bernal Diaz del Castillo. Copyright © 1956 by Farrar, Straus & Cudahy, renewed © 1984 by Farrar, Straus & Giroux. Reprinted by permission of Farrar, Straus & Giroux, Inc.

# THE SHANG DYNASTY

Meanwhile, halfway around the world, the **Shang Dynasty** had already experienced greatness and suffered decline. The Shang Dynasty, which was the first civilization in China, arose in the Valley of the Huang Ho (Yellow) River about 3500 ya. There, tens of thousands of Shang peasant farmers raised the crops and tended the animals that generated the surplus food used to support the occupational specialists who resided in urban centres such as the walled city of Yin (see Figure 10.13). At its peak, Yin's population included scribes who kept written records of Shang history; priests who predicted the future; engineers who designed magnificent monuments and buildings; artisans who excelled in metalwork and ceramics; and traders, administrators, and soldiers who were responsible for economic, political, and military matters. Above them all stood the Shang king, who lived in a palace, rode in a chariot, owned slaves, and was entitled to be buried in an elaborate tomb. About 3000 ya, however, after a dynasty of twelve monarchs had ruled in succession, the Shang were overrun by Zhou invaders from the west. Zhou people established their own dynasty, which subsequently gave way to the dynasties from which the modern state of China eventually emerged. See QQ 10.18.

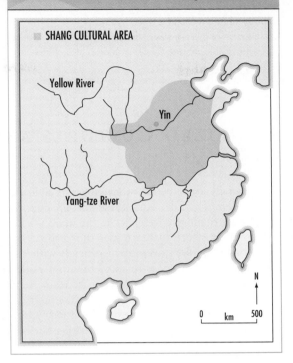

**FIGURE 10.13** The heartland of Shang civilization

SHANG CULTURAL AREA

Yellow River

Yin

Yang-tze River

N

0 km 500

## 10.18 DYNASTIC CHINA

K. C. Chang writes about the emotions that archaeologists experience in their work.

• • •

For the archaeologist dealing with data of the Shang and Chou [Zhou] periods tombs are those of kings with names instead of impersonal "persons of regal status"; bronze vessels provide silent testimony to ceremonies and processions . . . and weapons smell of blood they drank from enemies who rebelled in the East. The feeling and experience of identity with — not to mention admiration and respect for — his people as individuals as well as in collective groups transcend

The British Museum, Department of Oriental Antiquities, Collection Ref. #1936.11.18-1, Transparency #K6108

A bronze ritual vessel used to store goods for high-ranking Shang officials

# CIVILIZED CULTURES AND THE STATE

## THE NATURE OF STATES

Finally, there is one other aspect of civilized cultures that merits special consideration. Like ours, these ancient cultures featured a political system that anthropologists refer to as the **state** — a political system in which public policy is formulated under the auspices of an overriding political ideology that provides government officials with a monopoly on the use of physical force with respect to implementing policy decisions. Since the legitimate exercise of political power in states is vested in the few rather than in the many, and since those who exercise political power in states control access to life-sustaining resources such as food, scholars have long wondered how and why such an obviously inequitable political system arose.

## CONFLICT AND INTEGRATION THEORIES

Although many answers have been provided, for the sake of convenience these can be grouped into two broad categories, each of which contains a number of theories that attempt to account for the origin of the state. One category includes **conflict theories**. What these have in common is that they are all are based on the idea that the state was created in order to reduce or to eliminate social and economic conflict, either between individuals, between groups within society, or between societies themselves. The second category consists of **integrative theories**. Their distinguishing feature is that they are all based on the idea that the state was born of compromise, either to maintain or to accrue social and economic benefits that otherwise would not be in the offing. See Table 10.5.

## AN EMERGING CONSENSUS

However, given what anthropologists and others have learned about the ecological and historical circumstances that led to the formation of states in various parts of the world, it is now almost certain that both theoretical orientations have merit. In some situations states were likely born of conflict, in other circumstances they probably developed as a result of compromise, and in still other cases conflict and compromise both may have played

**TABLE 10.5** A synopsis of the major theories that have been proposed to account for the origin of the state

## CONFLICT THEORIES

### Individual conflicts

Theories of this sort maintain that the catalyst for the creation of the state was the social and economic disruption that marred interpersonal relationships in prestate societies. Social contract theories are examples. These hold that the state was formed when individuals who were living in an anarchical "state of nature" established a "social contract" with one another which permitted them to live together under the rule of law that flowed from the authority of the state.

### Intra-societal conflicts

These theories propose that states were formed to deal with the conflict that arose in prestate societies because some groups in such societies became wealthier and more powerful than others. One of the most famous theories of this kind was formulated by Marx and Engels. They argued that the state was formed by the rich to preserve their superior economic position *vis-a-vis* the underclasses, who would not be willing to provide the ruling class with the labour on which their superior economic position was based were it not for the repressive institutions of the state.

### Inter-societal conflicts

Inter-societal conflict theories hold that the state originated because of conflicts between societies. One such theory, known as conquest theory, proposes that the state arose in order to preserve the subjugation of less powerful societies by more powerful ones. Lest they be overthrown at some future date by those they had conquered, victorious societies formed the state to permanently control the vanquished.

## INTEGRATIVE THEORIES

### Circumscription theory

Circumscription theory maintains that the state was born of compromise. According to its proponents, that compromise occurred when the leaders of potentially dissident groups who lived in resource-rich homelands that were hemmed-in (circumscribed) by geographical and/or military barriers sought to maintain their *status quo* in the face of a military invasion. Instead of deciding to risk losing everything in battle, the leaders of those who were invaded agreed to act under the authority of the more powerful invaders. Although the invaders formed the state to consolidate their political power, they did not displace those they had conquered in return for their allegiance.

### Organizational benefits theory

Organizational benefits theory also maintains that the state was born of compromise. However, what its proponents argue is that the state arose when people who were otherwise relatively independent began to follow leaders who were willing to use their political and military authority to enhance not only their own social and economic well-being, but also the social and economic position of their followers. Some say that the leaders did so by supporting their followers in economic emergencies, others claim that the leaders rewarded their followers by giving them a share of the booty the state had acquired in war, and still others maintain that the leaders gained the trust of their followers by allowing them to use state owned and operated irrigation systems which increased their followers' wealth. There are also some who claim that followers were rewarded for their allegiance because of the "national pride" that the followers enjoyed on account of the massive monuments and tombs that were constructed to celebrate the achievements of the state and of its leaders. Whatever the case, according to organizational benefits theory, the followers had good reason to compromise their individual liberty in favour of obeying the dictates of the state.

Source: Based on information contained in: Elman R. Service, "Classical and Modern Theories of the Origins of Government." In Ronald Cohen and Elman R. Service, eds. *Origins of the State: The Anthropology of Political Evolution* (Philadelphia: Institute for the Study of Human Issues, 1978), pp. 21–34.

a role. In any event, what the majority of anthropologists currently believe is that a state is ultimately best regarded as an adaptive institution that is first and foremost a response to the problem of how best to govern a large and economically specialized population through the vehicle of public policy. And, while a ruling elite may promote the development of a state in order to satisfy its own political and economic interests, it is not only those who hold power who benefit from the effective operation of a state but also the majority of those who follow their lead. See QQ 10.19.

## 10.19 THE RISE OF THE STATE

Brian Hayden, Professor of Archaeology at Simon Fraser University, discusses the emergence of the state.

• • •

The emergence of the state can best be understood as the outgrowth of the efforts of a chiefly elite to pursue their own *material interests* and of their *practical manipulations* of access to goods and services that they are able to supply. Adequate resources, technology, and trade play critical roles in this development. Where this practical base offered elites opportunities to increase their power, elites also promoted other changes in social organization and values to consolidate the state organization. The spontaneous or planned introduction of new cognitive values or social relationships by themselves cannot adequately account for the emergence of the state, nor could they become widely accepted in the absence of practical reasons to adopt them. Ultimately, the development of the state significantly raised the standard of living of most people. In the most successful states, the population as a whole benefitted.

Source: From *Archaeology: The Science of Once and Future Things* by Brian Hayden. Copyright © 1993 by W.H. Freeman and Company. Reprinted with permission.

# KEY TERMS AND CONCEPTS

ARCHAIC CULTURE

LAURENTIAN ARCHAIC CULTURE

SHIELD ARCHAIC CULTURE

COMPOSITE TOOLS

MAGLEMOSIAN CULTURE (*mag-leh-moh'-zee-en* + *culture*)

MICROLITHS

NATUFIANS (*na-too'-fee-enz*)

NEOLITHIC REVOLUTION

ÇATAL HÜYÜK (*chah'-tahl* + *hoo'-yook*)

OASIS HYPOTHESIS

READINESS HYPOTHESIS

FERTILE CRESCENT

MARGINAL HABITAT HYPOTHESIS

CYBERNETICS MODEL

COEVOLUTIONARY MODEL

CIVILIZED CULTURES

PRIMARY CHARACTERISTICS

# SELECTED READINGS

## INTRODUCTION

Grayson, D. "Vicissitudes and Overkill: The Development of Explanations of Pleistocene Extinctions." In Shiffer, M., ed. *Advances in Archaeological Method and Theory*. New York: Academic Press, 1980

Martin, P. and Wright, H., Jr., eds. *Pleistocene Extinctions: The Search for a Cause*. New Haven: Yale University Press, 1967.

## MESOLITHIC CULTURE IN THE AMERICAS

Aldenderfer, M. "The Archaic Period in the South-Central Andes." *Journal of World Prehistory*, Vol. 3 (1989), pp. 117–158.

McGhee, R. *Ancient Canada*. Ottawa: Canadian Museum of Civilization, 1989.

Wright, J. *Ontario Prehistory: an Eleven Thousand-Year Archaeological Outline*. Ottawa: National Museums of Canada, 1972

## MESOLITHIC CULTURE IN EUROPE AND THE NEAR EAST

Belfer-Cohen, A. "The Natufian in the Levant." *Annual Review of Anthropology*, Vol. 20 (1991), pp. 167–186.

Clark, J. *Prehistoric Europe: The Economic Basis*. Stanford, California: Stanford University Press, 1952.

Koslowski, S. *The Mesolithic in Europe*. Warsaw: Warsaw University Press, 1973.

Price, T. "The Mesolithic of Northern Europe" *Annual Review of Anthropology*, Vol. 20 (1991), pp. 211–233.

Zvelebil, M., ed. *Hunters in Transition: Mesolithic Societies of Temperate Eurasia and their Transition to Farming*. Cambridge: Cambridge University Press, 1986.

## NEOLITHIC CULTURES

Ammerman, A., and Cavalli-Sforza, L. *The Neolithic Transition and the*

*Genetics of Populations.* Princeton: Princeton University Press, 1984.

MacNeish, R. "The Origins of New World Civilization." *Scientific American*, Vol. 211 (1964), pp. 29–37.

Mellart, J. *The Neolithic of the Near East.* New York: Scribner's, 1975.

Mellart, J. *Çatal Hüyük: A Neolithic Town in Anatolia.* New York: McGraw-Hill, 1967.

Moore, A. "Neolithic Societies in the Near East." *Advances in Archaeological Method and Theory,* Vol. 4 (1985), pp. 1–69.

Sarunas, M. and Kruk, J. "Neolithic Economy in Central Europe." *Journal of World Prehistory,* Vol. 3 (1989), pp. 403–446.

## Explaining The Neolithic

Bender, B. *Farming in Prehistory: From hunter-gatherer to food-producer.* New York: St. Martin's Press, 1975.

Boserup, E. *The Conditions of Agricultural Growth: The Economics of Agrarian Change Under Population Pressure.* Chicago: Aldine, 1965.

Braidwood, R. "The Agricultural Revolution." *Scientific American,* Vol. 203 (1960), pp. 130–148.

Childe, V. *Man Makes Himself.* New York: New American Library, 1951.

Cohen, M. *The Food Crisis in History.* New Haven: Yale University Press, 1977.

Flannery, K. "The Cultural Evolution of Civilizations." *Annual Review of Ecology and Systematics,* Vol. 3 (1972), pp. 399–426.

Flannery, K. "The Origins of Agriculture." *Annual Review of Anthropology,* Vol. 2 (1973), pp. 271–310.

Hayden, B. "Research and Development in the Stone Age: Technological Transitions Among Hunter-Gatherers." *Current Anthropology,* Vol. 22 (1981), pp. 519–548.

Iltis, H. "From Teosinte to Maize: The Catastrophic Sexual Mutation." *Science,* Vol. 222 (1983), pp. 886–894.

Streuver, S., ed. *Prehistoric Agriculture.* New York: Natural History Press, 1971.

Sauer, C. *Seeds, Spades, Hearths, and Herds.* Cambridge: MIT Press, 1969.

## Civilized Cultures

Butzer, K. "Civilizations: Organisms or Systems." *American Scientist,* September–October 1980, pp. 517–523.

Saggs, H. *Civilization Before Greece and Rome.* New Haven: Yale University Press, 1989.

## Egyptian Civilization

Davies, W. *Reading the Past: Egyptian Hieroglyphs.* Berkeley, University of California Press, 1987.

Hassan, F. "The Predynastic of Egypt." *Journal of World Prehistory,* Vol. 2 (1988), pp. 135–185.

Kemp, B. *Ancient Egypt: Anatomy of a Civilization.* London: Routledge and Kegan Paul, 1989.

Romer, J. *Ancient Lives: Daily Life in Egypt of the Pharoahs.* New York: Holt, Rinehart and Winston, 1984.

Trigger, B., *et. al. Ancient Egypt: A Social History.* Cambridge: Cambridge University Press, 1983.

Strudwick, N. *The Administration of Egypt in the Old Kingdom.* London: Routledge and Kegan Paul, 1985.

## THE RISE AND FALL OF CIVILIZED CULTURES

Sabloff, J. and Lamberg-Karlovsky, C., eds. *The Rise and Fall of Civilizations: Modern Archaeological Approaches to Ancient Cultures.* Menlo Park, California: Cummings Publishing Company, 1974.

Yoffee, N. and Cowgill, G., eds. *The Collapse of Ancient States and Civilizations.* Tucson, Arizona: University of Arizona Press, 1988.

## THE AZTECS

Bernal, I. *A History of Mexican Archaeology.* London: Thames and Hudson, 1980.

Clendinnen, I. *Aztecs: An Interpretation.* New York: Cambridge University Press, 1991.

Demarest, A. and Conrad, G. "Ideological Adaptation and the Rise of the Aztec and Inca Empires." In Leventhal, M. and Kolata, A., eds. *Civilization in the Ancient Americas.* Albuquerque, New Mexico: University of New Mexico Press, 1983, pp. 373–400.

Smith, M. "The Aztec Marketing System and Settlement Patterns in the Valley of Mexico: A Central Place Analysis. *American Antiquity,* Vol. 44 (1979), pp. 10–24.

## THE SHANG DYNASTY

Chang, K. *Shang Civilization.* New Haven: Yale University Press, 1980.

Gernet, J. *A History of Chinese Civilization.* Cambridge: Cambridge University Press, 1987.

Ho, P. *The Cradle of the East: An Inquiry into the Indigenous Origins of Techniques and Ideas of Neolithic and Early Historic China, 5000–1000 B.C.* Hong Kong: Chinese University of Hong Kong, 1975.

## CIVILIZATION AND THE STATE

Carniero, R. "The Circumscription Theory: Challenge and Response." *American Behavioral Scientist,* Vol. 31 (1988), pp. 497–511.

Fried, M. *The Evolution of Political Society: An Essay in Political Anthropology.* New York: Random House, 1967.

Haas, J. *The Evolution of the Prehistoric State.* New York: New York University Press, 1982.

Jones, G. and Kautz, R., eds. *The Transition to Statehood in the New World.* Cambridge: Cambridge University Press, 1981.

Service, E. *Origins of the State and Civilization: The Process of Cultural Evolution.* New York: W. W. Norton, 1975

Wright, H. "Recent Research on the Origin of the State." *Annual Review of Anthropology,* Vol. 6 (1977), pp. 379–397.

# CHAPTER ELEVEN

## RACE AND RACISM

*contents at a glance*

# INTRODUCTION

Although all living people are members of the same species, there are nonetheless differences among us. Aside from individual differences, one of the most striking contrasts in our appearance is the colour of our skin — the largest organ of the body. Its colour, which fascinates almost everyone, is determined by the interplay between environmental factors, such as direct sunlight, and two major pigments. One of these is **melanin**, a brown-black pigment that is produced in the lowest layer of the **epidermis** or outermost skin by special cells called **melanocytes**, and is then dispersed in granular form near the surface of the skin. Although all people generally possess the same number of melanocytes, in some individuals these cells secrete more melanin than in others, and in such persons skin colour tends to be darker. The red pigment of the haemoglobin in the red blood cells that are found in the **dermis** or innermost skin also has a bearing on skin colour; when relatively little melanin is dispersed near the surface of the skin the presence of this underlying red pigment imparts a pinkish hue to the skin. Another factor that influences the colour of the skin is **keratin**, a protein that is found in the dead, outer or corneal layer of the epidermis; it imparts a yellowish tinge to the skin. So, too, does the thickening of the outer layer of the epidermis, which is primarily responsible for producing copper-coloured skin. See QQ 11.1.

## 11.1 SKIN COLOUR

Stephen Molnar discusses the composition and the colour of the skin.

• • •

The human body is clothed in a protective covering of renewable, elastic skin consisting of . . . several structures organized into two major layers, the dermis and epidermis . . . . The innermost or *dermal* layer consists of thick collagenous fibres [composed of protein] and contains the blood vessels, nerves, hair follicles, and gland cells . . . [such as the sweat cells]. This layer is covered by a thinner protective sheath of *epidermis*, which protects the dermis from the outside environment, especially solar radiation. Epidermis is a tissue of four cellular layers that act together to provide the basic protection of the body as well as a place to synthesize vitamin D$_3$.

The lowest epidermal layer, *stratum germinativum* contains long columnar cells that make contact with the

dermis; within this layer are also specialized cells, the *melanocytes*, which synthesize the brown-black pigment granules, melanin. These granules are found scattered throughout the upper layers of the epidermis. The next layer is the *stratum granulosum*, where the cells have changed their shape and are more densely packed together providing a concentration of keratin, a protein synthesized by these cells. *Keratin* is a dense, relatively inert substance that is the major constituent of nails, hair, and the outer flaky layers of skin. There are two other layers: a thin translucent one, *stratum lucidum*, and a dense *corneum*, composed of fused cells now consisting almost entirely of keratin . . . .

The color of *Homo sapiens'* skin . . . is caused by several factors. In fair-skinned people the blood flowing through the innermost layer of the skin (dermis) contributes to the pinkish hue . . . . Keratin of the dead skin cells in the corneum has a yellowish tinge and also contributes to skin color as is demonstrated by the calluses on the palms of the hands and soles of the

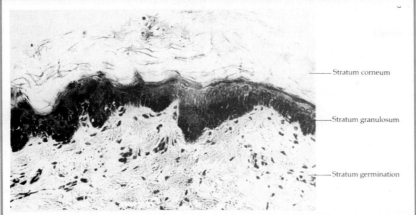

Human skin: a photograph of the epidermis as seen through a microscope. In this cross-section of dark skin, the concentration of melanin can be seen in the stratum granulosum.

From *Physical Anthropology* by Stein & Rowe, © 1989 McGraw-Hill, Inc.

Stratum corneum

Stratum granulosum

Stratum germination

feet. The lightness or darkness of the human skin, however, depends mainly on the amount of *melanin* made by the melanocytes.

Source: Stephen Molnar, *Human Variation: Races, Types, and Ethnic Groups*, third edition (Englewood Cliffs, New Jersey: Prentice-Hall, Inc., 1992), pp. 159–161.

Notwithstanding the fact that most people are unaware of the factors that produce the colour of our skin, or of those that are responsible for traits such as the amount, form, and colour of our hair; the width of our noses; the shape of our lips; and the colour of our eyes; in many people's minds such immediately recognizable differences are very important. Some maintain that they can be used to draw a clear distinction between the races or subgroups of humankind. There are also those who believe that the differential treatment of certain of these groups can be justified on scientific grounds. The first conclusion is hotly disputed; the second is mistaken.

## THE SO-CALLED THREE GREAT RACES

Many of us have heard of the so-called three great races of humankind; the **Caucasoid race**, the **Mongoloid race**, and the **Negroid race**. Many may also be aware that, in the past, each of these so-called races was thought to be unique because it was assumed that each possessed its own special bundle of distinctive, observable traits. The Caucasoid bundle reputedly included light-coloured skin; a medium to heavy amount of straight, wavy, or loosely curled body hair, either blond, red, or brown in colour; a narrow nose; thin to medium-thick lips; and light-coloured eyes. Meanwhile, the Mongoloid bundle supposedly included yellow- or copper-coloured skin; a sparse amount of medium to dark brown body hair with a straight shape; a medium-wide nose; medium-thick lips; and medium to dark brown eyes covered by a pronounced **epicanthic fold** — a fold of skin on the upper eyelids that drops down over one or both corners of the eyes and makes them appear to be almond-shaped. Finally, the Negroid bundle reputedly included dark-coloured skin; a sparse to medium amount of dark black, kinky or frizzly body hair; a wide nose; thick lips; and dark brown to black eyes (see Table 11.1). Since almost all cultures have a **folk taxonomy** of racial types, which is a classification of racial types that relies on customary beliefs to describe alleged races and to account for the differences among them, it is not surprising that the idea that the so-called three great races are distinct biological entities is one of humankind's most widely held convictions (see QQ 11.2). As we shall see, however, labels such as Caucasoid, Mongoloid, and Negroid are best regarded as social constructs rather than as scientific terms.

**TABLE 11.1** Alleged features of the so-called three great races

| FEATURE | CAUCASOID | MONGOLOID | NEGROID |
|---|---|---|---|
| Skin colour | light brown to white, pink, or ruddy | yellow or yellow-brown | dark brown to black |
| Beard and body hair | usually medium to heavy | less than Caucasoid or Negroid | medium to sparse |
| Hair form | wavy to straight, sometimes loosely curled | straight, coarse texture | woolly to frizzly |
| Hair colour | rarely black, all lighter shades | black | black |
| Nasal form | usually narrow | intermediate to Caucasoid and Negroid | usually broad |
| Lip shape | medium to thin | medium | usually thick |
| Eye colour | never black, all lighter shades | medium to dark brown | dark brown to black |

Based on information contained in: Richard H. Osborne, "The History and Nature of Race Classification." In Richard H. Osborne, ed. *The Biological and Social Meaning of Race* (San Francisco: W. H. Freeman and Company), 1971, p. 166.

## 11.2 FOLK TAXONOMIES

Virtually all cultures have a folk taxonomy of racial types; Klass and Hellman discuss that of the Cherokee Indians.

• • •

When the Cherokee Indians first came into contact with Europeans and Africans, they were astonished at how different in appearance human beings could be.

They wondered about these differences — and it wasn't long before a myth came into existence to explain them.

When the time came to make man, the myth begins, the Creator built an oven. Then he molded three figures, much like gingerbread men, out of dough. But the creator had no previous experience in making men, and had no idea how long the dough figures had to bake. The first one he took out too soon; it was underdone. Pale and unpleasant in color, "half-baked" if you will, and this was the ancestor of the white man.

The Creator waited a bit, then removed the second figure from the oven. This one was just right — light brown and pleasing to the eye of the Creator. From this figure all Indians are said to descend.

So pleased was the Creator with his second effort that he forgot to watch the oven, and when he finally pulled the third figure out it was too late. From this scorched and blackened figure, say the Cherokees, all black men are descended.

Source: M. Klass and H. Hellman, *The Kinds of Mankind: An Introduction to Race and Racism* (New York: J. B. Lippincott Company, 1971), p. 9.

# RACE AS A BIOLOGICAL CONCEPT

Given the existence of many such tripartite folk taxonomies around the world, it might seem safe to assume that, however they account for racial origins, the divisions they propose are an accurate representation of the facts. The proposed divisions may seem all the more accurate since the term race is not only part of our everyday speech, but also is entrenched in the technical languages of many biological sciences. In disciplines such as biology and zoology, for instance, the term **race** is generally used to identify sub-populations of a species that differ from one another in terms of the frequency of certain inherited traits — some of which manifest themselves in an organism's outward appearance — and this definition has been put to good use. Over the years, the identification of races has helped to refine the taxonomy of many plant and non-human animal species. This is especially true in cases where the subpopulations are separated by geographical or other barriers that prevent them from sharing genetic information, and that may ultimately give rise to new species in accord with neo-evolutionary theory. However, when it comes to humankind, scientifically valid biological entities that correspond to terms such as Caucasoid, Mongoloid, and Negroid, or to other lesser-known racial divisions, are notoriously hard to identify, and this means that the typologies that are based on such categories are of extremely limited scholarly value.

# PROBLEMS WITH HUMAN RACIAL TYPOLOGIES

## A FAULTY ASSUMPTION

One problem with the typologies is that they are based on a widespread but faulty assumption, namely, that divisions such as Caucasoids, Mongoloids, and Negroids represent what are popularly known as **pure-blooded races**, that is, biological populations with little physical variation within them, but with clear-cut differences between them. The idea is false. What biological anthropologists have discovered is that there is no such thing as a pure-blooded race. Instead, within and across so-called racial divisions, human variation occurs in a **cline** — a continuum with individuals arranged on an unbroken curve stretching from one extreme to the other for each of their genetically-based, observable traits. This finding has been confirmed by **anthropometry**, a research technique that involves using highly precise measuring instruments to determine the colouration of the skin, the width of the nose, the thickness of the lips, and so on among living humans. Since the turn of the century anthropometric studies have been undertaken throughout the world, and, based on the results, many anthropologists have concluded that clines, rather than races, best describe what has been discovered about human variation. See QQ 11.3.

## 11.3 CLINES

Frank Livingstone explains why cline is a more valuable scientific concept than race in describing human biological variation.

. . .

In the last decade there has been a remarkable increase in our knowledge of the complexities of human . . . variability. To an increasing number of anthropologists the concept of race seems to be losing its usefulness in describing this variability. In fact, for the human populations among which some of us have worked, it seems impossible even to divide these populations into races. At the same time a growing minority of biologists in general are advocating a similar position with regard to the usefulness of the concept of sub-species for classifying such diverse organisms as

grackles, martens, and butterflies . . . . Although there appears to have been a minimum of communication between anthropologists and biologists on this common problem, many of the arguments of the two groups are quite similar. It should be pointed out that the two similar positions on subspecific variation do not imply that there is no biological or genetic variability among the populations of organisms which comprise a species, but simply that this variability does not conform to the discrete packages labelled races or subspecies. For man the position can be stated in other words: There are no races, there are only clines . . . [which are measurable gradients of genetically-based traits].

Source: Frank B. Livingstone, "On the Nonexistence of Human Races." In Montagu, Ashley, ed. *The Concept of Race* (London: Collier-Macmillan, Limited, 1964), pp. 46–47.

## SKIN COLOUR

Variation in human skin colour, for example, occurs in a cline. Although it is certainly true that people who live alongside one another tend to have skin that is similar is colour, there are nonetheless differences among them. In fact, by measuring the light reflection of the skin, anthropologists have discovered that Caucasoids are not all the same shade of white, Mongoloids are not all the same shade of yellow, and Negroids are not all the same shade of black. Instead, although the frequency of dark skin is highest near the equator, in each of the continents associated with the so-called three great races — Europe with Caucasoids, Asia with Mongoloids, and Africa with Negroids — the skin colour of the people who live there occurs in a cline. In fact, research has shown that when it comes to traits such as skin colour, the overall variation within alleged human races is greater than the average difference between the divisions (see Figure 11.1). Internal consistency, in other words, is the exception rather than the rule. The idea that Caucasoids, Mongoloids, Negroids, and other such racial categories represent pure-blooded races is thus contradicted by the facts.

The idea that humankind can be divided into pure-blooded races has also been contradicted by another research finding. If there were simply variation in observable physical traits within proposed racial divisions but no overlap between the divisions, then the idea that humankind could be divided into pure-blooded races would make statistical sense. However, exactly the opposite is true. In the case of the so-called three great races, for instance, there is considerable overlap in the colour of the skin thought to be characteristic of each division, and this means that pure-blooded races do not exist among humankind even in a statistical sense.

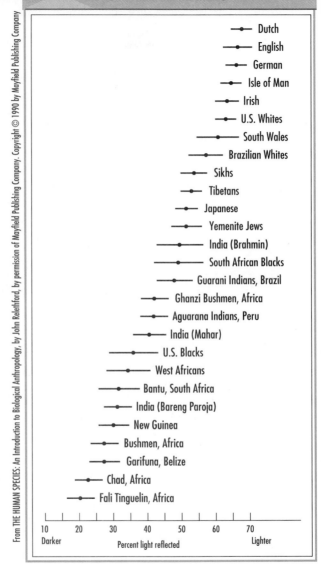

**FIGURE 11.1** Variation in human skin colour as measured by the amount of reflected light

From THE HUMAN SPECIES: An Introduction to Biological Anthropology, by John Relethford. Copyright © 1990 by Mayfield Publishing Company, by permission of Mayfield Publishing Company.

## EXCEPTIONS

Another problem with existing racial typologies is that millions of people cannot be clearly assigned to any of their categories. Although such schemes seemingly account for a large number of people from Europe, Asia, and Africa, many of the original inhabitants of those continents actually possess physical traits that include alleged features from two or more groups. In North Africa, for example, there are millions of people with wavy hair, narrow noses, and thin lips who otherwise appear to be Negroid, and in the southern part of the continent some Bushmen have an epicanthic fold. A similar situation exists in Asia, where the Ainu of northern Japan, who at first glance seem to be Mongoloid, have pale skin and are among the hairiest people in the world. Meanwhile, in India, there are millions of people who possess mostly so-called Caucasoid traits, except that their skin colour is dark brown or black. There are also exceptions in Europe, where the Saami or Lapps of Scandinavia look as much like the Inuit as they do like Europeans.

Nor does the scheme account for millions of people who do not live in Europe, Asia, or Africa. Among them are Australian Aborigines, most of whom have dark skin, which, when they are children, is sometimes accompanied by blond, wavy hair. The Polynesians in the South Pacific also possess what appear to be a combination of traits; some Caucasoid features, some Mongoloid ones, and some Negroid, and this means that assigning such people membership in a racial group is arbitrary. See QQ 11.4.

## 11.4 EXCEPTIONS TO THE RULE

James King considers the diversity that makes racial classifications suspect.

• • •

Not only are there numerous populations which do not fit neatly into . . . [the classification] system — the Polynesians, the Ainus, and the Australian Aborigines, for example — but each of the . . . groups can be subdivided, and the subdivisions can be made to appear or disappear with complete randomness depending on what character or characters one bases them on. Furthermore, there are intergrades among the . . .

groups: sometimes gradual clines like those between Caucasian and Mongolian extending from Europe through Siberia to Manchuria or between Caucasian and Negro from the Nile delta to the Sudan; sometimes sudden shifts as from North India into Tibet or from southern Algeria into the Niger region. Instead of accepting this situation for what it is — a complex of shifting biological equilibria — students have wasted huge amounts of energy trying to make it conform to some other pattern.

Source: James C. King, *The Biology of Race* (New York: Harcourt Brace Jovanovich, Inc., 1971), p. 12.

## ADMIXTURE AND GENE FLOW

Of course, it could be argued that the reason some people possess the physical traits of more than one group is because of interbreeding between the members of what were formerly pure-blooded races. However, the idea that admixture of this sort is responsible for what are presumably "mixed races" is contradicted by the facts. Consider the dark skinned people of India, who in many racial classifications are assigned membership in the so-called

Caucasoid race. On the surface, it might be suspected that the colour of their skin is due to admixture between Europeans and Africans. Yet there is no archaeological or historical evidence that the peoples of India are the result of admixture between these two populations.

Equally important, even in cases where certain human populations were isolated from others and there was no direct gene flow or exchange of alleles between them for thousands of years, there is absolutely no doubt that such populations exchanged genetic information before they became isolated. In other words, while it is certainly true that admixture between previously isolated human populations has been accelerated by modern transportation and communications, there is simply no reason to believe that such populations did not share genes until recent times, especially so considering the biological unity of humankind. The fact of the matter is that gene flow is and always has been part of the human condition, either directly via genetic exchange between neighbouring populations, or indirectly via genetic exchange between distant populations through intervening groups. See QQ 11.5.

## 11.5 GENETIC EXCHANGE

Theodosius Dobzhansky discusses the history of genetic exchange among humankind.

• • •

[The] continuous, sometimes slow, but unfailing gene flow between neighboring clans, tribes, [and] nations . . . upholds the biological and evolutionary unity of mankind. There may be no recorded case of a marriage of an Eskimo with, say, a Melanesian or a Bushman, but there are genetic links between all these populations via the geographically intervening groups. In contrast with distinct species, a beneficial change arising in any population anywhere in the world may become a part of the common biological endowment of all . . . mankind. This genetic oneness of mankind has been growing steadily since the development of material culture has made travel and communication between the inhabitants of different countries progressively more rapid and easier. What should be stressed, however, is that mankind has not become a meaningful biological entity just recently, since men began to travel often and far. The human species was such an entity even before it became recognizably human.

Lyn Hancock

▼

Inuit father and son

▲

Source: From *The Biological and Social Meaning of Race* by Theodosius Dobzhansky. Copyright © 1971 by W.H. Freeman and Company. Reprinted with permission.

# Genetics and Race

## Phenylthiocarbamide

Yet another problem with human racial typologies is that they tend to ignore differences between people that are less obvious than traits such as skin colour, hair form, body stature, and so on. The ability to taste the chemical compound **PTC (phenylthiocarbamide)** is an example. Geneticists have shown that the ability is based on a dominant allele (**T**), which is known as the "taster gene." Individuals who are homozygous recessive for the gene (**tt**) cannot taste PTC except in a highly concentrated solution, whereas those who are homozygous dominant (**TT**) or heterozygous (**Tt**) find the taste of PTC intensely bitter. At one time it was thought that this discovery might show genetic differences between races. However, this has not proved to be the case. Like skin colour, PTC tasting occurs in a cline.

## ABO Blood Groups

The distribution of **ABO blood groups** is arranged in a similar fashion. Many adults in the industrialized world know whether their blood is type "**A**," "**B**," "**O**," or "**AB**." What they may not know is that the ABO system is based on the same principles that govern the inheritance of traits such as the ability to taste PTC. There are three alleles in the ABO system: (**A**), (**B**), and (**o**), one of which is inherited from each parent. Both the (**A**) allele and the (**B**) allele are dominant over the (**o**) allele, but when the (**A**) allele and the (**B**) allele recombine, they act in a codominant fashion to form the blood type "**AB**."

**FIGURE 11.2** The frequency of gene A in the ABO system

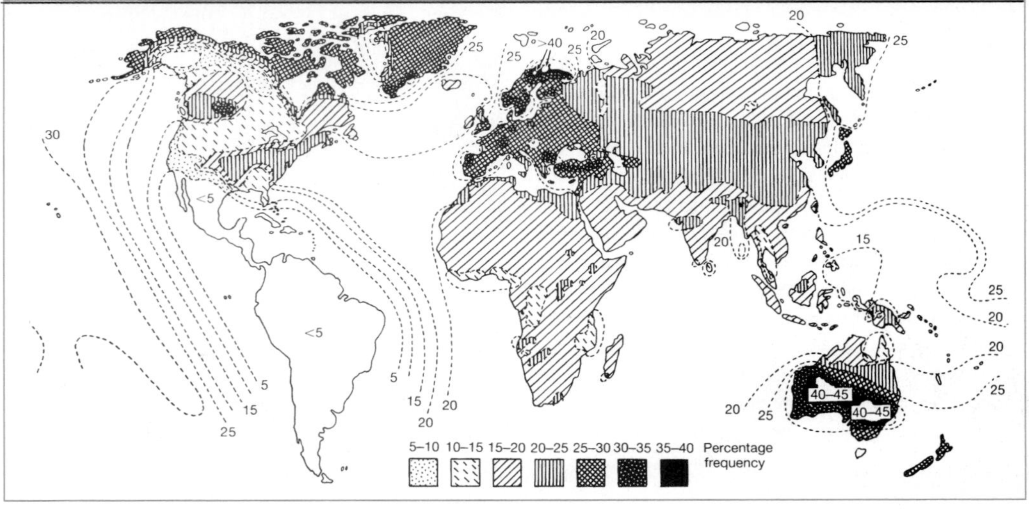
5–10
10–15
15–20
20–25
25–30
30–35
35–40
Percentage frequency
40–45
>40
<5

From *Human Biology*, 3/e, by Harrison, Tanner, Pilbeam & Baker, 1992. By permission of Oxford University Press.

Since the turn of the century, when the ABO system was discovered, simple procedures have been developed to test for blood type. As a result, it has been possible to determine the blood type of millions of people who were typically seen as Caucasoids, Mongoloids, or Negroids, or regarded as members of other alleged races. Presumably, if such groups were biologically distinct, then each should be characterized by one type of blood. However, what scientists have discovered is that while some cultural groups are marked by a relatively high frequency of one type of blood, no racial typology has ever been proposed that contains a division that is characterized by a single blood type (see Figure 11.2). In other words, there is no direct connection between the ABO blood group system and the divisions such typologies contain, and this too makes the classifications suspect. In fact, there is not one genetic marker that is exclusively associated with any alleged race. See QQ 11.6.

## 11.6 GENETICS AND RACE

L. L. Cavalli-Sforza looks at the relationship between genetics and race.

• • •

No person objects to the notion that all human beings belong to one single species. This means that all humans are potentially interfertile and that there is no limitation to mating between humans or having fertile progeny. This does not mean, however, that all human beings are genetically the same. In fact, work in recent years has shown that the diversity between individuals is much greater than we had previously anticipated. For almost every gene that has been investigated, there has been evidence that more than one allele exists, and frequently more than two . . . . [T]he poten-tial number of different individuals is [consequently] incredibly high . . . .

Such considerations ridicule completely the concept of a "pure" race. With so many differences between even a parent and a child, how can one divide the human population into clear-cut groups of individuals all of whom are very similar to one another . . . ? There may be cultures that are relatively "pure" in the sense of homogeneous, but today anyone who believes in the existence of biologically pure races in man [each with its own genetic constitution] must be considered a charlatan.

Source: L. L. Cavalli-Sforza, *Elements of Human Genetics*, second edition (Menlo Park, California: W. A. Benjamin, Inc., 1977), pp. 113–114.

# ELIMINATING THE CONCEPT OF RACE

## SOCIAL ISOLATES

In view of the abovementioned shortcomings, some biological anthropologists maintain that the concept of race should be eliminated from the language of anthropology. The opinion is not without merit. As already mentioned, in the life sciences the term *race* is used to identify biologically related groups that have the potential to evolve into new species. However, this is not the case among humankind. If anything, as a result of increased gene flow in the modern world, instead of becoming increasingly different with respect to their physical traits, people are becoming more and more alike. Besides, when it is used to refer to humankind, the term race is often confused with **social isolates** — groups of people who are isolated from one another because of geographical or cultural barriers, but who are by no means biologically unique. In everyday terms such isolates are frequently referred to as **ethnic groups**, which can be defined as subcultures within a larger society, and such groups are often misrepresented as races by those who fail to distinguish between human biology and human culture. The terms *race* and *ethnicity*, however, are not interchangeable. *Race* is used to identify subpopulations of a species that differ from one another in terms of the frequency of certain inherited traits; *ethnicity*, on the other hand, is a cultural concept. See QQ 11.7.

## 11.7 RACE AND ETHNICITY

Jean Elliot, Associate Professor of Sociology and Social Anthropology at Dalhousie University, and Augie Fleras, Assistant Professor of Sociology at the University of Waterloo, discuss the differences between "race" and "ethnicity."

• • •

We can define ethnicity as an organizing principle that distinguishes groups of individuals with a common sense of peoplehood from others on the basis of felt identification with select elements of the cultural past. Like race . . . this definition is focused on the process of classifying people into categories . . . . In the case of ethnicity, however, affiliation . . . is based on the criteria of perceived identification with others rather than on physical characteristics (race) . . . .

Other differences are evident. The race concept implies a relatively static and permanent classification imposed on the population. Racial typologies are constructed by those who justify such classifications primarily on the basis of preselected qualities. By contrast, ethnicity conveys a much more dynamic quality. Classification along ethnic lines reflects subjective properties and voluntaristic involvement according to the demands of a given situation . . . . Finally, the concept of ethnicity reflects a series of cultural differences that are regarded as worthy of retention by those

From *Ethnic Studies*, Vol. 25, No. 2, 1993, Canadian Ethnic Studies Association, Calgary

VOL. XXV, NO. 2, 1993

Cover of an issue of *Canadian Ethnic Studies/Etudes Ethniques au Canada* — A journal devoted to the study of ethnic groups and ethnicity

whose identity is bounded by these characteristics. Race, however, does not possess any intrinsic significance on its own.

Source: Jean Leonard Elliot and Augie Fleras, *Unequal Relations: An Introduction to Race and Ethnic Dynamics in Canada* (Scarborough, Ontario: Prentice-Hall, Canada, Inc., 1992), pp. 133–138.

## RACE AND INTELLIGENCE

Yet another reason for abandoning the concept of race is that it has been misused by people who claim that race determines intelligence. Research has shown that, on average, African-Americans ("Blacks") in the United States score lower on IQ (intelligence quotient) tests than do European-Americans ("Whites") from the same country. Does this mean, as some have claimed, that the Caucasoid race is intellectually superior to the Negroid race? Of course not. African-Americans and European-Americans are not racial groups; instead, they are social isolates. In any event, since African-Americans and European-Americans come from different cultural backgrounds and occupy different social, economic, and political positions in American society, and since both culture and social status are known to be extremely important in determining IQ test score results, attributing the difference between African-American and European-American scores to race is clearly mistaken.

In fact, in one study, when the IQ scores of northern African-Americans and southern European-Americans were compared, the northerners' scores were found to be statistically superior, thus showing that social and cultural factors are critical. And when questions on one "standard" IQ test were phrased in Black English Vernacular, which is the dialect of English that many African-Americans currently speak, the African-Americans who took the test scored consistently higher than their European-American counterparts, once again demonstrating the importance of social and cultural factors.

## THE RUSHTON CONTROVERSY

Similar criticisms can be levelled against proposals that claim that recent human evolution has produced progressively better types of people, and that some modern human populations are more "advanced" than others as a result. This is precisely what psychologist J. Philippe Rushton from the Department of Psychology at the University of Western Ontario argued in a paper that he read at the 1989 Annual Meeting of the American Association for the Advancement of Science. More specifically, what Rushton maintained was that Mongoloids first evolved about 41 kya, Caucasoids about 110 kya, and Negroids about 200 kya, and that because evolution, including recent human evolution, is "progressive," Orientals, on average, have larger brains and are more intelligent and culturally advanced than Whites, and Whites, on average, have larger brains and are more intelligent and culturally advanced than Blacks. See Figure 11.3.

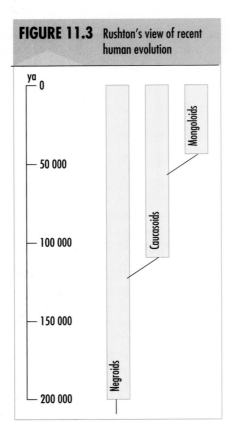

**FIGURE 11.3** Rushton's view of recent human evolution

ya 0

Mongoloids

Caucasoids

Negroids

50 000

100 000

150 000

200 000

However, as many Canadian scholars have pointed out, Rushton's argument suffers from a number of serious flaws. For one thing, as already mentioned, it is highly doubtful that Mongoloids, Caucasoids, and Negroids represent distinct biological races. For another, even if one assumes that the classification is accurate, there is simply no evidence to support the view that the so-called three great races of humankind evolved in any particular sequence. Finally, although Rushton did not claim that the environment plays no role in shaping intelligence and culture, his contention that Orientals are superior to Whites, and Whites to Blacks, largely because of inherited differences in each group's genetic potential also flies in the face of the facts. See QQ 11.8.

## 11.8 SZATHMARY ON RUSHTON

Emöke Szathmary, Professor of Anthropology, Honorary Professor of Zoology, and Dean of the Faculty of Social Science at the University of Western Ontario, comments on the flaws in Rushton's paper.

• • •

A fundamental question concerning Professor Rushton's paper is the appropriateness of the evolutionary framework within which he analyzes the data he has assembled. Professor Rushton says that the three human races can be ranked in terms of their antiquity. According to him, Negroids evolved first, around 200 000 years ago, followed by Caucasoids appearing about 110 000 years ago, and then Mongoloids, appearing about 41 000 years ago. Professor Rushton postulates that more recently evolved groups are more "advanced" than others. He says this explains the patterns he observes in the rankings of the three races on a host of morphological and behavioral variables. However, what explanation is there, if Professor Rushton is using an inaccurate evolutionary scenario, based on a misunderstanding of what has been published?

What evidence is there for this ranked ordering of the evolution of human races? None. To begin with, human population biologists know that the number of human races is in itself open to dispute. Some have argued that there are no races, only clines . . . . Most others will allow that the minimum estimate is three . . . but the maximum exceeds 100 . . . . It all depends on what traits one chooses to define a "race," and the choice is dependent on the type of problem one addresses.

Regarding the antiquity of the "races" (if we allow that there are at least three) . . . [i]t is worth noting that there are many considerations, theoretical and statistical, that enter into the choice of methods and the choice of data used to reconstruct biological history. One cannot choose a "correct" scenario without considering the methods and data used to determine it. Proteins and enzymes are thought by some to provide a better reconstruction of what happened in human history . . . [than other markers]. Nevertheless, even the lines of descent based on proteins and enzymes do not allow the classification of modern human "races" into "least evolved," "more evolved," and "most evolved." The evidence they provide does not support Professor Rushton's claim that Negroids evolved 200 000 years ago, Caucasoids 110 000 years ago, and Mongoloids, 41 000 years ago.

As for superiority and inferiority — any geneticist finds this notion amusing. Each is dependent on the environmental context within which the "superior" or "inferior" trait (not population) evolved. Change the environment, and one may change the labelling attributed to the trait. While I do not think that natural selection has favoured the emergence of human intellectual ability, it is worth remembering that intelligence is not necessarily indicated by brain size or cranial capacity. Neanderthals on average had larger cranial capacities than do modern humans, and this is even more obvious if we correct cranial capacity for body size (Neanderthals were a short people). They may have been gifted for all we know, but none remain to debate the point with us.

Source: Emöke J. E. Szathmary, "A 'fundamental question' concerning Rushton paper." *Western News*, 9 February 1989, p. 4.

# HOW SHOULD HUMAN VARIATION BE STUDIED?

## A FOCUS ON ADAPTATION

Despite the fact that many anthropologists agree that races cannot and should not be defined on the basis of observable physical traits, it remains true that such traits are well worth studying, especially as genetically-based adaptations to specific environmental conditions. Consider skin colour. In the tropics, in areas without forests, the dispersion of melanin in the epidermis of black-skinned individuals helps to shield their bodies from the ultraviolet radiation of the sun. Although such radiation triggers the conversion of dehydrocholesterol (a cholesterol derivative) into calciferol or vitamin D, which helps the body to absorb calcium and discourages the onset of bone diseases such as rickets (soft bones) and osteomalacia (weakened bones), the deep penetration of ultraviolet radiation is known to cause skin cancer. Under the intense rays of the tropical sun, dark-coloured skin may consequently help to prevent skin cancer, and, at the same time, allow just enough vitamin D to be produced to encourage the proper development of bones. However, in Europe, where the sun's rays are less intense, dark-coloured skin could lead to a deficiency of vitamin D and increase the probability of rickets and osteomalacia. There, in other words, light-coloured skin would be advantageous. See QQ 11.9. This has led to the hypothesis that the first hominids had dark-coloured skin, but, as they moved north of the equator, natural selection favoured light-coloured skin. See Figure 11.4.

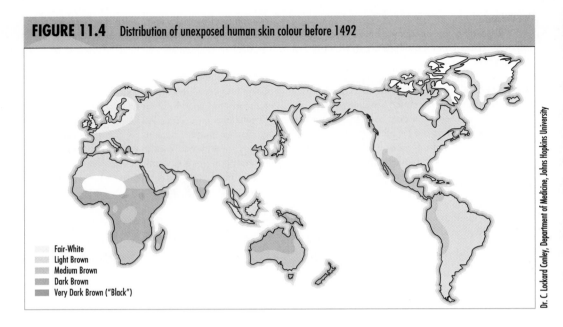

**FIGURE 11.4**   Distribution of unexposed human skin colour before 1492

Fair-White
Light Brown
Medium Brown
Dark Brown
Very Dark Brown ("Black")

Dr. C. Lockard Conley, Department of Medicine, Johns Hopkins University

However, as interesting as this hypothesis may be, some scholars believe that there may be no direct link between natural selection and human skin colour. Variation in human skin colour may well be the result of neutral mutations or the by-product of natural selection. Of course, this does not mean that the adaptive significance of genetically-based traits such as skin colour should be ignored, only that they should be studied with caution, keeping in mind that the origin of a trait and its function are not necessarily the same. See QQ 11.9.

## QQ 11.9 NATURAL SELECTION AND HUMAN SKIN COLOUR

Walter Quevedo, Jr., Thomas Fitzpatrick, and Kowichi Jimbow discuss the relationship between natural selection and human skin colour.

• • •

A number of theories have been proposed to account for the existence of cutaneous melanin pigmentation [or skin pigmentation attributable to melanin] and its variation throughout the human species. They hold that cutaneous melanin pigmentation is or has been associated with (1) resistance to sunburn, premature aging and cancer of the skin, (2) camouflage of man in evasion of predators and in hunting, (3) thermoregulation [or regulation of body temperature] by enhancement of absorption of solar radiation or loss of metabolic heat, (4) regulation of vitamin D synthesis by influences on the penetration of U[ltra] V[iolet] R[adiation] into skin, (5) protection of light-sensitive substances in skin from photodestruction, (6) sensitivity to frostbite and (7) aggressive display behavior patterns. These theories have linked cutaneous melanin with human fitness, thus emphasizing that skin pigmentation in man has been carefully shaped by natural selection throughout human evolution . . . .

[However,] there is a widening consensus that evolutionary mechanisms do not hold life on as tight a leash as previously thought. As a corollary, not all traits have been built either directly (adaptation) or indirectly (preadaptation) by natural selection for the roles that they currently play . . . . Many traits may represent the outward expression of evolutionarily neutral genetic processes of the developmental by-products of natural selection operating on particularly complex components of bodily form and function. Therefore, although biochemistry and electron microscopy may provide the investigator with more understanding of the mechanism of melanin pigmentation in man, it does not necessarily follow that variations in fine details of the melanogenic mechanism among human populations are invariably of profound evolutionary significance.

Source: Walter C. Quevedo, Jr., Thomas B. Fitzpatrick, and Kowichi Jimbow, "Human Skin Color: Origin, Variation and Significance." *Journal of Human Evolution* Vol. 15, No. 1 (1985), pp. 52–54.

## BREEDING POPULATIONS

In addition to looking at the adaptive significance of human variation, biological anthropologists have also begun to focus their attention on **breeding populations**, that is, on populations whose members mate with one another rather than with individuals in other groups, and thus represent groups in which the transmission and significance of selected biological characteristics may be scientifically studied. Although formerly isolated populations such as the American Indians and Australian Aborigines are both examples, there are also many contemporary human breeding populations. Such populations are characterized by **assortative mating**, in which individuals select and reject potential mates on the basis of shared biological and cultural characteristics. Canadian Jews are an example. Census data indicate that as recently as the 1970s Jews from Toronto were about five times more likely to marry Jews than non-Jews because of their common cultural heritage. See QQ 11.10.

## 11.10 ASSORTATIVE MATING AMONG THE JEWS OF TORONTO

Evelyn Kallen, Professor of Social Science and Anthropology at York University, looks at the factors responsible for assortative mating among the Jews of Toronto.

• • •

The tendency to confine intimate relationships to fellow Jews is rooted not only in the Judaic prohibition against intermarriage but also in the related notions of collective responsibility . . . . In the shtetl [or European Jewish ghettos of former times], obligations to fellow Jews were customarily scaled in terms of degree of kinship; hence one's primary obligation was to immediate . . . family, then close relatives, then friends, local Jewish community, and eventually all Jews everywhere. The traditional Jewish obligation to put "family first"

From *The Jewish Community in Canada, Vol. 1: A History,* by Stuart Rosenberg, 1970, McClelland & Stewart, Toronto

▼

Some of the leading male members of the Toronto Jewish community in 1925

▲

remains operative among contemporary Orthodox, Conservative, and Reform Jews in Toronto . . . .

The vast majority of Jewish adults in all three groups profess to feel a "common bond," a sense of "kinship" with fellow Jews . . . . Most Jewish adults feel that Jews should give preference to fellow Jews, as opposed to non-Jews, in some, if not all, social

contexts . . . . Many suggest that they "should not" give preference to fellow Jews but that they "have to": "We must . . . because if *we* don't look after our own, *who will?*"

Source: Evelyn Kallen, *Spanning the Generations: A Study in Jewish Identity* (Don Mills, Ontario: Longman Canada Limited, 1977), pp. 92–93.

## THE ADVANTAGES OF STUDYING BREEDING POPULATIONS

Although the study of human breeding populations is reminiscent of earlier attempts to understand how and why biological variation occurs among humankind, it has at least two major advantages when compared to traditional racial studies. One is that a focus on breeding populations eliminates the problem of forcing each and every person into a fixed number of predetermined racial categories that are notoriously hard to define. Another benefit is that the new approach reflects social reality, for the fact of the matter is that the members of some modern social isolates not only condone and encourage assortative mating on philosophical grounds, but also practise it in their everyday lives. What remains to be determined is the number of human breeding populations that exist today, how these are formed and change over time, and the degree to which their members exchange genetic information with individuals from other groups.

## TAY-SACHS DISEASE

Although the study of human breeding populations is a comparatively recent development, there have been some very promising initial results. Take the case of **Tay-Sachs disease (TSd)**, a hereditary illness that causes deterioration of the brain and ultimately results in death, usually within the first two years of life (see QQ 11.11). Studies have shown that TSd is due to a recessive allele (**r**), which means that homozygous dominant (**RR**) and heterozygous (**Rr**) individuals are not afflicted. However, if two heterozygous individuals mate, then the chances are one in four that their offspring will manifest the disease since: (**Rr x Rr**) = **RR** or **Rr** or **Rr** or **rr**.

Equally important, research has also shown that the descendents of Ashkenazic Jews from Eastern Europe are especially prone to producing children who suffer from TSd. Under the circumstances, it is advantageous for

### QQ 11.11 TAY-SACHS DISEASE

In 1887, Dr. Bernard Sachs, an American neurologist, described the case of an infant who was stricken with a strange and fatal condition.

. . .

The following is the history of the case: The little girl, S., who was but two years old at the time of death . . . was born at full term, and appeared to be a healthy child in every respect; its body and head were well proportioned, its features beautifully regular. Nothing abnormal was noticed until the age of two to three months, when the parents observed that the child was much more listless than children of that age are apt to be; that it took no notice of anything, and that its eyes rolled about curiously . . . . The child would ordinarily lie upon its back, and was never able to change its position; muscles of head, neck, and back so weak that it was not able either to hold its head straight or to sit upright. It never attempted any voluntary move-

ments . . . . The child as it grew older gave no signs of increasing mental vigor. It could not be made to play with any toy, did not recognize people's voices, and showed no preference for any person around it. During the first year of its life, the child was attracted by the light, and would move its eyes, following objects drawn across its field of vision; but later on absolute blindness set in . . . .

Hearing seemed to be very acute . . . the slightest touch and every sound were apt to startle the child . . . . The child never learned to utter a single sound; if left to itself it would occasionally make a low gurgling noise . . . .

During last summer (1886), the child grew steadily weaker, it ceased to take its food properly, its bronchial troubles increased, and finally, pneumonia setting in, it died August, 1886.

Source: Bernard Sachs, "On Arrested Cerebral Development, with Special Reference to its Cortical Pathology." *Journal of Nervous and Mental Disease* Vol. 14, Nos. 9 and 10 (1887), pp. 543–545.

descendents of Ashkenazic Jews, who continue to practise assortative mating, to possess this information. Fortunately a simple and inexpensive test has been developed to identify the presence of the defective allele before reproduction, so that genetic counselling can be employed to advise people who need be concerned about the probability of producing a Tay-Sachs child. See Figure 11.5.

**FIGURE 11.5** Sample questionnaire used in prenatal genetic screening for inherited disorders (female respondent)

---

### SAMPLE PRENATAL GENETIC SCREEN*

Name_____ Patients _____ Date _____

1. Will you be 35 years or older when the baby is due?                                        Yes _____  No _____

2. Have you, the baby's father, or anyone in either of your families ever had any of the following disorders?
   - Down syndrome (mongolism)                                                              Yes _____  No _____
   - Other chromosomal abnormality                                                         Yes _____  No _____
   - Neural tube defect, ie. spina bifida (meningomyelocele or open spine), anencephaly     Yes _____  No _____
   - Hemophilia                                                                            Yes _____  No _____
   - Muscular dystrophy                                                                    Yes _____  No _____
   - Cystic fibrosis                                                                       Yes _____  No _____

   If yes, indicate the relationship of the affected person to you or to the baby's father: _____

3. Do you or the baby's father have a birth defect?                                          Yes _____  No _____

   If yes, who has the defect and what is it? _____

4. In any previous marriages, have you or the baby's father had a child, born dead or alive,
   with a birth defect not listed in question 2 above?                                       Yes _____  No _____

   If yes, what was the defect and who had it? _____

5. Do you or the baby's father have any close relatives with mental retardation?             Yes _____  No _____

   Indicate the cause, if known: _____

6. Do you, the baby's father, or a close relative in either of your families have a birth defect, any familial disorder,
   or a chromosomal abnormality not listed above?                                           Yes _____  No _____

   If yes, indicate the condition and the relationship of the affected person to you or to the baby's father: _____

7. In any previous marriages, have you or the baby's father had a stillborn child or three or more first-trimester
   spontaneous pregnancy losses?                                                            Yes _____  No _____

   Have either of you had a chromosomal study?                                              Yes _____  No _____

   If yes, indicate who and the results: _____

8. If you or the baby's father are of Jewish ancestry, have either of you been screened for Tay-Sachs disease?   Yes _____  No _____

   If yes, indicate who and the results: _____

9. If you or the baby's father are black, have either of you been screened for sickle cell trait?   Yes _____  No _____

   If yes, indicate who and the results: _____

10. If you or the baby's father are of Italian, Greek, or Mediterranean background, have either of you been tested for β-thalassemia?   Yes _____  No _____

    If yes, indicate who and the results? _____

11. If you or the baby's father are of Philippine or Southeast Asian ancestry, have either of you been tested for a-thalassemia?   Yes _____  No _____

    If yes, indicate who and the results: _____

12. Excluding iron and vitamins, have you taken any medications or recreational drugs since being pregnant or since your
    last menstrual period? (include nonprescription drugs.)                                 Yes _____  No _____

    If yes, give name of medication and time taken during pregnancy: _____

* Any patient replying "YES" to questions should be offered appropriate counseling. If the patient declines further counseling or testing, this should be noted in the chart. Given that genetics is a field in a state of flux, alterations to this form will be required periodically.

From Technical Bulletin #108 (Sept/87) of Antenatal Diagnosis of Genetic Disorders, American College of Obstetricians & Gynecologists

## SICKLE CELL DISEASE

Genetic counselling is also employed to advise individuals about the blood disorder that is known as sickle cell disease — a debilitating and often fatal form of anemia that is caused by the presence of a pair of mutated recessive alleles governing the production of haemoglobin. Among those who are not affected by this disease the red blood cells are round, but among those who suffer from the disease the red blood cells are characterized by a crescent or sickle-like shape that obstructs the flow of oxygen to the various parts of the body causing severe weakness and pain.

Fortunately, the genetic basis of the disease is well understood. People who are homozygous for normal haemoglobin are not affected by sickle cell disease. Nor are heterozygous individuals, although their genetic make-up may produce a mild form of anemia. However, among individuals who are homozygous for the sickling allele, the disease begins to manifest itself between two and four years of life, and, if left untreated, can lead to death before 20 years of age. Most important, by examining the shape of a person's red blood cells physicians can forewarn them about the probability of producing children who will suffer from the disease since, as is the case with TSd, there is a one in four chance that "healthy" carriers will produce stricken offspring. Such information is especially important to African-Americans; in their breeding population it has been estimated that one in 300 is born with the disease.

# RACISM

Whatever value the study of human variation may have, there will never be any anthropological support for the doctrine of racism, which, as already mentioned, is the belief that people whose physical features are allegedly or actually different from our own are biologically inferior and can consequently be mistreated.

## GENOCIDE

One of the most infamous expressions of racism was manifested in Germany under Hitler's Nazi regime (1933–1945). Hitler argued that the German people represented the **Aryan race** — a pure-blooded "master race" of people reputedly descended from Nordic stock whose members possessed white skin, light-coloured hair and eyes, muscular bodies, and superior intelligence. Although no such race exists or has ever existed in a biological sense, so-called Nazi racial experts spun a web of theories about the superiority of the so-called Aryan race and proclaimed that non-Aryans such as Jews, Gypsies, and Blacks were lesser beings, fit only to serve the master race. During World War II Hitler and his supporters took it upon themselves to eliminate these undesirables, especially the Jews. Throughout Nazi-held territories, and most notoriously in the death factories at Auschwitz, Treblinka, and Buchenwald, six million Jews were subject to **genocide** — the deliberate and systematic extermination of a group of people on account of their alleged biological and behavioral traits — all in order to rid the world of those who Hitler believed, if allowed to live, might pollute the (fictitious) Aryan race (see Figure 11.6). The term **holocaust**, which literally means being completely consumed by fire, is well-suited to describe the destruction of European Jewry by the Nazis. See QQ 11.12.

## SLAVERY

Another institutionalized form that racism takes is **slavery**, in which some people are treated by others as property often because of their observable physical traits. Although slavery was once practised by many of the world's peoples, including the ancient Egyptians, the Aztecs, and the classical Greeks and Romans, it reached its peak between the sixteenth and the nine-

## 11.12 THE HOLOCAUST

Nora Levin examines the holocaust — the destruction of European Jewry by the Nazis.

• • •

In Hitler-Germany, a highly developed people devised the rationale and methodology for exterminating six million human beings — over a million of them children — and for converting them into fat for soap, hair for mattresses and bone for fertilizer. For the Nazis, Jews became part of the non-human universe — the objects of functional exploitation, undifferentiated from other non-human matter in nature, and requiring the same detachment. This new formulation enabled mass murderers to think of themselves as technicians following orders and to call mass murder "special treatment". . . .

In the atmosphere of this period . . . many Germans succumbed to the cult of race . . . . The "Aryan" myth, which Hitler used with great cunning, was an offshoot of this obsession with race. The use of "Aryan" resulted from a confused meaning of the word . . . . [L]inguists had used it earlier to denote a family of related languages, including German, Greek, Latin, Celtic, Slavic and Sanskrit. They also theorized that there must once have been an original, single, primitive Aryan people. But now the racialists identified this peo-

The Bettmann Archive

▼

Emaciated, dead Jewish people killed by the Nazis
(probably at Auschwitz)

▲

ple with the tall, blond, longheaded Germans and concluded that the Germans must be *the* Aryans par excellence — the purest, strongest and noblest of all peoples, the greatest of the culture-bearers.

Source: Nora Levin, *The Holocaust: The Destruction of European Jewry, 1933–1945* (New York: Thomas Y. Crowell Company, 1968), pp. xii–12.

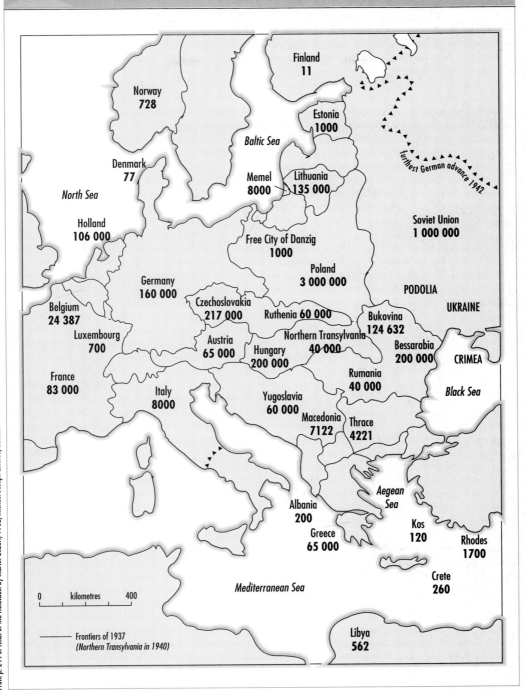

**FIGURE 11.6**  Jews murdered between 1 September, 1939 and 8 May, 1945: an estimate

Finland
11

Norway
728

Estonia
1000

Baltic Sea

Denmark
77

Memel
8000

Lithuania
135 000

furthest German advance 1942

North Sea

Holland
106 000

Free City of Danzig
1000

Soviet Union
1 000 000

Germany
160 000

Poland
3 000 000

PODOLIA

Belgium
24 387

Czechoslovakia
217 000

Ruthenia 60 000

Bukovina
124 632

UKRAINE

Luxembourg
700

Austria
65 000

Northern Transylvania
40 000

Bessarabia
200 000

CRIMEA

Hungary
200 000

France
83 000

Italy
8000

Yugoslavia
60 000

Rumania
40 000

Black Sea

Macedonia
7122

Thrace
4221

Aegean
Sea

Albania
200

Kos
120

Greece
65 000

Rhodes
1700

Crete
260

Mediterranean Sea

0   kilometres   400

Frontiers of 1937
(Northern Transylvania in 1940)

Libya
562

From p. 244 of Atlas of the Holocaust by Martin Gilbert, 1982, Michael Joseph Limited, London

teenth centuries, when Europeans enslaved millions of people whose skin colour was different from their own.

The first to be victimized were the Native peoples of Central and South America who were enslaved by Portuguese and Spanish invaders in order to work on plantations and in mines in European colonies in the New World. But inhumane treatment and European diseases decimated these slaves, causing the Portuguese and Spanish to turn to Africa as the source of slaves.

**TABLE 11.2** The year in which slavery was legally abolished in selected countries

| COUNTRY | YEAR WHEN SLAVERY WAS LEGALLY ABOLISHED | COUNTRY | YEAR WHEN SLAVERY WAS LEGALLY ABOLISHED |
|---|---|---|---|
| Afghanistan | 1923 | Ivory Coast | 1848 |
| Antigua | 1833 | Jamaica | 1834 |
| Austria | 1826 | Korea | 1905 |
| Argentina | 1853 | Madagascar | 1896 |
| Australia | 1833 | Mali | 1905 |
| Bahamas | 1834 | Mexico | 1850 |
| Brazil | 1888 | Nepal | 1926 |
| Cameroon | 1952 | Netherlands | 1860 |
| Canada | 1833 | New Hebrides | 1833 |
| Ceylon | 1844 | Norway | 1902 |
| Chad | 1848 | Pakistan | 1879 |
| Chile | 1833 | Peru | 1933 |
| China | 1929 | Philippines | 1913 |
| Congo | 1891 | Rwanda | 1923 |
| Cuba | 1886 | Rhodesia | 1891 |
| Cyprus | 1824 | Russia | 1861 |
| Dahomey | 1905 | Saudi Arabia | 1962 |
| Denmark | 1848 | South Africa | 1838 |
| Equador | 1851 | Spain | 1878 |
| Ethiopia | 1942 | St. Vincent | 1833 |
| France | 1848 | St. Lucia | 1834 |
| Germany | 1945 | Sudan | 1899 |
| Ghana | 1874 | Tunisia | 1891 |
| Guatemala | 1824 | United Arab Republics | 1896 |
| Hungary | 1848 | United Kingdom | 1833 |
| India | 1943 | United States | 1865 |
| Iran | 1928 | Venezuala | 1854 |
| Iraq | 1924 | West Indies | 1838 |

Based on information contained in: Mohamed Awad, *Report on Slavery* (New York: United Nations, 1966); John Kells Ingram, "Slavery" *Encyclopaedia Britannica*, Vol. 25 (1911), pp. 216–227; Franklin W Knight, "Slavery" *Encyclopaedia Americana, Vol. 25* (1989), pp. 19–24; and Orlando Patterson, "Slavery" *Collier's Encyclopedia*, Vol. 21 (1989), pp. 71–76.

The Netherlands, France, and especially Britain also engaged in the practice of slavery. By the early part of the seventeenth century Britain had gained control of the lucrative African slave trade from its European competitors, and it was then that the capture and sale of Black slaves started to accelerate. The principal destinations for slaves were Britain's American colonies along the Atlantic seaboard. There, Blacks imported from Africa initially were allowed to earn their freedom by working as indentured servants, who were contractually bound to serve as apprentices for a set term. Within a short time, however, the colonial legislatures passed a series of laws that recognized and affirmed the institution of slavery. The outcome was that an untold number of men, women, and children were taken from their African homelands and transported on slave ships to the New World, mainly to the American south, where those who survived the voyage were sold like cattle in open markets to European-American entrepreneurs who used them to work on sugar, cotton, and tobacco plantations. The practice was condoned until the nineteenth century when most of the world's countries finally outlawed slavery (see Table 11.2).

## SEGREGATION

In the meantime, another racist tactic had arisen: **segregation** — the enforced separation of people on account of their observable physical traits. This practice was particularly widespread in the United States, where, even after slaves were granted their freedom by *The Emancipation Proclamation* of 1864, the legal segregation of African-Americans and European-Americans was maintained until the 1960s. The social and psychological damage caused by this longstanding division of the American population was profound; it created rifts in the country that have yet to be healed. See QQ 11.13.

 ## 11.13 THE IMPACT OF SEGREGATION

On 16 April 1963, Martin Luther King, Jr., wrote the following letter from a Birmingham jail.

. . .

My Dear Fellow Clergymen . . .

We have waited for more than 340 years for our constitutional and God-given rights . . . . Perhaps it is easy for those who have never felt the stinging darts of segregation to say, "Wait." But when you have seen vicious mobs lynch your mothers and fathers at will and drown your sisters and brothers at whim; when you have seen hate-filled policemen curse, kick and even kill your black brothers and sisters; when you see the vast majority of your twenty million Negro brothers smothering in an airtight cage of poverty in the midst of an affluent society; when you suddenly find your tongue twisted and your speech stammering as you seek to explain to your six-year-old daughter why she can't go to the public amusement park that has been advertised on television . . . and see ominous clouds of inferiority beginning to form in her little mental sky, and see her beginning to distort her personality by developing an unconscious bitterness toward white people . . . when you are humiliated day in and day out by nagging signs reading "white" and "colored"; when your first name becomes "nigger," your middle name becomes "boy" (however old you are) and your last name becomes "John," and your wife and mother are never given the respected title "Mrs."; when you are harried by day and haunted by night by the fact that you are a Negro, living constantly at tiptoe stance, never quite knowing what to expect next, and are plagued with inner fears and outer resentments; when you are forever fighting a degenerating sense of "nobodiness" — then you will understand why we find it difficult to wait. There comes a time when the cup of endurance runs over, and men are no longer willing to be plunged into the abyss of despair. I hope, sirs, you can understand our legitimate and unavoidable impatience . . . .

Yours for the cause of Peace and Brotherhood,
Martin Luther King, Jr.

Source: Martin Luther King, Jr., "Letter from a Birmingham Jail." In Blaustein, Albert P. and Zangrando, Robert L., eds. *Civil Rights and the American Negro: A Documentary History* (New York: Washington Square Press, Inc., 1968), pp. 502–509.

Joseph Louw, Life Magazine © Time Warner

The assassination of Martin Luther King, Jr., 4 April 1968

## APARTHEID

The United States is not the only nation to have adopted segregation as official government policy. The Republic of South Africa is another. There, as in the United States, segregation was preceded by slavery. Only a few years after they had settled at the Cape of Good Hope in 1652, White Dutch farmers began to import slaves from West Africa. They justified the practice on religious grounds. Verses 23 and 24 in the *Book of Joshua* state that some "are accursed and will for ever be serfs, as wood-cutters and water-carriers in the house of . . . God," and this, the Dutch settlers claimed, was proof that God had intended Blacks to be slaves.

Much later, after the Dutch colony had fallen into British hands, it was made a Dominion of Great Britain under the terms of the 1910 *South Africa Act*. However, despite the fact that slavery had been abolished in the British Empire in 1834, the new South African government began to pass a series of laws that elevated Whites to a superior position with respect to the other residents of the country. These laws were the basis of **apartheid** — the legal segregation of Whites, Asians, Coloureds, and Blacks, which became official government policy in 1948. During the election campaign that year, Daniel François Malan, who was subsequently elected Prime Minister, stated that the policy of his National Party was based on "two fundamental principles, of separation and trusteeship. As the words themselves indicate," he said, "this means in no way oppression of non-Europeans . . . . [We acknowledge] their right of existence . . . but everything in their own spheres and under the sovereignty and leadership of the Europeans."

The spheres of separation embodied in the legislative acts passed before, during, and after Malan's regime were all-encompassing. What they did was to guarantee the political and economic domination of a small White minority over a much larger population of non-Europeans, especially Blacks, who were told where to live, where to work, and even where to bury their deceased. Apartheid, however, is now dead. On 1 February 1991 President F. W. de Klerk announced in parliament that his government was prepared "to repeal the racially discriminatory legislation which have become known as the cornerstone of apartheid." More recently, on 23 March 1992, in a "Whites only" national referendum, an overwhelming majority (68.6%) voted in favour of continuing ongoing discussions with Black South African leaders aimed at developing a constitution that will guarantee Blacks the right to participate fully in the political process. Since then, President de Klerk and Nelson Mandela, leader of the African National Congress, have been awarded the Nobel Peace Prize "for laying the foundation for a new democratic South Africa." Mandela said that "[t]he Nobel Prize is a tribute to all South Africans." De Klerk added that the award "will bring a message to all South Africans that the world wants to achieve everlasting peace." See QQ 11.14.

## 11.14 PREAMBLE TO THE *FREEDOM CHARTER*

In 1956, the African National Congress adopted the *Freedom Charter*.

. . .

We, the people of South Africa, declare for all our country and the world to know:

That South Africa belongs to all who live in it, black and white, and that no government can justly claim authority unless it is based on the will of the people;

That our people have been robbed of their birthright to land, liberty and peace by a form of government founded on injustice and inequality;

That our country will never be prosperous or free until all our people live in brotherhood, enjoying equal rights and opportunities;

That only a democratic state, based on the will of all the people, can secure to all their birthright without distinction or colour, race, sex or belief;

And therefore, we, the people of South Africa, black and white together — equals, countrymen and brothers — pledge ourselves to strive together, sparing nothing of our strength and courage, until . . . democratic changes . . . have been won.

Source: African National Congress, "The *Freedom Charter*." In McCartan, Greg ed. *Nelson Mandela Speeches 1990* (New York: Pathfinder Press, 1990), p. 67.

# RACISM IN CANADA

## BANNING THE POTLATCH

Canada, too, has been marred by racism, especially with respect to the ways in which the Aboriginal peoples of Canada have been mistreated by both their fellow citizens and the Canadian government. Consider how the Canadian government reacted to the **potlatch** — a ceremony in which Indian peoples on the west coast of Canada and the northern United States gave away items of value to publicly validate a change in social status.

Although the Native peoples who held the celebrations were technically hunters and gatherers, the mainstay of their economy was fishing. And because seafood was abundant, they almost always produced a surplus of food. They also produced an abundance of material items such as blankets and canoes, which were accumulated by an elite group of nobles who were the leaders of their societies. The rank of each noble depended on the honorary titles and crests that he was entitled to possess. Each time an honour was inherited, which allowed the noble to enjoy specific social and economic prerogatives, a potlatch would be held in which the wealth that had been accumulated was given to those persons who publicly confirmed the change in the recipient's status by attending the ceremony. The guests included people from other villages and frequently people from other tribes, who were told about the scope and nature of the inheritance and who received gifts as witnesses to the proceedings that were commensurate with the prestige that flowed from the inheritance. In addition, although potlatching was not undertaken for economic purposes, it would sometimes have had a beneficial economic result as well; whenever resources were scarce the redistribution of wealth helped to ensure the survival of those who were in need. See QQ 11.15.

Kwakiutl potlatch,
late nineteenth century
— gift boxes are
shown in the
foreground

Milwaukee Public Museum

# 11.15 THE POTLATCH

Philip Drucker explains the cultural significance of the potlatch.

• • •

The ceremonial at which the various prerogatives intimately associated with social status were assumed was called in Chinook jargon, which was the *lingua franca* of the Northwest Coast, the "potlatch." Each major cultural division, from the Tlingit to the Lower Chinook, had its own variations in procedure and detail of this performance, but the function was everywhere the same. The potlatch brought to expression basic principles involved in social status and also served as a major force for social integration . . . . [E]ach . . . [noble] in the social unit was born with an inherent right to use group properties of major or minor importance, but he could not exercise these rights — in other words, assume his proper status — until his title to them had been formally announced and validated. This formal announcement and public validation was accomplished during and by the potlatch. The heir presumptive to a chieftainship would be presented formally to a group of guests at such an affair. His relationship to the incumbent chief would be explained and he would be given a name or the right to use some crest specifically related to the position he would eventually occupy . . . . The guests who heard these claims announced, and recognized their validity, were regarded as witnesses to the proceedings. As such, they were rewarded and their subsequent good will was insured by giving them feasts and gifts. While at times the demonstration of privileges or the giving away of material goods might appear to overshadow the essential announcement and validation of rights and status, this last-named function was the essence and basic goal of the whole performance.

Source: Philip Drucker, *Indians of the Northwest Coast* (New York: American Museum of Natural History, 1955), pp. 131–132.

To its great discredit, the Canadian government passed a law in 1894 that made potlatching a crime and threatened participants with imprisonment. Twenty-eight years later, after a number of unsuccessful attempts to prosecute alleged offenders, jail terms were imposed on forty-five Kwakiutl Indians who had participated in a potlatch at Village Island, British Columbia, in 1921. Whatever the government's logic — at the time it claimed that potlatching was economically wasteful and unchristian — the jail terms had the desired effect. Until the law was finally changed in 1951, potlatching was forced underground.

## POLITE RACISM

While we may take comfort from the fact that most Canadians now consider racist tactics such as banning the potlatch to be unacceptable, it would be a serious mistake to assume that racism in Canada is a thing of the past. Some scholars maintain that racism continues to dominate the relationship between the Aboriginal peoples of Canada and peoples of European descent, and that the current social and economic position of Aboriginal peoples in Canadian society demonstrates the accuracy of this conclusion.

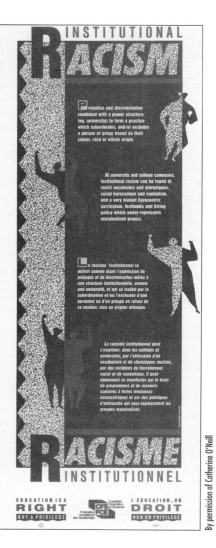

Poster decrying institutional racism

By permission of Catherine O'Neill

Nor should it be forgotten that while federal and provincial laws have been passed to ensure that racism does not lead to blatant discrimination, prejudice against African-Canadians, Asian-Canadians, and other visible minorities is part and parcel of our everyday lives. In addition, those who are subject to racism in Canada can and do sometimes behave in racist ways themselves. In fact, despite Canada's current reputation as a tolerant country, all that may have changed in recent years is that blatant discrimination has given way, at least in part, to **polite racism** — a form of racism that is more covert than overt, but equally real. See QQ 11.16.

We can all work towards eliminating both the overt and the covert racist practices that continue to divide us. In the meantime, it is worthwhile to keep in mind that one of the main lessons that anthropology teaches is that there is no scientific basis for the view that one group of people is inherently superior to others because of either its biological or cultural traits. Such a view is simply an invention of people and of governments who are trying to increase their political and economic power at the expense of their fellow human beings by pretending that social stereotypes are valid biological and cultural facts.

## QQ 11.16 RACISM IN CANADA

Maxwell Yalden, Chief Commissioner of the Canadian Human Rights Commission, speaks out in favour of "no tolerance for intolerance."

• • •

"Racism's back," announced the November 1991 issue of The Economist. While some may question whether it ever goes away, the present upsurge of xenophobia around the world undoubtedly touches a nerve in Canada as well. The call for more pluralistic politics and practical respect for cultural and religious diversity is real and strong. But so are the latest ethnic conflicts and the sometimes violent backlash against minorities. The kinder, gentler society being proclaimed by our neighbour to the south has been Canada's official philosophy for decades now. This is no doubt as much the result of a complex history as it is of humanitarian vision. But Canada has never completely freed itself, any more than any other country, from the corrosive effects of ingrained assumptions of racial superiority. We need to come to terms more directly with the reality that our nation is culturally a very mixed bag, and that that in itself is more of a blessing than a curse . . . .

[O]ur journey to becoming a tolerant society is incomplete. This . . . [is] borne out by the persistent promotion of racial hatred by extremist organizations across the country — through rallies, telephone messages, newspapers, books and pamphlets. The defacing of prayer halls, cemeteries and other religious sites happens not only in Europe or Asia, it happens in towns and cities in Ontario, Alberta or Quebec.

Before we dismiss these as fringe phenomena, we need to recognize that the operating premise of hate-groups does not change: it is that scapegoats for society's ills are found most readily among those who can be distinguished from the mainstream. These 'strangers in our midst' are portrayed as standing in the way of a traditional society with an established hierarchy of power. It is true that at many levels, both official and unofficial, Canada has set its face against ethnic and religious intolerance, but it is equally true that the virus exists virtually everywhere and only a steady determination can keep it at bay.

Source: Maxwell Yalden, *Canadian Human Rights Commission Annual Report 1991* (Ottawa: Minister of Supply and Services Canada, 1992), pp. 30–34.

# KEY TERMS AND CONCEPTS

MELANIN  (*meh'-la-nin*)

EPIDERMIS

MELANOCYTES

DERMIS

KERATIN  (*ker'-a-tin*)

CAUCASOID RACE  (*kah'-kah-zoyd + race*)

MONGOLOID RACE  (*mohn'-gohl-oyd + race*)

NEGROID RACE  (*nee'-groyd + race*)

EPICANTHIC FOLD  (*ep-i-kan'-thik + fold*)

FOLK TAXONOMY

RACE

PURE-BLOODED RACES

CLINE

ANTHROPOMETRY

PTC (PHENYLTHIOCARBAMIDE)
(*fee-nil-thy-oh-kahr'-bih-myd*)

ABO BLOOD GROUPS

SOCIAL ISOLATES

ETHNIC GROUPS

BREEDING POPULATIONS

ASSORTATIVE MATING

TAY-SACHS DISEASE (TSD)  (*tay-saks + disease*)

ARYAN RACE

GENOCIDE

HOLOCAUST  (*hoh'-loh-kahst*)

SLAVERY

SEGREGATION

APARTHEID  (*a-pahrt'-hayt*)

POTLATCH

POLITE RACISM

# Selected Readings

## Introduction

United Nations Educational, Scientific, and Cultural Organization (UNESCO). "Statement on Race and Racial Prejudice." *International Social Service Journal*, Vol. 20 (1968), pp. 93–97.

## The So-Called Three Great Races

Haller, J. *Outcasts from Evolution: Scientific Attitudes of Racial Inferiority, 1859–1900*. Urbana: University of Illinois Press, 1975.

Huntington, E. *The Character of Races*. New York: Arno, 1977 (reprint of 1925 edition).

Stanton, W. *The Leopard's Spots: Scientific Attitudes toward Race in America, 1815–1859*. Chicago: Phoenix Books, 1960.

## Race as a Biological Concept

Stepan, N. *The Idea of Race in Science: Great Britain, 1800–1960*, London: Macmillan, 1982.

## Problems with Human Racial Typologies

Montagna, W. "The Evolution of Human Skin Color." *Journal of Human Evolution*, Vol. 14 (1985), pp. 3–22.

Nurse, G., Weiner, J., and Jenkins, F. *The Peoples of Southern Africa and their Affinities*. Oxford: Oxford University Press, 1985.

Szathmary, E. and Reed, T. "Caucasian Admixture in Two Ojibwa Indian Communities in Ontario." *Human Biology*, Vol. 44 (1972), pp. 655–671.

## Genetics and Race

Mourant, A. *Blood Relations: Blood Groups in Anthropology*. Oxford: Oxford University Press, 1983.

Mourant, A., Kopec, A., and Domaniewska-Sobezak, K. *The Distribution of Human Blood Groups and Other Polymorphisms* (second edition). Oxford: Oxford University Press, 1976.

Provine, W. "Geneticists and the Biology of Race Crossing." *Science*, Vol. 182 (1973), pp. 790–796.

Race, R. and Sanger, R. *Blood Groups in Man* (sixth edition). Oxford: Blackwell Scientific Publications, 1975.

## Eliminating the Concept of Race

Anderson, A. and Frideres, J. *Ethnicity in Canada: Theoretical Perspectives*. Toronto: Butterworths, 1981.

Evans, B. and Waites, B. *IQ and Mental Testing: An Unnatural Science and Its Social History*. London: Macmillan, 1981.

Li, P., ed. *Race and Ethnic Relations in Canada*. Toronto: Oxford University Press, 1990.

Littlefield, A., Lieberman, L., and Reynolds, L. "Redefining race: The Potential Demise of a Concept in Physical Anthropology." *Current Anthropology*, Vol. 23 (1982), pp. 641–655.

Loehlin, J., Lindzey, G., and Spuhler, J. *Race Differences in Intelligence*. San Francisco: W. H. Freeman, 1975.

Montagu, A. *Man's Most Dangerous Myth: The Fallacy of Race* (fifth edition). New York: Oxford University Press, 1974.

## How Should Human Variation Be Studied?

Beet, E. "The Genetics of Sickle Cell Trait in a Bantu Tribe." *Annals of Eugenics*, Vol. 14 (1949), pp. 279–284.

Blum, H. "Does the Melanin Pigment of Human Skin Have Adaptive Value?" *Quarterly Review of Biology*, Vol. 35 (1961), pp. 50–63.

Dagenais, D., Courville, L., and Dagenais, M. "A Cost-Benefit Analysis for the Quebec Network of Genetic Medicine." *Social Science Medicine*, Vol. 20 (1985), pp. 601–607.

Damon, A., ed. *Physiological Anthropology*. New York: Oxford University Press, 1975.

Emery, A. and Pullen, I. *Psychological Aspects of Genetic Counseling*. New York: Academic Press, 1984.

Hanna, J. and Brown, D. "Human Heat Tolerance: Biological and Cultural Adaptations." *Yearbook of Physical Anthropology*, Vol. 22 (1979), pp. 164–181.

Jinks, D., *et. al.* "Molecular Genetic Diagnosis of Sickle Cell Disease Using Dried Blood Specimens on Blotters Used for Newborn Screening." *Human Genetics*, Vol. 81 (1989), pp. 363–366.

Kaback, M., Zeiger, R., Reynolds, L., and Sonneborn, M. "Approaches to the Control and Prevention of Tay-Sachs Disease." *Progress in Medical Genetics*, Vol. 10 (1974), pp. 103–134.

Loomis, W., "Skin-Pigment Regulation of Vitamin-D Biosynthesis in Man." *Science*, Vol. 157 (1967), pp. 501–506.

Perutz, M. "Hemoglobin Structure and Respiratory Transport." *Scientific American*, Vol. 239 (1978), pp. 92–125.

Reed, S. *Counseling in Medical Genetics* (third edition). New York: Alan R. Liss, 1980.

Veatch, R. "Ethical Issues in Genetics." *Progress in Medical Genetics*, Vol. 10 (1974), pp. 223–264.

## Racism

Frederickson, G. *The Black Image in the White Mind: The Debate on Afro-American Character and Destiny, 1817–1914*. New York: Harper & Row, 1971.

Goffman, E. *Stigma: Notes on the Management of Spoiled Identity*. Englewood Cliffs, New Jersey: Prentice-Hall, 1963.

Gossett, T. *Race: The History of an Idea in America*. New York: Schocken, 1965.

Gutman, I., ed. *Encyclopedia of the Holocaust* (4 volumes). New York: Macmillan, 1990.

Harmsen, H. "The German Sterilization Act of 1933." *Eugenics Review*, Vol. 46 (1955), pp. 227–232.

Kelves, D. *In the Name of Eugenics: Genetics and the Uses of Human Heredity*. New York: Knopf, 1985.

Lifton, R. *The Nazi Doctors: Medical Killing and the Psychology of Genocide*. New York: Knopf, 1986.

Mosse, G. *Toward the Final Solution: A History of European Racism.* New York: Howard Fertig, 1978.

Müller-Hill, B. *Murderous Science: Elimination by Scientific Selection of Jews, Gypsies, and Others, Germany, 1933–1945.* Oxford: Oxford University Press, 1988.

Poliakov, L. *The Aryan Myth: A History of Racist and Nationalist Ideas in Europe.* New York: Basic Books, 1970.

Proctor, R. *Racial Hygiene: Medicine Under the Nazis.* Cambridge, Massachusetts: Harvard University Press, 1988.

Wagley, C. and Harris, M. *Minorities in the New World.* New York: Columbia University Press, 1958.

## RACISM IN CANADA

Bode, P. "Simcoe and the Slaves." *The Beaver,* Vol. 73 (1993), pp. 17–19.

Bolaria, B., and Li, P. *Racial Oppression in Canada* (second edition). Toronto: Garamond Press, 1988.

Bolt, M. and Long, J., eds. *The Quest for Justice: Aboriginal Peoples and Aboriginal Title.* Toronto: University of Toronto Press, 1985

Driben, P. and Trudeau, R. *When Freedom is Lost: The Dark Side of the Relationship between Government and the Fort Hope Band.* Toronto: University of Toronto Press, 1983.

Hughes, D. and Kallen, E. *The Anatomy of Racism: Canadian Dimensions.* Montreal: Harvest House, 1974.

Miller, J. *Skyscrapers Hide the Heavens: A History of Indian-White Relations in Canada.* Toronto: University of Toronto Press, 1989.

Parillo, V. *Strangers to These Shores: Race and Ethnic Relations in Canada.* New York: MacMillan, 1990.

Trigger, B. *Natives and Newcomers.* Kingston and Montreal: McGill and Queen's University Press, 1985.

# GLOSSARY

## a

**ABO blood groups**, blood types A, B, O, and AB, which are determined in accord with Mendelian principles of inheritance. Alleles A and B are both dominant with respect to allele O, but when alleles A and B recombine they act in a codominant fashion to form blood type AB.

**absolute or chronometric age**, an age that is usually expressed in terms of years before the present, but that can also be expressed in terms of other standard intervals of time such as centuries, millennia, and so on.

**Acheulian** (*a-shoo'-lee-en*), large oval-shaped stone cores and flakes, characteristically double-sided or bifacial, that were made during the lower palaeolithic.

**Acheulian hand-axe**, an almond-shaped or teardrop-shaped lower palaeolithic stone implement between 12 centimetres and 15 centimetres long with sharp edges around the perimeter.

**Adapidae** (*a-dap'-i-day*), an extinct family of Eocene primates that possessed lemur-like traits and that may be the stock from which lemurs evolved.

**adapted**, suited to survive and reproduce in a particular environment.

**adaptive radiation**, the diversity that results when the members of a rapidly evolving population become adapted to a wide variety of ecological niches that are available for them to exploit.

**Aegyptopithecus** (*ee-jip-tuh-pith'-i-kus*), an extinct genus of Oligocene primates that possessed anthropoid-like traits and that may be the stock from which Old World monkeys, apes, and humans evolved.

**Age of Mammals**, the period during which mammals started to flourish, beginning about 65 mya.

**albinism**, a hereditary disorder caused by a pair of recessive alleles that features a lack of pigmentation in the skin, eyes, and hair.

**alleles** (*a-leelz*), alternate forms of a particular gene.

**allopatric model** (*al-oh-pa'-trik + model*), the model that calls attention to the evolution of new species that arise from a parent population whose subunits become separated by physical barriers; the natural history of almost all species conforms to this model.

**Altiatlasius** (*al-tee-at-la'-see-us*), a recently discovered 60-million-year-old mammal from North Africa that possessed primate-like teeth and that may be the stock from which all subsequent primates evolved.

**altruistic hypothesis**, the hypothesis that the behaviour of Neandertals was sometimes governed, as is our own, by an unselfish concern for the welfare of others of their kind.

**Ambrona** (*um-broh'-nah*), an alleged *H. erectus* kill site located in north-central Spain.

**American Sign Language** (ASL), the sign language that many deaf persons in Canada and the United States use to communicate; linguistic researchers have taught apes such as the chimpanzee and the gorilla to communicate with humans via ASL.

anagenesis *(an'-a-gen'-i-sis)*, one species evolving directly into another without giving rise to any side branches.

anatomically modern humans, the common designation for the most intelligent of the wise humans, ourselves.

anatomy, the structural makeup (organs, etc.) of the body.

ancestral traits, traits that reflect a relatively longstanding adaptive pattern within the evolutionary line comprised of an ancestor and its descendants.

angiosperm radiation theory, the theory that the adaptive radiation of the first primates was promoted by the presence of flowering plants with protected seeds (angiosperms) because these provided the first primates with foods such as flowers, fruits, gums, nectars, and seeds that they could readily exploit since they possessed the visual acuity and hand-eye coordination that were necessary to feed on small food objects in a comparatively dark ecological niche.

Anthropoidea *(an'-throh-poy'-dee-a)*, the primate suborder that includes monkeys, apes, and humans.

anthropoids *(anth'-roh-poydz)*, the common designation for the subdivision of the primates that includes monkeys, apes, and humans.

anthropology, the study of humankind, all over the globe in the past as well as the present.

anthropometry, a research technique that involves using highly precise measuring instruments to determine the colouration of the skin, the width of the nose, the thickness of the lips, and so on among living humans.

Anthropomorpha *(an-throh-poh-mor'-fah)*, the taxonomic category in which

Linnaeus grouped humans and the other primates of which he was aware.

apartheid *(a-pahrt'-hayt)*, the legal segregation of Whites, Asians, Coloureds, and Blacks in the Republic of South Africa, which became official government policy in 1948.

Apidium *(a-pid'-ee-um)*, an extinct genus of Oligocene primates that possessed platyrrhine-like traits and that may be the stock from which New World monkeys evolved.

aquatic theory of human evolution, the theory that the gap in the hominoid-hominid fossil record between 8 mya and 4 mya occurred because the hominoids that gave rise to hominids began to spend the majority of their time in the water during the latter part of the Miocene. The theory also holds that it was the aquatic environment that fostered the development of upright posture.

arboreal habitats *(ahr-bor'-ee-uul + habitats)*, a variable group of tropical forest environments in which survival is enhanced by hands and feet that can grasp branches and limbs; acute vision; good hand-eye coordination; and a brain that facilitates the ability to learn, store, and act on information that is acquired from others.

arboreal quadrupeds *(arboreal + kwah'-droo-pedz)*, animals that move slowly on all fours up the trunks of trees and then scurry likewise but more rapidly along branches.

arboreal theory, the theory that the first primates were the descendants of insectivores that were better equipped to live, move, and feed in the trees than their competitors because natural selection had endowed them with traits such as short faces, small snouts, close-set eyes, stereoscopic vision, clawless digits, grasping hands and feet, small snouts, and teeth that were especially well-suited to feeding on fruits and leaves in the upper canopy of the tropical forest.

archaeological culture, a prehistoric culture.

archaeological record, what remains of ecofacts and artifacts from archaeological sites around the world.

archaeologists *(ahr'-kee-ahl'-uh-jists)*, cultural anthropologists who focus their attention on the cultural achievements of prehistoric peoples.

Archaic culture, a mesolithic culture that arose about 8 kya in the southwestern and eastern regions of North America where it had evolved from a Palaeo-Indian base, and then proceeded to expand until about 3 kya when it gave way to subsequent prehistoric American Indian cultural traditions.

archaic *H. sapiens*, an extinct variety of *H. sapiens* reminiscent of both *H. erectus* and *H. sapiens* who lived in Africa, Europe, and Asia between about 400 kya and about 130 kya.

artifacts, the remnants of manufactured prehistoric items ranging from weapons to art. Archaeologists rely on such items, which they recover from archaeological sites, to draw inferences about the behaviour of those who manufactured the items.

artificial selection, the practice that plant and animal breeders employ when they allow only certain organisms in a population to mate in order to develop specific traits in subsequent generations.

Aryan race, a pure-blooded "master race" of people reputedly descended from Nordic stock whose members had white skin, light-coloured hair and eyes, muscular bodies, and a superior intelligence. No such race exists or has ever existed in a biological sense; the Aryan race is a political construct, not a scientific one.

assemblage, an integrated collection of subassemblages from a prehistoric site that archaeologists use to draw inferences about the various cultural activities that were undertaken at the site.

assortative mating, the practice of selecting and rejecting potential mates on the basis of shared biological and cultural characteristics.

Aurignacian culture *(aw-rig-nay'-shen + culture)*, an upper palaeolithic culture that persisted in Europe between 34 kya and 30 kya.

australopithecines *(ahs-tray-loh-pith'-i-seenz)*, the extinct species that comprised the genus *Australopithecus*.

*Australopithecus aethiopicus (australopithecus + ee-thee-oh-pikus')*, an extinct species of especially ruggedly built hominids that lived in East Africa about 2.5 mya; the black skull is an example.

*Australopithecus afarensis (australopithecus + af-ahr-en'-sis)*, an extinct species of lightly built hominids that lived in East Africa between about 4 mya and about 3 mya; Lucy is an example.

*Australopithecus africanus (australopithecus + af-ra-kan'-us)*, an extinct species of lightly built hominids that lived in South Africa between about 3 mya and about 2 mya; the Taung child is an example.

*Australopithecus boisei (australopithecus + boy'-zee-eye)*, an extinct species of especially ruggedly built hominids that lived in East Africa between about 3 mya and about 2 mya; Zinj is an example. The designation is employed by those who regard *boisei* as a species rather than as a subspecies of *Australopithecus*.

*Australopithecus robustus (australopithecus + roh-bus'-tus)*, an extinct species of ruggedly built hominids that lived in South Africa between about 2 mya and about 1.5 mya.

*Australopithecus robustus boisei (australopithecus + roh-bus'-tus + boy'-zee-eye)*, an extinct subspecies of especially ruggedly built hominids that lived in East Africa between about 3 mya and about 2 mya; Zinj is an example. The designation is employed by those who regard *boisei* as a subspecies rather than as a species of *Australopithecus*.

Aztec civilization, a civilized culture that flourished in what is now Mexico from about 1325 A.D. until 1521 A.D.

# b

band societies, small, politically autonomous groups that likely contained between 50 to 75 persons who were related through blood and through marriage, and who occupied comparatively large territories where they were generally self-sufficient with respect to food, clothing, and shelter.

bear cult, a religious observance in which bears are worshipped; European Neandertals may have worshipped bears.

Beringia *(beh-rin'-jee-uh)*, a 1000-kilometre-wide to 2000-kilometre-wide land bridge between Siberia and Alaska that was exposed frequently during the past 40 000 years because of the formation of glaciers in the high altitude and high latitude regions of the world. It is likely that the first Americans crossed from the Old World to the New via Beringia.

Big Game Hunting Tradition, the first undisputed upper palaeolithic culture in the New World; its practitioners

specialized in hunting big game such as mammoth and bison.

biological anthropology, the branch of anthropology that is concerned with human biology in the broadest possible sense.

bipedal, walking upright on two feet.

black skull, the common designation for KNM-WT 17000, who is classified as *Australopithecus aethiopicus*.

blade technique, a method of making stone tools in which flakes at least twice as long as wide were struck from a core previously prepared for that purpose.

borers, icepick-like implements that were used to punch holes in leather and other soft materials.

brachiators *(bray'-kee-ay-torz)*, animals that travel by using their hands to swing underneath branches from one branch to another in the trees.

*Branisella (bran-i-sel'-uh)*, an extinct genus of Oligocene primates that possessed platyrrhine-like traits and that may be the stock from which New World Monkeys evolved.

breeding populations, populations whose members mate with one another rather than with individuals in other groups.

Broca's area, an area in the left cerebral cortex of the brain towards the front that is necessary for uttering syllables and for comprehending spoken words.

bronze age cultures, cultures that arose about 5 kya in the Near East and featured implements made from bronze; such cultures are generally considered to be civilized cultures.

brow ridge, a thick bone projection that protrudes outward over the eyes.

burins *(bur'-inz)*, chisel-like implements that were used for shaving, trimming, and engraving wood, bone, and ivory.

# c

Callitrichidae *(kal-i-trik'-i-day)*, the family of New World monkeys that includes marmosets and tamarins.

calls, characteristic sounds that non-human primates utter in the wild in order to convey information.

cancer, a group of diseases that is produced when a single somatic cell or a group of such cells begins to multiply and spread in an uncontrolled fashion.

carbon-14 method, an absolute dating method that is used to estimate the age of carbon-bearing organic remains such as bone, wood, charcoal, hair, skin, shell, and seeds from the recent past, generally between about 500 years old and about 50 000 years old.

carbon isotopes, carbon elements with the same atomic number but with different atomic weights; carbon 12 and carbon 14 are examples.

carnivores, meat eaters.

cartouche *(kahr-toosh')*, a special ring with a knot at the base inside of which the name of an Egyptian pharaoh was enclosed.

Çatal Hüyük *(chah-tahl + hoo'-yook)*, a neolithic town in southern Turkey that was occupied between about 8250 ya and about 7400 ya.

catarrhine nose *(ka̱'-ti-reen + nose)*, a nose that features narrow nostrils that open downward.

Catarrhini *(ka̱-ti-ree'-nee)*, the primate infraorder that includes Old World monkeys, apes, and humans.

catastrophism *(ka-tas'-troh-fiz-e̱m)*, Baron Georges Cuvier's mistaken idea that the remains of extinct organisms are simply examples of species originally created by God, but then destroyed by miraculous catastrophic events such as the flood described in *Genesis.*

Caucasoid race *(kah'-kah-zoyd + race)*, the common designation for persons with light-coloured skin; a medium to heavy amount of straight, wavy, or loosely curled body hair, either blond, red, or brown in colour; a narrow nose; thin to medium-thick lips; and light-coloured eyes. The so called Caucasoid race is not a race in the biological sense of the term; it is best regarded as a social construct.

Cebidae *(seh'-bi-day)*, the family of New World monkeys that includes squirrel monkeys, capuchins, howler monkeys, woolly monkeys, spider monkeys, and woolly spider monkeys.

cells, the units that make up the bodies of all living things.

Cercopithecinae *(sur-koh-pith'-i-sin-ay)*, the family of Old World monkeys that includes frugivores such as guenons, mangabeys, baboons, drills, mandrills, and macaques.

Cercopithecoidea *(sur-koh-pith-i-koy'-dee-a̱)*, the primate superfamily that includes Old World monkeys.

Châtelperronian culture *(sha-tel-pur-oh'-nee-en + culture)*, an upper palaeolithic culture that persisted in Europe between 32 kya and 28 kya.

chinampas *(tchee-nahm'-pahs)*, small raised fields surrounded by water that the Aztecs made by dragging floating mats of vegetation from open water to marshy locations where these were anchored into place with stakes and cypress roots and then dressed with fertile muck from below.

chordata *(kor'-da̱-ta̱)*, a phylum in the animal kingdom that includes all animals whose bodies are supported, at some stage in their lives, by a notochord. All primates are assigned to this phylum.

**Christian fundamentalists**, Christians who insist on a literal interpretation of the contents of the Bible.

**chromosomal mutations**, a change in the structure of an organism's chromosomes; such mutations affect multiple alleles.

**chromosomes** *(kroh'-moh-sohmz)*, the thread-like structures on which alleles and hence DNA are located. Each species has its own characteristic number of chromosomes; humans, for instance, possess 46 that are arranged into 23 pairs.

**civilized cultures**, cultures that are characterized by urban centres, occupational specialization, taxation of surplus wealth, distinct social classes, centralized government, large-scale public works, long distance trade, standardized monumental artwork, writing, and scientific achievements.

**cladogenesis** *(klad'-uh-gen'-i-sis)*, the evolution of one species into multiple, branching species.

**classic Neandertals**, the common designation for the Neandertals who lived in Western Europe during the latter part of the Pleistocene.

**cline**, a continuum with individuals arranged on an unbroken curve stretching from one extreme to the other for each of their genetically based, observable traits.

**clovis point** *(kloh'-vis + point)*, a lance-shaped stone projectile between 1.5 centimetres and 4 centimetres wide and about 13 centimetres long with flutes or grooves extending about halfway up from the base to the tip on opposite faces.

**codominant**, alternate alleles of the same gene whose effects are both expressed in the heterozygous state.

**coevolutionary model**, the model that portrays the development of agriculture as the result of mutually beneficial relationships that were established between humans and plants, with humans acting as the unintentional agents of natural selection.

**Colobinae** *(koh-loh'-bin-ay)*, the family of Old World monkeys that includes folivores such as guerezas and langurs.

**Combe-Grenal** *(kom-bu' gre-nal)*, a series of rock shelters in a box canyon in southwestern France where Neandertals once lived.

**comparative method**, a research strategy that is based on analyzing similarities and differences in a comprehensive way.

**composite tools**, implements consisting of several parts that were joined together into a single unit.

**conflict theories**, a class of theories that hold that the state was created to alleviate conflict, either between individuals, between groups within society, or between societies themselves.

**core**, a lump of stone that was fashioned into one or more implements.

**core chopper**, a stone chopper made by removing a relatively small number of flakes from a core; such choppers were likely used for skinning animals, cutting meat, and primitive woodworking.

**creation story**, the rendition of creation contained in the Book of Genesis.

**crepuscular** *(kree-pus'-kyoo-lur)*, species that are active at dawn and again at twilight.

**cross-fertilization**, the fertilization of the sex cell of one organism with the sex cell of another.

**culture**, the learned patterns of thought and behaviour that are characteristic of either humankind as a whole, the members of a particular society, or the members of a particular subgroup in a society.

cultural anthropology, the branch of anthropology that is concerned with humankind's cultural achievements.

cultural relativism, analyzing cultural differences in anthropological terms or making no negative value judgements about cultures with the exception of those practices that are racist or ethnocentric.

culture shock, the feeling of alienation that is associated with being a foreigner in a cross-cultural setting. This feeling sometimes comes to the fore when anthropologists undertake fieldwork in cultures other than their own.

cusps, bump-like projections on the crowns of molar teeth.

cybernetics model, the model that portrays the neolithic revolution as a lengthy process governed by the feedback that people received from deliberate experiments that they undertook to increase their food supply.

# d

Dabban industry (dah-bawn + industry), an upper palaeolithic cultural industry that persisted in northwestern Africa between 40 kya and 13 kya.

Darwin's finches, the thirteen species of finches that Darwin encountered on the Galápagos Islands.

degradation (deg-reh-day'-shun), Georges Louis de Buffon's idea that the natural history of modern life forms could be traced back to several pure original ancestors from which they had "degenerated" when the original ancestors spread into new territory and encountered new environmental conditions.

demotic (dee-maw'-tik), a form of ancient Egyptian writing that was written with connected letters and used primarily to record secular events.

dendrochronology (den'-droh-krawn-ahl'-uh-jee), a technique that is used to determine the absolute age of wooden beams and similar products, especially from the American Southwest, whose parent trees were felled as early as 7400 ya.

denticulates (den-tik'-yoo-layts), saw-like stone implements that were used to cut and shred wood.

DNA (deoxyribonucleic acid) (dee-ox'-ee-ry-bow-noo'-klay-ik + acid), the hereditary material that governs how each and every living thing is constructed.

derived traits, traits that reflect a relatively recent adaptation to a specific environment.

dermis, the innermost layer of the skin.

dietary hypothesis, the hypothesis that the skulls, jaws, and teeth of the gracile australopithecines indicates that they were omnivores, and that the corresponding features of the robust australopithecines indicates that they were herbivores.

digits, fingers and toes.

diurnal (dy-ur'-nal), species that are active by day.

Dollo's law, the contention that evolutionary change in irreversible.

dominance hierarchy, a social arrangement in which those with the highest rank possess special rights and responsibilities.

dominant, an allele that is present in the underlying genetic make-up (genotype) of an organism, and that is expressed in the organism's outward appearance (phenotype) in both the homozygous and heterozygous state.

**drills**, bit-like drilling implements that were used to create holes in hard materials.

**dryomorphs** *(dry'-oh-morfs)*, the common designation for an extinct group of forest-dwelling Miocene hominoids from East Africa.

**Dunkers**, a small religious sect of Old German Baptist Brethren whose members migrated from Germany to the farmlands of eastern Pennsylvania in the 18th century.

**Dyukhtai Tradition** *(dy-yuhk'-ty + tradition)*, an upper palaeolithic cultural tradition that persisted in northeastern Siberia between about 18 kya and about 14 kya.

*e*

**ER 1470** (East Rudolf + 1470), an almost complete *H. habilis* cranium from East Lake Turkana in Kenya that is about 1.9 million years old.

**ecofacts**, what are left of prehistoric natural resources such as minerals, plants, and animals. Archaeologists rely on such items, which they recover from archaeological sites, to draw inferences about the behaviour of those who used the resources.

**Egyptian civilization**, a civilized culture that flourished along the banks of the Nile River between about 5000 ya and about 2000 ya (*c.* 3 168 B. C. – 332 B. C.).

**Egyptian hieroglyphs**, an ancient form of Egyptian writing that conveys meaning via semograms and phonograms.

**Eoanthropus dawsoni** *(ee-an'-throh-pus + daw'-soh-nee)*, the taxonomic designation originally assigned to Piltdown.

**epicanthic fold** *(ep-i-kan'-thik + fold)*, a fold of skin on the upper eyelids that droops down over one or both corners of the eyes and makes them appear to be almond-shaped.

**epidermis**, the outermost layer of the skin.

**ethnic groups**, the common designation for human social isolates.

**ethnocentrism**, evaluating the patterns of culture that prevail in one culture on the basis of standards that prevail elsewhere.

**ethnographic record** *(eth'-naw-gra-fik + record)*, the compendium of anthropological descriptions of contemporary cultures and those of the recent past.

**ethnographic research**, gathering information about cultures on a first-hand basis through fieldwork.

**ethnographies** *(eth-naw'-gra-feez)*, anthropological descriptions of particular cultures.

**ethnohistorical research**, gathering information about cultures by examining historical records.

**ethnologists** *(eth-nawl'-uh-jists)*, cultural anthropologists who study contemporary cultures and those of the recent past.

**evolution**, either biological change through time, the process by which new species arise from old ones, or the theory that Charles Darwin proposed.

**evolution of species**, how species change over time, so much so that new species inevitably arise.

*f*

**Fayum** *(fay-yoom')*, a region in Egypt that contains the remains of many Oligocene primates.

**fertile crescent**, an arc of fertile land extending from the Levant to Iraq.

**fieldwork**, a research method that brings anthropologists into direct contact with the subjects and objects they study.

**first family**, the fossil remains of thirteen *A. afarensis* discovered in Ethiopia near the location where Lucy was found.

**fission**, the process whereby small groups of individuals slowly drift away from somewhat larger social units.

**fission-track method**, an absolute dating technique that is used to estimate the age of volcanic rocks that are between about 100 000 years old and about three million years old.

**flake knife**, a stone flake struck from a core that was likely used for butchering and sharpening sticks.

**fluorine test**, a dating technique that is used to establish the relative age of bone that is buried in a moist sediment; the technique is based on the fact that bone absorbs fluorine from the groundwater.

**folivores** (*foh'-li-vorz*), species that feed mainly on leaves.

**folk taxonomy**, a classification of racial types that relies on customary beliefs to describe alleged races and to account for the differences among them.

**foramen magnum**, the large hold at the base of the skull where the spinal cord enters the brain.

**fossil record**, what remains of organisms that are no longer alive.

**fossils**, the remains or traces of the remains of organisms that lived in the past.

**frugivores** (*froo'-gi-vorz*), species that feed mainly on fruit.

**gallery forests**, relatively luxuriant forest that line the banks of waterways.

**gametes** (*ga'-meets*), sex cells.

**geological time scale or calendar**, a time scale that divides the earth's history into a number of eras, periods, and epochs, each interval characterized by its own depositional features.

**gene flow**, the infusion of the alleles from one gene pool in a population into another.

**gene pool**, the total number of alleles in a population.

**genes**, chemical blueprints that determine the structure and function of the various parts of an organism's body.

**genetic drift**, fluctuations in the allele frequencies in the gene pool of a population that are caused by chance events. Genetic drift produces genetic change in a population.

**genocide**, the deliberate and systematic extermination of a group of people on account of their alleged biological and behavioural traits.

**genotype** (*jee'-noh-typ*), an organism's underlying genetic make-up.

*Gigantopithecus* (*jy-gan-toh-pith'-i-kus*), an extinct genus of late Miocene and middle Pleistocene hominoids that may have included the largest primates that ever lived.

**glaciers**, huge sheets of ice and debris.

*Gondwana* (*gahn-dwah'-nuh*), a former supercontinent that was composed of what are now the land masses in the southern hemisphere.

**gracile australopithecines**, the common designation for *Australopithecus africanus*.

gradualism, the idea that evolution is a slow, gradual, and continuous process in which one species shades almost imperceptibly into the next as a result of incremental biological change.

Gravettian culture (gra-ve'-tee-en + culture), an upper palaeolithic culture that persisted in Europe between 30 kya and 20 kya.

Great Chain of Being, the doctrine that nature is arranged like a ladder, with inanimate matter, plants, and non-human animals located on the bottommost rungs, angels and God on the uppermost, and humankind midway between, part body and part spirit.

great restrictive law, Thomas Malthus' idea that while the number of individuals in a population increases at a geometric rate their food supply increases at a arithmetic rate. It was this idea that led Malthus to coin the expression "struggle for existence".

Great Rift Valley, a long trough stretching from Syria down through eastern Africa for thousands of kilometres to Mozambique. The Great Rift Valley is where the first hominids lived.

grooming, an interactive behaviour in which one animal picks dirt and parasites from another's hair with its hand and teeth to initiate social contact.

## h

Hadar (hah-dahr'), a site in the Afar Triangle in northeastern Ethiopia where Lucy and the first family were discovered.

haemoglobin (hee-moh-gloh'-bin), a complex iron-bearing protein found in human red blood cells that plays an essential role in enabling those cells to transport oxygen from the lungs to the various cells of the body.

hammerstone, a stone used to alter the shape of a core.

Hardy-Weinberg theorem, the proposition that, under certain conditions, the genotypic frequencies that prevail in a population will remain inherently stable through time.

herbivores, vegetarians.

heredity, the process by which biological characteristics originate and are transmitted from one generation to the next.

heterozygous (he-ter-oh-zy'-gus), possessing alternate alleles of the same gene.

hieratic (hy-ra'-tik), a form of ancient Egyptian writing that was written with connected letters and used primarily to record religious doctrine.

holistic (hoh-lis'-tik), a view that focuses on the whole rather than on the parts of the whole.

holocaust (hoh'-loh-kahst), the destruction of European Jewry by the Nazis (1933–1945).

home bases, selected campsites where early hominids may have manufactured stone tools, shared food, and otherwise interacted with one another on a regular basis.

home range, the territory that a primate community generally travels within and defends.

Hominidae (haw-min'-i-day), the primate family that is represented exclusively by humans.

hominids (haw'-min-idz), the common designation for the members of primate family Hominidae.

**Homininae** (*haw-min'-i-nay*), the primate subfamily that includes gorillas, chimpanzees, and humans.

**hominization** (*haw-min-i-zay'-shun*), the process by which hominids became increasingly human.

**Hominoidea** (*haw-min-oy'-dee-a*), the primate superfamily that includes apes and humans.

**hominoids** (*haw'-min-oydz*), the common designation for the members of the primate superfamily Hominoidea.

**Homo**, the primate genus that includes ourselves and the immediate forerunners of our kind.

**Homo erectus** (*homo + eh-rek'-tus*), an extinct species of humankind that originated in Africa about 1.8 mya and subsequently migrated to Asia and perhaps Europe.

**Homo ergaster** (*homo + ur-gas'-tur*), an extinct species of humankind, much like *H. habilis*, except that its members possessed less robust faces, smaller brains, and smaller teeth.

**Homo habilis** (*homo + ha-bil'-us*), an extinct species of humankind that lived in Africa between about 2.2 mya and about 1.5 mya.

**Homo neandertalensis** (*homo + nee-an'-der-tahl-en-sis*), an extinct species of humankind whose members flourished in Africa, Asia, and Europe between about 80 kya and about 40 kya.

**Homo sapiens**, the species that includes humans such as ourselves.

**Homo sapiens neandertalensis**, a subspecies of *H. sapiens* that supporters of the preneandertal theory believe lived in the Near East and North Africa.

**Homo sapiens sapiens**, the subspecies of *H. sapiens* that is represented by contemporary humankind.

**homozygous** (*hoh-moh-zy'-gus*), possessing identical alleles of the same gene.

**Hylobatidae** (*hy-loh-bat'-i-day*), the primate family that includes gibbons and siamangs.

**hyperrobust australopithecine**, the common designation for an extinct species of especially ruggedly built hominids that lived in East Africa between about 3 mya and about 2 mya.

*i*

**Ice Age**, the common designation for the Pleistocene epoch (*c.* 1.8 mya – 11 kya).

**ice-free corridor**, a long trough that stretched through the surrounding ice fields from the northwestern corner of North America down through what are now British Columbia and Alberta into Montana. It is believed that the first Americans moved south through this corridor after they entered the New World.

**immutability of species**, the principle that species only can reproduce exact replicas of their own kind; the principle is based on a literal interpretation of the contents of the Bible.

**inheritance of acquired characters**, Jean Baptiste de Lamarck's mistaken idea that offspring inherit the biological features that their parents develop during their lives.

**insectivora** (*in-sek-ti-vor'-uh*), the mammalian order that includes modern insect-eating animals such as moles, shrews, and Old World hedgehogs.

**insectivores** (*in-sek'-ti-vorz*), species that feed mainly on insects.

integrative theories, a class of theories that hold that the state was born of compromise, either to maintain or to accrue social and economic benefits that otherwise would not be in the offing.

interglacial period, an intervening period of warm weather between glacial advances that caused glaciers to melt.

iron age cultures, cultures that arose about 4 kya in Asia and featured implements made from iron; such cultures are generally considered to be civilized cultures.

# k

karyotypes (ker'-ee-oh-typs), photographic representations of chromosomes.

KNM-WT 15000 (Kenya National Museums-West Turkana + 15000), an almost complete skeleton of a 1.6 million year old *H. erectus* boy who was about 12 years old at the time of his death; the specimen is also known as the "Turkana boy".

KNM-WT 17000 (Kenya National Museums-West Turkana + 17000), a well-preserved 2.5 million year old cranium with an afarensis-like brain volume and a massively rugged, boisei-like shape. The specimen is also known as the black skull.

keratin (ker'-a-tin), a protein found in the dead, outer or corneal layer of the dermis that imparts a yellowish tinge to the skin.

keystone herbivore hypothesis, the hypothesis that Ice Age peoples in the Americas, in Europe, and in Australia hunted many large herbivores to the point of extinction, and that their demise, combined with climatic change, adversely affected the habitat of medium-size herbivores condemning them to a similar fate.

kill sites, places where big game was slaughtered, butchered, and made ready to carry back to temporary base camps.

Kingdom Animalia, the taxonomic designation for the animal kingdom, which includes all organisms whose body cells are surrounded by salt water, whose nourishment is acquired by ingesting their food, and whose anatomy and physiology makes them capable of moving about. All primates are assigned to this kingdom.

knuckle-walking, resting the weight of the lower body on the soles of the feet and the weight of the upper body on the knuckles while moving forward. Chimpanzees and gorillas frequently move by knuckle-walking.

# l

Lake Mungo, a 25 000-year-old to 30 000-year-old upper palaeolithic site located in the southeastern corner of Australia.

Lamarckism (la-mahr'-kiz-em), Jean Baptiste de Lamarck's mistaken idea that species are capable of changing their physical features in order to become better suited to the environment.

language, words and rules for their use.

Laurasia (lor-ayzh'-uh), a former supercontinent that was composed of what are now the land masses in the northern hemisphere.

Laurentian Archaic culture, a mesolithic culture that persisted from about 6 kya until about 3 kya in the Lower Great Lakes region of southern Ontario and in adjacent areas.

Laetoli *(ly-toh'-lee)*, an early hominid site in Tanzania not far from Olduvai Gorge that was made famous because of the 3.5-million-year-old fossil footprints that were found there.

Lemuriformes *(lee-mur'-i-formz)*, the prosimian infraorder that includes the lemurs and the indriids of Madagascar and the lorises of Africa and Asia.

leopard hypothesis, the hypothesis that leopards rather than australopithecines were responsible for the bones found in Swartkrans and other South African caves.

Levallois method *(li-val-wah' + method)*, a method of making stone tools that involved removing standardized flakes from a parent core that, after it had been worked, resembled a small tortoise shell.

linguistic anthropologists, cultural anthropologists who study the relationship that exists between communication and culture.

living floors, sites that contain substantial numbers of Oldowan stone cores and flakes and animal bones.

lower palaeolithic cultures, old stone age cultures that arose about 2.5 mya and persisted until about 75 kya.

Lucy, the oldest, most compete Australopithecine fossil discovered to date.

*m*

▼

Magdalenian culture *(mag-duh-lin'-ee-en + culture)*, an upper palaeolithic culture that persisted in Europe between 18 kya and 11 kya.

Maglemosian culture *(mag-leh-moh'-zee-en + culture)*, a mesolithic culture that flourished from about 9500 ya until about 7700 ya in the peat-swamp forest country of northern Europe.

Magnetic Polarity Time Scale, a time scale that divides the earth's history into a number of intervals on the basis of reversals in magnetic polarity.

mammalia, the class of vertebrates whose members have hair, are warm blooded, and suckle their young. All primates are assigned to this class of animals.

man-the-hunter scenario, the idea that the origin of human society can be traced back to a group of dominant and aggressive male hunters who attracted females to themselves because of their superior subsistence skills.

marginal habitat hypothesis, the hypothesis that the neolithic revolution arose in response to people outstripping the carrying capacity of their homelands in selected marginal regions of the world.

meiosis *(my-oh'-sis)*, the way that gametes or sex cells divide.

melanin *(meh'-la-nin)*, a brown-black pigment that is produced in the lowest layer of the outermost skin and then dispersed in granular form near the surface of the skin.

melanocytes, the cells that produce melanin.

mesolithic cultures *(meh'-soh-lith'-ik + cultures)*, middle stone age cultures that first appeared about 11 kya.

Métis *(may'-tee)*, a Canadian culture whose members trace their biological roots back to European fur traders and Indian women.

microliths, small stone blades one centimetre to five centimetres long that were set into pieces of wood, bone, and antler in order to fabricate composite tools.

microsyopids *(my-kroh-sy'-oh-pidz)*, a group of primate-like Palaeocene mammals that were apparently well-suited to an arboreal habitat and that may have given rise to *Altiatlasius*.

middle palaeolithic cultures, old stone age cultures that arose about 75 kya and persisted until about 40 kya.

mineralization, the process in which the hard parts of dead plants and animals dissolve completely but are replaced with minerals.

mitochondrial DNA *(my-toh-kawn'-dree-uul + DNA)*, the DNA molecules that are situated in the minuscule, granular, mitochondrial bodies inside a cell but outside its nucleus.

Mitochondrial Eve *(my-toh-kawn'-dree-uul + Eve)*, a hypothetical representative of a small 200 000-year-old African gene pool from which all modern human populations may have been derived.

mitosis *(my-toh'-sis)*, the way in which somatic or body cells divide.

molecular clock, an analytical device that is used to estimate when living species diverged from a common ancestor and began to evolve along separate lines.

Mongoloid race *(mohn'-gohl-oyd + race)*, the common designation for persons with yellow or copper-coloured skin; a sparse amount of medium to dark brown body hair with a straight shape; a medium-wide nose; medium-thick lips; and medium to dark brown eyes covered by a pronounced epicanthic fold. The so-called Mongoloid race is not a race in the biological sense of the term; it is best regarded as a social construct.

monkey trial, the anti-evolution trial of John T. Scopes that was held in Dayton, Tennessee, in 1925.

Mousterian *(moo-steer'-ee-en)*, a late Pleistocene toolkit that is associated primarily with Neandertals. Neandertal culture is also frequently referred to as Mousterian.

mutation, a change in the hereditary instruction that alleles provide.

*n*

Natufians *(na-too'-fee-enz)*, a mesolithic people who lived in what are now Israel, Jordan, and Syria between about 12 500 ya and about 10 500 ya.

natural selection, the concept that Darwin employed to describe the elimination of individuals from a population because of differential survival and reproduction rates in the population.

Neandertal-phase theory, the theory that Neandertals were the first *H. sapiens* — the evolutionary end product of a straight line that ran from *H. erectus* through the transitional forms to ourselves.

Neandertals *(nee-an'-der-tahlz)*, the common designation for a descendant of *H. erectus* who may have appeared as early as early as 130 kya but who flourished nearer the present, between about 80 kya and about 40 kya.

*Necrolemur (nek-roh-lee'-mur)*, an extinct tarsier-like Eocene primate.

Negroid race *(nee'-groyd + race)*, the common designation for persons with dark-coloured skin; a sparse to medium amount of dark black, kinky or frizzly body hair; a wide nose; thick lips; and dark brown to black eyes. The so-called Negroid race is not a race in the biological sense of the term; it is best regarded as a social construct.

neo-Darwinism, an evolutionary theory that is based on Darwinian and Mendelian principles.

**neolithic cultures**, new stone age cultures that first appeared about 8 kya.

**neolithic revolution**, a food producing revolution in which people began to depend on domesticated rather than on wild species for their livelihoods.

**nitrogen dating**, a dating technique that is used to establish the relative age of bone that is buried in a moist sediment; the technique is based on the fact that the nitrogen burden of such bone decreases over time.

**nocturnal**, species that are active at night.

**notches**, sharp-edged stone flakes with angular indentations that were used as knives for cutting meat and hides.

*Notharctus* (*noh-thahrk'-tus*), an extinct lemur-like Eocene primate.

**notochord**, a flexible rod or cartilage that runs lengthwise along the back from the rump to the neck where it leads into a hollow nerve cord that is connected to the brain.

**nuclear DNA**, the DNA molecules that are located in the nucleus of a cell.

**numeric changes**, the addition or the deletion of a complete chromosome.

## O

**oasis hypothesis**, the hypothesis that the neolithic revolution arose in the Near East as a result of the hot and dry climate that prevailed there at the end of the Pleistocene.

**obsidian hydration** (*awb-si'-dee-en + hydration*), a technique that is used to estimate the absolute age of volcanic glass (obsidian) upwards of 800 000 years old that was once widely used to manufacture prehistoric cutting and scraping implements.

**Old Man of La Chapelle** (*old man of + lah + shah-pel'*), the common designation for a Neandertal specimen discovered in a cave near the village of La Chapelle-aux-Saints, in southern France. The specimen was mistakenly reconstructed as a brute by palaeontologist Pierre Marcellin Boule.

**Oldowan tools** (*ol'-doh-wahn + tools*), core choppers and flake knives found in East Africa that may have been manufactured by *Australopithecus* or *Homo*.

**Olduvai Gorge** (*ol'-doo-vy + gorge*), a site in Tanzania that contains the skeletal and cultural remains of many extinct hominids.

**omnivores**, species that feed on both meat and vegetable foods.

**Omomyidae** (*oh-moh-my'-i-day*), an extinct family of Eocene primates that possessed tarsier-like traits and that may be the stock from which tarsiers evolved.

**opposable big toe**, a big toe whose underside can be placed against the sole of the foot.

**opposable thumb**, a thumb whose underside can be placed against the palm of the hand.

**osteodontokeratic culture** (*ah'-stee-oh-don-toh-keh'-ra-tic + culture*), a hypothetical australopithecine culture that featured a toolkit with weapons that were made from bones, teeth, and horns.

## P

**palaeoanthropologists** (*pay'-lee-oh + anthropologists*), biological anthropologists who study the antecedents of humankind.

**Palaeo-Indians**, the first substantial Indian population in the New World.

**palaeolithic cultures** *(pay'-lee-oh-lith'-ik + cultures)*, old stone age cultures that first appeared about 2.5 mya.

**Pangea** *(pan-jee'-uh)*, a former super-continent that was composed of what are now all of the continents of the world.

**panmixis** *(pan-miks'-us)*, random mating in a population.

**peppered moth**, the common name for *Biston betularia.*

**permineralization**, the process in which the pores or spaces in the hard parts of dead plants and animals are filled in with minerals such as silica, pyrite, and calcite that are absorbed from the local groundwater.

**petrification**, the process in which the hard parts of dead plants and animals are replaced either in whole or in part with minerals.

**phenotype** *(fee'-noh-typ)*, an organism's outward appearance.

**PTC** (phenylthiocarbamide) *(fee-nil-thy-oh-kahr'-bih-myd)*, a chemical compound that some people cannot taste but that others find intensely bitter.

**phonograms**, written signs that represent specific sounds in speech.

**physiology**, the operations of the organs of the body.

**Piltdown**, the common designation for the infamous fossil forgery that was once accepted as the missing link between apes and humans.

*Pithecanthropus erectus (pith-i-kan'-throh-pus + erectus)*, the taxonomic designation that Eugène Dubois assigned to the *H. erectus* fossils that he discovered in Java.

**plate tectonics**, the slow but continual movement of the semirigid plates on which continents and oceans rest.

**platyrrhine nose** *(pla'-ti-reen + nose)*, a nose that features widely separated nostrils that open to the side.

**Platyrrhini** *(pla-ti-ree'-nee)*, the primate infraorder that includes New World monkeys.

**plesiadapiformes** *(plee-zee-uh-dap'-i-formz)*, an extinct infraorder of Palaeocene mammals whose position in the natural history of the primates is hotly disputed.

*Plesiadapis (plee-zee-uh-dap'-is)*, an extinct genus of the infraorder plesiadapiformes.

**point mutations**, mutations that affect a single gene.

**polite racism**, a form of racism that is more covert than overt.

**polygenesis**, the idea that the world's species have evolved from several original ancestors rather than one.

**polygenic**, physical features that are produced by multiple genes.

**Pongidae** *(pon'-ji-day)*, the primate family that includes orangutans, gorillas, and chimpanzees.

**Ponginae** *(pon'-ji-nay)*, the primate subfamily that is represented exclusively by orangutans.

**population biologists**, biological anthropologists who study the outward appearance and the genetic traits of contemporary human populations.

**population genetics**, the study of the way that genetic information is exchanged among individuals who can and do interbreed.

**postorbital bar**, a bony ring that surrounds and protects the eyes of living primates.

**potassium-argon (K-Ar) method**, a dating technique that geochronologists use to calculate the absolute age of compacted volcanic ash called tuffs that are between about one million years old and about five million years old, as well as rocks in the same age range that were formed through volcanic activity.

potlatch, a ceremony practiced by a wide variety of Indian peoples on the west coast of Canada and the northern United States in which items of value were given away in order to publicly validate a change in social status.

prebiotic soup, the conglomeration of organic molecules in the atmosphere that were washed into oceans and rivers where they may have been transformed into the first living organisms.

prehensile tail, a grasping tail that is used like a hand.

prehistoric, the interval between the time that humankind developed the capacity for culture to the beginning of written history.

preneandertal theory, the theory that *H. erectus* gave rise to transitional forms who then evolved into an early Neandertal population called third interglacial forms, and that these forms then evolved, on the one hand, into the progressive Neandertals of Asia and North Africa who afterwards evolved into *H. sapiens*, and, on the other, into the classic Neandertals of Western Europe who ultimately became extinct.

presapiens theory, the theory that *H. erectus* gave rise to two distinct populations: one that led from *H. erectus* to transitional forms such as Swanscombe and then on to ourselves (*H. sapiens*), and another that led from H. erectus to transitional forms such as Steinheim and then on to *H. neandertalensis*.

primary characteristics, the social features of civilized cultures, including urban centres, occupational specialization, taxation of surplus wealth, distinct social classes, and centralized government.

primary rain forests, rain forests with fully mature trees, dense canopies, and dark understoreys.

primate pattern, an amalgam of social and biological traits that, on the whole, is especially well-suited to arboreal habitats.

primates *(pry'-mayts)*, the group (order) of animals that includes prosimians and anthropoids.

primatologists, biological anthropologists who study the similarities and the differences between modern humankind and the other surviving primates.

Principle of Recombination, the principle that the single set of alleles contributed by the male and the female gametes recombine when an offspring is formed.

Principle of Segregation, the principle that alternate alleles of the same gene separate during meiosis and become housed in different gametes.

Principle of Stratigraphy, the principle that those items discovered at the bottom of an undisturbed collection of prehistoric remains are older than those found nearer the surface.

Principle of Superposition, the principle that the formations that cover the earth's core have been deposited in successive layers or strata.

*Proconsul (pro-kahn'-suul)*, an extinct group of dryomorphs that may include the last common ancestor of apes and humans.

progressive Neandertals, the common designation for the Neandertals who lived in Asia and North Africa.

prosimians *(proh-sim'-ee-anz)*, the subdivision of the primates that includes lemurs, indriids, lorises, and tarsiers.

*Prosimii (proh-sim'-ee-eye)*, the primate suborder that includes lemurs, indriids, lorises, and tarsiers.

proto-World, the hypothetical parent language of *H. sapiens*.

provisioning hypothesis, the hypothesis that natural selection favoured bipedalism because it allowed male hunters to carry the meat that they killed back to females and their infants who remained behind while the males were away hunting.

punctuated equilibrium, the idea that the evolution of species is characterized by saltations.

pure-blooded races, biological populations with little variation in their own physical traits, but with clear-cut differences between them.

*Purgatorius (pur-guh-tor'-ee-us)*, an extinct genus of Palaeocene mammals that may include the first primates.

# *q*

Qafzeh cave *(kahf'-tzeh + cave)*, a cave site in Israel in which the oldest skeletal remains of fully modern humans were discovered; the remains are about 92 000 years old.

quadrumanual arborealists, animals that use both their hands and their feet to grasp and suspend themselves from branches and trees.

# *r*

race, subpopulations of a species that differ from one another in terms of the frequency of certain inherited traits.

racism, the belief that people whose physical features are allegedly or actually different from our own are biologically inferior and can consequently be mistreated.

rafting hypothesis, the idea that the ancestors of New World monkeys originated in Africa and came to the New World across the Atlantic either on floating rafts composed of driftwood and living plants or else on natural floating islands.

ramamorphs *(ra'-ma-morfs)*, the common designation for an extinct group of woodland and savanna-dwelling Miocene hominoids from Africa, Asia, and Europe.

*Ramapithecus (ra-ma-pith'-i-kus)*, a group of extinct ramamorphs that were once thought to be the first hominids.

*ras* gene, a gene that may be responsible for upwards of 15 percent of 100 or so types of cancer from which human beings suffer.

readiness hypothesis, the hypothesis that the neolithic revolution arose in the fertile crescent in the Near East when the mesolithic peoples who lived there had learned enough to domesticate the plants and animals in their immediate environment.

recessive, an allele that is present in the underlying genetic make-up (genotype) of an organism, but that is expressed in the organism's outward appearance (phenotype) only in the homozygous state.

regional continuity or multiregional model, the model that portrays the transition from *H. erectus* to *H. sapiens* as a gradual process beginning about 1.5 mya when *H. erectus* spread from Africa to other parts of the world where they gave rise to several regionally distinct *H. erectus* populations that independently evolved into *H. sapiens* in a step-like fashion in the regions that they occupied.

relative age, an age that indicates whether one item is older than another, although not by how much.

replacement theory, the theory that transitional forms in Africa gave rise to a *H. sapiens* population that spread into the regions where the Neandertals lived and replaced the Neandertals when the Neandertals became extinct.

replication, the process in which the intertwined strands of the DNA molecule unwind as the hydrogen bonds between the bases dissolve, and the unpaired bases on both strands attract new complementary bases that are present as raw materials in the cell.

robust australopithecines, the common designation for *Australopithecus robustus*.

Rosetta Stone, a massive black basalt slab that bears a trilingual inscription written in hieroglyphic, demotic, and Greek characters that honours the Egyptian monarch Ptolemy V.

## S

sagittal crest *(saj'-i-tal + crest)*, a ridge of bone that runs along the midline of the top of the skull from the front to the back. The sagittal crest helps to support the muscles that are used in chewing.

Sahul *(sah-ool')*, a continental shelf that united what are now Australia, New Guinea, and Tasmania about 50 kya.

saltations *(sal-tay'-shunz)*, relatively long periods of biological stability interrupted by the comparatively sudden appearance of new forms.

savanna, tropical grasslands.

savanna theory of human evolution, the theory that the transition from hominoids to hominids took place on the savanna or grasslands of Africa.

scavenging hypothesis, the hypothesis that natural selection favoured bipedalism because it was the most efficient and effective way to locate dead animals, to fend off competitors that were intent on eating the same animals, and to secure access to other sources of food when scavenging failed.

scientific creationists, those who insist that the creation story should be incorporated into the science curriculum of public schools and taught in science classes.

scrapers, small tools with a steeply bevelled working edge that were used to clean hides and carve wood, bone, and antler.

secondary characteristics, the institutional features of civilized cultures, including large-scale public works, long distance trade, standardized monumental artwork, writing, and scientific achievements.

secondary rain forests, rain forests that contain maturing trees, discontinuous canopies, and green understoreys.

seed-eating hypothesis, the hypothesis that natural selection favoured bipedalism because it facilitated two-handed feeding on small food objects on the grasslands of Africa.

segregation, the enforced separation of people on account of their observable physical traits.

self-fertilizing, sexual reproduction without a sexual partner.

semograms, written symbols that convey meaning such as the symbols that were used in part to fashion Egyptian hieroglyphs.

**sex chromosomes**, the chromosomes that house the alleles that are responsible for determining the sex of an organism.

**sexual division of labour**, the cultural practice of assigning different tasks to men and to women on the basis of their sex.

**sexually dimorphic**, species that are characterized by distinctive non-reproductive differences between males and females.

**Shang dynasty**, a civilized culture that flourished in China between about 3500 ya and about 3000 ya.

**Shield Archaic culture**, a mesolithic culture that persisted from about 6500 ya until about 2500 ya in the region stretching from the Keewatin District in the Northwest Territories to Cape Breton in Nova Scotia.

**sickle cell disease**, a debilitating and often fatal form of anemia that is caused by the presence of a pair of mutated recessive alleles governing the production of haemoglobin.

**sickle cell trait**, a mild form of anemia that is caused by the presence of a single mutated recessive allele governing the production of haemoglobin; the sickle cell trait can also provide protection against malaria.

**silverback**, an adult male gorilla at least eleven years old to which unrelated adolescent females and mature females become attached.

*Sinanthropus pekinensis (sin-an'-throh-pus + pee'-kin-en'-sis)*, the taxonomic designation that Davidson Black assigned to the *H. erectus* fossils that were discovered at Zhoukoudian.

**single origin or Noah's Ark model**, the model that represents the transition from *H. erectus* to *H. sapiens* as a relatively recent event that took place about 200 kya, when a regionally distinct African *H. erectus* population gave rise to a *H. sapiens* population that became differentiated in Africa and then spread from there throughout the world where modern regional features eventually became entrenched.

**sites**, places where prehistoric materials are buried.

*Sivapithecus (shee-va-pith'-i-kus)*, an extinct genus of ramamorphs that likely gave rise to the orangutan.

**slavery**, the practice in which people are treated by others as property often because of their observable physical traits.

*Smilodectes (smy-loh-dek'-teez)*, an extinct lemur-like Eocene primate.

**social isolates**, groups of people who are isolated from one another on account of geographic or cultural barriers, but who are by no means biologically unique.

**soft hammer technique**, a method of making stone tools in which flakes were carefully removed from a core by striking it with a "soft" hammer fashioned from either wood, bone, or antler.

**Solutrean culture** *(suh-loo'-tree-en + culture)*, an upper palaeolithic culture that persisted in Europe between 20 kya and 18 kya.

**somatic**, pertaining to the body.

**speciation**, the process by which one species arises from another.

**species**, a biological population whose members can interbreed with reproductive success.

**state**, a political system in which public policy is formulated under the auspices of an overriding political ideology that provides government officials with a monopoly on the use of physical force with respect to implementing policy decisions.

Steinheim *(shtyn'-hym)*, the common designation for the fossil skull of a 250 000-year-old to 200 000-year-old archaic H. sapiens discovered in a gravel pit in Steinheim, near Stuttgart, Germany.

stereoscopic vision, depth perception or the ability to see objects in three dimensions.

stone age cultures, cultures that arose in Africa about 2.5 mya and featured implements made of stone.

struggle for existence, the competition that takes place between organisms for limited resources that are critical to their survival. The expression was coined by Thomas Malthus.

subassemblage, a collection of ecofacts and artifacts that are indicative of a specific activity that a particular group of prehistoric people undertook in the past.

surface surveys, inspecting the ground for signs of human activity, which is usually done by walking or riding over the landscape in search of ancient material that is scattered on the surface.

Swanscombe *(swanz'-kohmb)*, the common designation for the fossil skull of a 250 000-year-old to 200 000-year-old archaic *H. sapiens* discovered in a gravel pit in Swanscombe, England.

sympatric model *(sim-pa̱'-trik + model)*, the model that calls attention to the evolution of new species that arise from a parent population whose subunits become separated by behavioural barriers. Although the natural history of few if any species conform to the model, it is nonetheless an important theoretical possibility.

taphonomy *(taf-on'-oh-mee)*, the study of what happens to the remains of animals between death and fossilization.

tarsal bones, ankle bones.

Tarsiiformes *(tar'-si-formz)*, the prosimian infraorder that is currently represented by four species of tarsiers that live in the primary and secondary tropical rain forests of Southeast Asia.

Taung child *(tah-oong' + child)*, the common designation for the first australopithecine fossil discovered.

taxonomy, a system of biological classification in which organisms are arranged first into large groups and then into increasingly smaller units on the basis of a rigorous and scrupulous analysis of the similarities and differences in their biological traits.

Tay-Sachs disease (TSd) *(tay-saks + disease)*, a hereditary illness that causes deterioration of the brain and ultimately results in death, usually within the first two years of life.

technical vocabulary, the terms and concepts that comprise the technical language of a specialized area of knowledge.

Tenochtitlán *(tay-notch-tee-tlahn')*, the capital city of the Aztecs; its ruins lie buried beneath modern Mexico City.

Terra Amata *(terra + ah-mah'-tah)*, an alleged seasonal camp that was established sometime between 300 kya and 200 kya in what is now the city of Nice on the French Riviera.

terrestrial quadrupeds, animals that habitually move about with all fours on the ground.

thermoluminescence *(thur'-moh-loom-in-es'-ens)*, a technique that is used to estimate the absolute age of fired pottery and similar wares that are upwards of 10 000 years old.

third interglacial forms, an early Neandertal population that supporters of the preneandertal theory believe gave rise to both progressive and classic Neandertals.

Three Age System, Christian Thompson's arrangement of prehistoric cultures into stone, bronze, and iron age cultures.

Torralba *(tor-ahl'-bah)*, an alleged *H. erectus* kill site located in north-central Spain.

transitional forms, the common designation for archaic *H. sapiens*.

Triple Alliance, an alliance made between the Aztecs and two rival Mexican states that the Aztecs controlled.

Trisomy 21 *(try'-soh-mee + 21)*, a hereditary disorder that is characterized by mental retardation, a short body with stubby fingers and toes, a round face, and a large, protruding tongue that makes coherent speech difficult.

troops, primate communities.

tuffs, compacted volcanic ash.

Tupaiidae *(too-py'-i-day)*, the non-primate family that includes tree shrews.

## u

uniformitarianism *(yoo'-ni-for-mi-ter'-ee-an-iz-em)*, the idea that age-old forces such as earthquakes and erosion are constantly changing the world.

upper palaeolithic cultures, old stone age cultures that arose about 40 kya and persisted until about 11 kya.

uranium dating, a dating technique that is used to establish the relative age of bone that is buried in a moist sediment; the technique is based on the fact that bone absorbs uranium from the groundwater.

## v

Venus figurines, small and delicate statues of pregnant women with exaggerated breasts and buttocks but lacking facial features.

vertical clingers and leapers, animals that hop short distances and make highly precise leaps from time to time between distantly spaced branches or trees.

vertebrata, the subphylum of the animal kingdom that includes chordates that have a true spine or backbone. All primates are included in this subphylum of animals.

visual predation theory, the theory that primate-like eyes and limbs are best regarded as adaptive devices that allowed those that originally possessed them to best survive by preying on insects that inhabited the slender vines and branches that formed the comparatively dark understorey of the tropical forest.

## w

woodland forests, forests bordered by tropical grasslands that feature deciduous trees interspersed with grasses and bushes.

## x

xenografts *(zee'-noh + grafts)*, cross-species organ transplants.

## y

Y-5 pattern, the cusp pattern on the lower molars of apes and humans; the pattern features five cusps on the top of the crown arranged in the shape of a sideways Y with the arms of the letter facing the cheek.

## z

Zhoukoudian *(zhoo'-kood-yen)*, a famous *H. erectus* cave site near Beijing, China.

Zhoukoudian cranium 5, a *H. erectus* skull from Zhoukoudian that shows imprints of the same neural connections that make it possible for modern humans to receive, interpret, and send linguistic information in the form of spoken language.

*Zinjanthropus boisei (zin-jan'-throh-pus + boy'-zee-eye)*, a taxonomic term that was formerly used to identify the first hyper-robust australopithecine fossil discovered.

zygote *(zy'-goht)*, a single cell that is formed by the union of a female gamete and its male counterpart.

# INDEX

## Aa

ABO blood groups, 390–91

Absolute age (chronometric age), 159

Absolute dating techniques (chronometric dating techniques), 159

Acheulian, 250–52, 313–14

Acquired characters, 90

Adams, Richard, 369

Adapidae, adapids, 173

Adaptation
  by Darwin's finches, 99
  by early mammals, 154–55
  by prosimians, 48–49
  by hominoids to an aquatic environment, 206

Adaptive radiation, 100

Adenine, 123

Admixture, 388, 389

*Aegyptopithecus*, 177

Africa, 61, 178, 248
  archaeological discoveries in, 332
  palaeoanthropological discoveries in, 2–3
  primatological studies in, 176–77

African-Americans, 393

African National Congress, 407

Age of Mammals, 154–55

Age of Reptiles, 166–67

Agriculture
  in Europe, 350–52
  in the Far East, 350, 355
  in the Near East, 346, 348, 352, 353, 355
  in the New World, 355–56
  in the Old World, 363, 367

Ainu (Japan), 317, 388

Albinism, 131

Aldred, Cyril, 363

Algeria, 54

Alland, Alexander Jr., 148

Allele, Alleles, 122, 124
  codominant, 144
  dominant, 128
  HbA, 144
  HbS, 144
  recessive, 128

Allen, Harry, 327

Allopatric model of speciation, 146

Altamira (Spain), 294

*Altiatlasius*, 172–73

Altruistic hypothesis, 318–19

Ambrona (Spain), 258–59

American Civil Liberties Union (ACLU), 108

American Sign Language (ASL), 67–69

Animals, 37

Amino acids, 124

Anagenesis, 147–48

Anatolia (Turkey), 348

Anatomically modern humans (*Homo sapiens sapiens*), 292–98
  biological features of, 292–93
  cultural attributes of, 293–95, 297
  fossil specimens of, 292
  origin of, 292

Anatomy, 34

Ancestral traits, 171

Andersson, John, 241–42, 244

Andrews, P., 267–68

Angiosperm radiation theory of primate origins, 169–70

## Bb

Baboons, 54
Baby Fae, 53–54
Bakker, Robert, 166
Band societies, 334
Barbary Ape (*Macaca Sylvana*), 54–55
Barid,Patricia, 143–44
Bar-Yosef, Ofer, 347
Barlow, Nora, 96–97
*Beagle, H.M.S.*, 96
Bear cult, 317
Beijing (formerly Pekin or Peking, China), 241
Belfer-Cohen, Anna, 347
Berckhemer, F., 269
Berdan, Frances, 370, 371
Bering Strait, 296
Beringia, 296
Berman, Carol, 38
Bible, 87–88, 110
Big game hunting tradition, 257–58, 330
  *See also* Palaeo-Indians.
Bigfoot, 183
Binford, Lewis, 257–58, 320–21, 354–55
Biological anthropology, 2–6
  goals of, 2
  subfields of, 2–6
Bipedalism, 69–70
  evolution of, 207–08
  origin of, 206
Black, Davidson, 242–44
Black, Debra, 140
Black skull (KNM-WT 17000), 217–18
Blade technique, 297, 298
Blood, 382, 391, 401
Boas, Franz, 28
Bohlin, Birger, 243-244

Bolivia, 175
Bonobo (*Pan paniscus*, pygmie chimpanzee), 63
Book of Genesis, 110
Borgaonkar, Digamber, 142
Borneo (Indonesia), 58
Borers, 281
Boule, Pierre Marcellin, 277–79, 318–19
Bow and arrow, 344
Bowler, Jim, 327
Brace, C. Loring, 282
Brachiation, 56–57
Braidwood, Robert, 353–54
Brain, C.K., 210–12
Brain, 39. *See also* cranial capacity.
*Branisella*, 175
Brazil, 18, 49
Breeding population, 397–98
Breeding seasons, 38
Broadbeck, May, 21
Broca's area, 249
Bronze Age cultures, 311
Broom, Robert, 198–201, 244
Brow ridge, 70
Brown, Frank, 247
Brues, Alice, 132
Brünn (Czech Republic), 119
Bryan, William Jennings, 108–09
Buchenwald (Germany), 401
Buffon, George Louis de, 89–90, 94
Burckhard, Richard Jr., 92
Burial, 315, 316, 322, 326, 342
Burial cult, 342
Burins, 281
Burton, Frances, 55
Butchery, 325
Butzer, Karl, 236–37

# Cc

## Dd

## Ee

## *Mm*

# Oo

# Pp

## Qq

## Rr

## *Ss*

## *Tt*